Sustainable Cities in a Changing Climate

Sustainable Cities in a Changing Climate

Enhancing Urban Resilience

Edited by

Sami G. Al-Ghamdi
King Abdullah University of Science and Technology (KAUST)
Saudi Arabia

This edition first published 2024
© 2024 by John Wiley & Sons Ltd

All rights reserved. No part of this publication may be reproduced, stored in a retrieval system, or transmitted, in any form or by any means, electronic, mechanical, photocopying, recording or otherwise, except as permitted by law. Advice on how to obtain permission to reuse material from this title is available at http://www.wiley.com/go/permissions.

The right of Sami G. Al-Ghamdi to be identified as the author of the editorial material in this work has been asserted in accordance with law.

Registered Offices
John Wiley & Sons, Inc., 111 River Street, Hoboken, NJ 07030, USA
John Wiley & Sons Ltd, The Atrium, Southern Gate, Chichester, West Sussex, PO19 8SQ, UK

For details of our global editorial offices, customer services, and more information about Wiley products visit us at www.wiley.com.

Wiley also publishes its books in a variety of electronic formats and by print-on-demand. Some content that appears in standard print versions of this book may not be available in other formats.

Trademarks: Wiley and the Wiley logo are trademarks or registered trademarks of John Wiley & Sons, Inc. and/or its affiliates in the United States and other countries and may not be used without written permission. All other trademarks are the property of their respective owners. John Wiley & Sons, Inc. is not associated with any product or vendor mentioned in this book.

Limit of Liability/Disclaimer of Warranty
While the publisher and authors have used their best efforts in preparing this work, they make no representations or warranties with respect to the accuracy or completeness of the contents of this work and specifically disclaim all warranties, including without limitation any implied warranties of merchantability or fitness for a particular purpose. No warranty may be created or extended by sales representatives, written sales materials or promotional statements for this work. This work is sold with the understanding that the publisher is not engaged in rendering professional services. The advice and strategies contained herein may not be suitable for your situation. You should consult with a specialist where appropriate. The fact that an organization, website, or product is referred to in this work as a citation and/or potential source of further information does not mean that the publisher and authors endorse the information or services the organization, website, or product may provide or recommendations it may make. Further, readers should be aware that websites listed in this work may have changed or disappeared between when this work was written and when it is read. Neither the publisher nor authors shall be liable for any loss of profit or any other commercial damages, including but not limited to special, incidental, consequential, or other damages.

Library of Congress Cataloging-in-Publication Data
Names: Al-Ghamdi, Sami G, editor.
Title: Sustainable Cities in a Changing Climate: Enhancing Urban Resilience / edited by Sami G. Al-Ghamdi.
Description: Hoboken, NJ : Wiley, 2024. | Includes index.
Identifiers: LCCN 2023035113 (print) | LCCN 2023035114 (ebook) | ISBN 9781394201549 (hardback) | ISBN 9781394201518 (adobe pdf) | ISBN 9781394201525 (epub)
Subjects: LCSH: Sustainable urban development. | Climate change mitigation. | City planning.
Classification: LCC HT241 .C5955 2024 (print) | LCC HT241 (ebook) | DDC 307.76—dc23/eng/20231101
LC record available at https://lccn.loc.gov/2023035113
LC ebook record available at https://lccn.loc.gov/2023035114

Cover Design: Wiley
Cover Image: © Photo of the KAUST campus

Printed and bound by CPI Group (UK) Ltd, Croydon, CR0 4YY

C9781394201549_291123

Contents

List of Contributors *xiii*
About the Editor *xv*
Preface *xvii*
Abbreviations *xix*

Part I Climate Change and The Built Environment: Foundations and Implications *1*

1 Understanding Climate Change Fundamentals: Exploring the Forces Shaping Our Planet's Future *3*
Introduction *3*
  Recent Climate Change is Anthropogenic *5*
  Spatial Distribution of Global Warming *6*
Modes of Climate Variability *6*
Find, Read, and Process Climatic Data *8*
  Climate Models (GCMs and RCMs) *8*
  Pathways and Scenarios *10*
  Observations and Reanalysis *10*
  Visualizing and Processing Climatic Data *12*
Conclusion *15*
References *15*

2 Advancing Urban Resilience and Sustainability Through the WRF-Urban Model: Bridging Numerical Modeling and Real-World Applications *17*
Introduction *17*
Nexus Between Urbanization and Climate Change *18*
Urban Modeling Through WRF-Urban Model *19*
  Overview of the WRF-Urban Model *20*
  Applications of the WRF-Urban Model *20*
Relevant Case Studies *21*
  Case Study 1: Urban Climate Modeling in Singapore Using WRF-Urban *21*
  Case Study 2: Summertime Air Conditioning Electric Loads Modeling in Beijing, China, Using WRF-Urban *21*

Case Study 3: Coastal-Urban Meteorology Study in the Metropolitan Region of Vitória, Brazil, Using the WRF-Urban Model *22*
Limitations of the WRF-Urban Model *22*
Ways Forward for Improvement *23*
Conclusions *24*
References *25*

3 Assessing and Projecting Climatic Changes in the Middle East and North Africa (MENA) Region: Insights from Regional Climate Model (RCM) Simulations and Future Projections *29*
Introduction *29*
Methodology *31*
GCMs vs. RCMs in Simulating MENA Temperature and Precipitation *32*
RCMs Performance in Simulating MENA Climatic Changes *34*
Projected Future Changes Over MENA-CORDEX *35*
Conclusion *36*
References *38*

4 Building for Climate Change: Examining the Environmental Impacts of the Built Environment *39*
Introduction *39*
Embodied Carbon Emission in Building Environment *40*
Embodied Carbon Emission for Selected Building Materials *40*
Embodied Carbon Emission of Limestone Quarrying *41*
Embodied Carbon Emission from Cement and Concrete Manufacturing *42*
Embodied Carbon from Asphalt Production and Construction *44*
Embodied Carbon Emission of Steel Production *45*
Embodied Carbon Mitigation Strategies *46*
MS1: Using Materials with a Lower Embodied Carbon *46*
Precast Hollow-Core Slabs *48*
Steel Framework System *48*
Use of Unfired Brick *48*
Ethylene Tetrafluoroethylene *49*
MS2: Reducing, Reusing, and Recovering—Heavy Building Materials *49*
MS3: Improvement in Design Phase and Efficient Construction *49*
MS4: Carbon Sequestration *51*
MS5: Extending the Building's Life *51*
Operation Carbon Emissions in Building Environment *51*
Operation Carbon Mitigation Strategies *52*
Efficient HVAC Systems in Buildings *53*
Renewable Resources Integration *53*
Strategy for Water Use *54*
Use of Lighting *54*
Conclusion *55*
References *56*

5 Unveiling the Nexus: Human Developments and Their Influence on Climate Change *61*
Introduction *61*
Life Cycle Assessment for Environmental Impact *63*
ReCiPe Impact Category: Climate Change *64*
Energy Sector Impact on Climate Change *65*
  Case Study 1: Electricity Generation in Turkey *65*
  Case Study 2: Coal Power Plant with Carbon Capture Technology in Czech Republic *67*
  Case Study 3: Solar Power with Energy Storage *68*
Emissions Savings from Energy Sector *69*
  Energy Efficiency Increase *70*
  Wind and Solar Plant Installation *71*
  Keep Running the Nuclear Plants *72*
Freshwater Sector Impact on Climate Change *72*
  Case Study 1: Water Supply in Singapore *72*
  Case Study 2: Seawater Desalination in South Africa *73*
  Case Study 3: Multistage Flash Desalination in Qatar *73*
Emission Savings from Water Sector *74*
  Groundwater Management *74*
  Energy Management in Water System *75*
  Smart Wastewater Treatment Technology *75*
Concluding Remarks *75*
References *76*

Part II Quantifying Resilience and Its Qualities *79*

6 Assessing Resilience in Urban Critical Infrastructures: Interdependencies and Considerations *81*
Introduction *81*
Individual Network Resilience *83*
  Transportation Network Resilience *84*
  Electrical Network Resilience *84*
  Water Network Resilience *85*
Case Study About Individual System Resilience: Transportation Resilience During Mega Sport Events *86*
Infrastructures Interdependencies and Resilience *88*
Case Study About Interdependent Systems Resilience *90*
Conclusion *92*
References *93*

7 Assessing Infrastructure Resilience: Approaches and Considerations *97*
Introduction *97*
Complex Networks *98*

Types of Graphs *98*
      Directed and Undirected Graphs *99*
      Weighted and Unweighted Graphs *99*
   Main Applications in Resilience Assessment *100*
      Betweenness Centrality *100*
      Graph Percolation *101*
      Strengths and Limitations of Complex Networks *101*
   Simulation Approaches *101*
      System Simulation *102*
      Agent-Based Modeling *103*
      GIS-Based Approaches *103*
      Strengths and Limitations of Simulation Approaches *103*
   Other Approaches *104*
      Statistical Approaches *104*
      Optimization Approaches *104*
   Conclusion *105*
   References *105*

8 Enhancing Buildings Resilience: A Comprehensive Perspective on Earthquake Resilient Design *111*
   Introduction *111*
   Structural Resilience Representation *112*
   Performance-Based Design (PBD) *114*
   Supporting Systems *115*
      Supporting Systems Within the Building *116*
      Beyond the Building Limits *116*
   Conclusion *117*
   References *118*

9 Enhancing Built Environment Resilience: Exploring Themes and Dimensions *121*
   Introduction *121*
      Uncertainty *122*
      Risk Identification and Assessment *123*
      Resilience Capacities *123*
      Resilience Components *124*
   Types of Resilience *124*
      Ecological and Engineering Resilience *125*
      Community and Social Resilience *127*
      Specified and General Resilience *128*
      Critical Infrastructure Resilience *128*
      Technical Systems, Products, and Production Resilience *129*
   Resilience Dimensions and Capitals *129*
   Resilience Measuring *130*
   Conclusion *133*
   References *134*

10 Unveiling Urban Resilience: Exploring the Qualities and Interconnections of Urban Systems *139*

Introduction *139*
Urban Resilience to Climate Change *140*
  Climate Change Impacts on Built Environment Systems *140*
  Temperature Rise *144*
  Sea Level Rise (SLR) *144*
  Interacting Stresses *144*
  Major Uncertainties and Interrelations *146*
Resilience Qualities *146*
  Reflectivity *146*
  Robustness *147*
  Redundancy *147*
  Flexibility *147*
  Resourcefulness *148*
  Rapidity of Recovery *148*
  Inclusivity *148*
  Integration *148*
Interrelation of Resilience Qualities *149*
Conclusion *149*
References *150*

11 Quantifying Urban Resilience: Methods and Approaches for Comprehensive Assessment *155*

Introduction *155*
Urban Resilience *156*
  Resilience Strategies *156*
  Urban and Community Resilience Assessment *157*
Resilience Assessment Approaches *159*
  Qualitative Resilience Assessment *160*
    Conceptual Frameworks *161*
    Semiquantitative Indices *163*
  Quantitative Resilience Assessment *163*
  General Resilience Approaches (Measures) *164*
    Deterministic Performance-based Approach *165*
    Probabilistic Performance-based Approach *165*
  Structural-based Models *165*
    Optimization Models *165*
    Simulation Models *165*
    Fuzzy Logic Models *166*
Frameworks and Tools for Measuring Resilience *166*
Conclusion *177*
References *177*

Part III Resilient Urban Systems: Navigating Climate Change and Enhancing Sustainability 183

12 Building Climate Resilience Through Urban Planning: Strategies, Challenges, and Opportunities 185
Introduction 185
Understanding Climate Change Impacts on Urban Areas 186
Urban Planning Strategies for Mitigating Climate Change Impacts 188
   Transit-Oriented Development (TOD) 188
   Fifteen Minutes City (FMC) 190
   Compact Cities 190
   Sustainable Land Use and Development Policies 191
   Low-Impact Development (LID) 191
   Sponge Cities 192
   Green Infrastructure and Urban Greening Initiatives for Cool Cities 193
   Waste Management and Recycling Systems, Public Participation, and Education 194
Risk Assessment and Adaptation in Urban Planning 195
Case Studies of Successful Climate-Responsive Urban Planning 200
Challenges and Opportunities 202
Major Key Points 203
Conclusion 204
References 204

13 Integrating Green–Blue–Gray Infrastructure for Sustainable Urban Flood Risk Management: Enhancing Resilience and Advantages 207
Introduction 207
   Green Infrastructure (GI) 208
   Gray Infrastructure (GRAI) 209
Green–Blue–Gray Infrastructure Combination 209
   Benefits of Combining Green–Blue–Gray Infrastructure (GBGI) Systems 209
   Green–Blue–Gray Infrastructure (GBGI) for Flood Risk Management 210
   Environmental Impacts of Floods and Green Climate Change Adaptation 210
   Regional Progress in GBGI Nexus Research 211
   Flood Risk Management Resilience 212
Conclusion 221
References 221

14 Enhancing Energy System Resilience: Navigating Climate Change and Security Challenges 227
Introduction 227
Adapting the Theory of Resilience to Energy Systems 229
Why Incorporate Resilience into Energy Systems? 234
What are the Threats to the Energy System? 235
Domains of Resilience Approaches to Energy Systems 237

Resilience Enhancement Approaches for Energy Systems 240
    System Hardening 240
    Distributed Generation 240
    Energy Storage 241
    Smart Grid Technology 241
    Enhancing Energy Efficiency 242
    Make Climate Resilience a Central Part of Energy System Planning 242
  Conclusion 243
  References 245

15 Building Resilient Health Policies: Incorporating Climate Change Impacts for Sustainable Adaptation 251
  Introduction 251
  Climate Change Impacts on Public Health 253
    Infectious Diseases 254
    Air Pollution 255
    Extreme Events 256
  Considerations in Health Policy Development 256
    Reducing Carbon Emissions 256
    Medical Interventions 257
    Healthy Lifestyle 257
    Monitoring 257
    Proactive Approaches 258
    Strengthening Institutions 258
  Conclusion 259
  References 259

16 Enhancing Resilience: Surveillance Strategies for Monitoring the Spread of Vector-Borne Diseases 263
  Introduction 263
  Vector-Borne Diseases 265
    Environmental Factors and Vector-Borne Diseases 265
    Climate Change Impacts on Vector-Borne Diseases 266
  Surveillance Strategies 266
    Monitoring of Human Cases 268
    Identification of Pathogen Species 269
    Distribution and Behavior of Vectors 269
    Climatic and Environmental Changes 270
    Control Measures 270
    Policy Development 270
  Conclusion 271
  References 271

Glossary 277
Index 281

List of Contributors

Nisreen Abuwaer
King Abdullah University of Science and Technology (KAUST)
Thuwal
Saudi Arabia

Sami G. Al-Ghamdi
King Abdullah University of Science and Technology (KAUST)
Thuwal
Saudi Arabia

Mohammed Al-Humaiqani
Hamad Bin Khalifa University (HBKU)
Doha
Qatar

Salah Basem Ajjur
Hamad Bin Khalifa University (HBKU)
Doha
Qatar

Fama N. Dieng
Hamad Bin Khalifa University (HBKU)
Doha
Qatar

Muhammad Imran Khan
Hamad Bin Khalifa University (HBKU)
Doha
Qatar

Mohammed G. Madandola
Hamad Bin Khalifa University (HBKU)
Doha
Qatar

Mehzabeen Mannan
Hamad Bin Khalifa University (HBKU)
Doha
Qatar

Mohammad Zaher Serdar
Hamad Bin Khalifa University (HBKU)
Doha
Qatar

Furqan Tahir
King Abdullah University of Science and Technology (KAUST)
Thuwal
Saudi Arabia

Safi Ullah
King Abdullah University of Science and Technology (KAUST)
Thuwal
Saudi Arabia

About the Editor

Prof. Sami G. Al-Ghamdi is a distinguished professor specializing in sustainable built environment and climate change resilient infrastructure at King Abdullah University of Science and Technology (KAUST). He earned his PhD in civil and environmental engineering from the University of Pittsburgh in 2015, an MSc in civil and construction engineering from Western Michigan University in 2010, and a BSc in architecture and building science from King Saud University in 2005. Prof. Al-Ghamdi is also a LEED-accredited professional, specializing in green building design and construction, certified by the US Green Building Council.

As a founding faculty member of KAUST's Climate and Livability Initiative (CLI), Prof. Al-Ghamdi's passion lies in conducting multidisciplinary research to develop innovative solutions that address climate change mitigation, optimize energy, water, and material consumption, and enhance the overall quality of life. His research group has focused on seven core objectives: reducing the contribution to global climate change, promoting individual human health, advocating for local sustainable and regenerative material cycles, protecting and restoring water resources, fostering a domestic green economy through technology entrepreneurship, enhancing community quality of life, and preserving and enhancing biodiversity and ecosystem services.

Demonstrating academic excellence, Prof. Al-Ghamdi boasts an impressive track record, having supervised and mentored numerous postdoctoral scholars, PhD students, and MSc students. His efforts have been recognized and rewarded, securing significant external competitive funding for research projects centered on resiliency, climate change, and related fields. Notably, he has been honored with several prestigious awards, including the Qatar Sustainability Awards in various years (research category) and the 2018 National Program for Conservation and Energy Efficiency Award for the Best Green Sustainable Initiative. Moreover, his dedication to research excellence is evident through the publication of a substantial number of refereed, Scopus-indexed papers.

Prof. Al-Ghamdi's contributions to the field have been invaluable in promoting sustainable development and advancing the domain of sustainable built environment. His commitment to innovative research and holistic approaches in tackling the challenges of climate change has had a far-reaching impact, not only in academia but also in practical applications and the wider community. As the editor of "Sustainable Cities in a Changing Climate: Enhancing Urban Resilience," Prof. Al-Ghamdi's expertise and leadership have brought together his group of experts, practitioners, and researchers to create a comprehensive and timely resource for shaping the future of resilient and sustainable cities.

Preface

Cities are the vibrant epicenters of human civilization, serving as hubs of innovation, economic growth, and cultural exchange. However, as we navigate the challenges of the twenty-first century, cities face unprecedented pressures, particularly in the face of climate change. Rising temperatures, extreme weather events, sea level rise, and other climate-related impacts pose significant threats to urban environments and the well-being of their inhabitants. To ensure a sustainable and prosperous future, it is imperative that we enhance the resilience of our cities.

"Sustainable Cities in a Changing Climate: Enhancing Urban Resilience" is a comprehensive and timely exploration of the vital intersection between urban development, climate change, and resilience. In this book, we bring together a diverse group of experts, practitioners, and researchers who share their insights, experiences, and innovative approaches to tackle the complex challenges of climate change in urban settings.

Divided into three main parts, this book takes readers on a journey through the foundations of climate change and its implications for the built environment, the quantification of resilience and its qualities, and the strategies for navigating climate change and enhancing sustainability in urban systems. It provides a holistic and multidisciplinary perspective that spans climate science, urban planning, infrastructure resilience, energy systems, healthcare, and more.

In Part I, "Climate Change and the Built Environment: Foundations and Implications," we lay the groundwork for understanding climate change and its impacts. From the basics of climate science and greenhouse gases to advancements in climate modeling and observations, readers gain a solid foundation in the science of climate change. We delve into the role of urbanization in shaping local hydroclimate and explore regional climate modeling to assess future projections. Additionally, we examine the impacts of the built environment and human developments on climate change, emphasizing the need for sustainable practices and efficient resource management.

Part II, "Quantifying Resilience and Its Qualities," delves into the assessment and enhancement of resilience in urban environments. We explore the interdependencies of critical infrastructures, ranging from transportation and water systems to buildings' structural resilience. The chapters provide insights into quantifying infrastructure and urban systems' resilience qualities, emphasizing the importance of holistic approaches and considering multiple dimensions of resilience. We also delve into the assessment methods and frameworks used to evaluate the resilience of physical infrastructures, providing valuable tools for resilience planning and decision-making.

In Part III, "Resilient Urban Systems: Navigating Climate Change and Enhancing Sustainability," we turn our attention to the critical domains of energy and healthcare systems within urban environments. These chapters focus on developing and implementing resilience strategies to mitigate climate change impacts in these key sectors. From integrating green–blue–gray infrastructure for flood risk management to enhancing energy system resilience and incorporating climate change considerations into health policies, we explore innovative approaches to enhance sustainability and adapt to a changing climate.

Throughout this book, our contributors emphasize the importance of collaboration, interdisciplinary approaches, and community engagement in building resilient cities. They highlight the need for policymakers, urban planners, researchers, and practitioners to work together to develop evidence-based solutions and implement transformative changes.

As the editor, I am deeply grateful to all the authors who have contributed their knowledge and expertise to this book. Their dedication and passion for enhancing urban resilience have made this collaborative effort possible. I also extend my gratitude to the readers for their interest in this important topic and their commitment to building sustainable cities in the face of climate change.

I hope that "Sustainable Cities in a Changing Climate: Enhancing Urban Resilience" serves as a valuable resource for all stakeholders involved in shaping the future of our cities. May this book inspire and empower readers to take action, innovate, and create urban environments that are not only resilient but also sustainable, inclusive, and equitable.

Together, let us embark on this journey toward a future where our cities thrive, adapt, and flourish in the face of a changing climate.

Prof. Sami G. Al-Ghamdi, Editor
King Abdullah University of Science and Technology (KAUST)

Abbreviations

These abbreviations provide readers with a convenient reference to commonly used terms throughout the book, allowing for easier comprehension and efficient communication of ideas.

ABM	agent-based-modelling
AHP	analytic hierarchy process
AIACC	assessments of impacts and adaptation of climate change
ASPIRE	Atlas of social protection indicators of resilience and equity
BEdZED	Beddington zero energy development
BEM	building energy model
BEP	building effect parameterization
BI	blue infrastructure
BMPs	best management practices
BRACED	building resilience and adaptation to climate extremes and disasters
BREEAM	building research establishment environmental assessment method
BRT	bus rapid transit
CBA	cost–benefit analysis
CCS	carbon capture and storage
CHP	combined heat and power
CI	critical infrastructure
CN	complex networks
CO_2	carbon dioxide
CoBRA	community-based resilience analysis
COP26	26th UN climate change Conference
COP	conference of the Parties
CP	collapse prevention
CSP	concentrated solar power
DUCT	digital urban climate twin
DRM	disaster risk management
DRR	disaster risk reduction
EAD	expected annual damage
EHEs	extreme heat events
EIA	environmental impact assessment

EIP	extrinsic incubation period
EPA	Environmental Protection Agency
EWF	energy–water–food
EWS	early warning system
FCM	fuzzy cognitive map
FMC	fifteen minutes city
FTOPSIS	fuzzy technique for order preference by similarity to ideal solution
FVI	Flood Vulnerability Index
GBGI	Green–Blue–Gray Infrastructure
GBI	Green–Blue Infrastructure
GCMs	global climate models
GDP	gross domestic product
GHG	greenhouse gas
GHGI	greenhouse gas inventory
GI	green infrastructure
GIS	Geographic Information System
GPS	Global Positioning System
GRAI	gray infrastructure
HGBGI	Hybrid Green–Blue–Gray Infrastructure
HVAC	Heating, Ventilation, And Air Conditioning
ICT	information and communication technology
IO	immediate occupancy
IPCC	Intergovernmental Panel on Climate Change
ISS	International space station
LCC	life cycle cost
LCCA	life cycle cost analysis
LCLUC	land cover and land use change
LEED	Leadership in Energy and Environmental Design
LID	low impact development
LS	life safety
MCA	multi-criteria analysis
MCDA	multi-criteria decision analysis
ME	Middle East
MENA	Middle East and North Africa
MEP	Mechanical, Electrical, and Plumping
MSE	Mega Sport Event
MSERRI	Mega Sport Event Road Resilience Index
NBS	nature-based solutions
NDCs	Nationally Determined Contributions
NGO	nongovernmental organization
PBD	Performance Based Design
PCA	Principal component analysis
PES	Payment for Ecosystem Services
PNW	Pacific Northwest
PPP	public–private partnership

PSP	participatory scenario planning
PTSD	post-traumatic stress disorder
PV	photovoltaic
R&D	research and development
RCP	representative concentration pathway
REDI	Resilience-Based Earthquake Design Initiative
RELI	Resilient Design Rating System
RRA	rapid risk assessment
RS	remote sensing
SCP	Sponge City Program
SDGs	sustainable development goals
SIA	social impact assessment
SUDS	sustainable urban drainage systems
SVI	social vulnerability index
SWMM	Storm Water Management Model
SWOT	strengths, weaknesses, opportunities, and threats
TDM	transportation demand management
TOD	transit-oriented development
TSS	total suspended solid
UCCR	urban climate change resilience
UCM	urban Canopy model
UHC	Universal health coverage
UHI	urban heat island
UN	United Nations
UNFCCC	United Nations Framework Convention on Climate Change
UNDP	United Nations Development Program
USAID	United States Agency for International Development
USGBC	U.S. Green Building Council
VBDs	vector-borne diseases
WASH	water, sanitation, and hygiene
WHO	World Health Organization
WRF	Weather Research and Forecasting
WSUD	water-sensitive urban design
WWC	Waterway Corridors

Part I

Climate Change and The Built Environment: Foundations and Implications

Part 1 of the book lays the groundwork for understanding climate change and its implications for the built environment. It begins with Chapter 1, where readers are introduced to the basics of climate change, including the role of greenhouse gases, global warming potential, and the scientific evidence of human-induced global warming. The chapter also explores natural climate variability and provides an overview of cutting-edge improvements in climate models and observations.

Chapter 2 delves into the WRF-Urban model and its significance in promoting urban resilience and sustainability. It highlights the impact of urbanization on local hydroclimate, emphasizing the need for advanced modeling techniques to improve real-time weather prediction and understand urban land surface processes.

In Chapter 3, the focus shifts to the MENA-CORDEX domain, where climate change simulations from regional climate models (RCMs) are evaluated and projected. The chapter presents a step-by-step methodology for assessing the trends in surface air temperature and precipitation, enabling a high-resolution evaluation of temperature and precipitation projections for the MENA region through the twenty-first century.

Chapter 4 examines the impacts of the built environment on climate change, emphasizing the importance of considering both embodied and operational carbon emissions. It highlights the need for sustainable practices such as utilizing low-embodied carbon materials, improving construction efficiency, and exploring carbon sequestration techniques to mitigate building-related emissions.

Finally, Chapter 5 focuses on the impact of human developments on climate change, specifically the production and consumption of energy and water and our reliance on fossil fuels. It underscores the need to use energy and water more efficiently to successfully tackle climate change issues.

Together, these chapters provide a comprehensive understanding of climate change, regional climate modeling, the role of the built environment in contributing to climate change, and the importance of sustainable practices in mitigating its effects. This knowledge sets the stage for the subsequent sections of the book, which will delve into strategies and approaches for enhancing climate resilience in the built environment.

1

Understanding Climate Change Fundamentals: Exploring the Forces Shaping Our Planet's Future

Salah Basem Ajjur[1,2] *and Sami G. Al-Ghamdi*[1,3,4]

[1] *Division of Sustainable Development, College of Science and Engineering, Hamad Bin Khalifa University, Qatar Foundation, Doha, Qatar*
[2] *Department of Earth, Environmental and Planetary Sciences, Brown University, Providence, RI, USA*
[3] *Environmental Science and Engineering Program, Biological and Environmental Science and Engineering Division, King Abdullah University of Science and Technology (KAUST), Thuwal, Saudi Arabia*
[4] *KAUST Climate and Livability Initiative, King Abdullah University of Science and Technology (KAUST), Thuwal, Saudi Arabia*

Introduction

Climate change is a global problem with severe consequences for humans and the environment. During the last decades, climatic observations and records have provided unequivocal evidence of increasing hazardous risks and tipping points that could leave future generations with a less livable planet. The implications of these tipping points include many irreversible actions for all species on Earth. Previous literature has demonstrated that poor and marginalized countries are the most affected by climatic change, which increases global inequalities [1]. Therefore, it is necessary to understand the basics of climate change and the role everyone can play in adaptation and mitigation measures. This chapter aims to provide such knowledge to researchers from outside the climate change community.

The chapter begins by giving a brief, but comprehensive definition of some specific terms, which should be interpreted in the context of this book. The terms are atmosphere, greenhouse gases (GHGs), Global Warming Potential (GWP), CO_2 equivalent (CO_2-eq) emission, aerosols, and carbon budget. Understanding these terms is essential in climate change studies. The next lines provide the scientific evidence that demonstrates the human cause of recent global warming and show the spatial distribution of global warming. The global modes of climate variability are defined after that. To include necessary knowledge in climate change studies, this chapter summarizes the cutting-edge improvements in global climate models, observations, reanalysis datasets, and relevant programs needed to visualize and process climatic data. The chapter is concluded by highlighting the main points and outlooks.

Sustainable Cities in a Changing Climate: Enhancing Urban Resilience. First Edition.
Edited by Sami G. Al-Ghamdi.
© 2024 John Wiley & Sons Ltd. Published 2024 by John Wiley & Sons Ltd.

The atmosphere is the gaseous envelope that surrounds the Earth. It has five layers: the inner to the outer, troposphere (contains 50% of the Earth's atmosphere), stratosphere, mesosphere, thermosphere, and exosphere. The atmosphere acts like a barrier to protect the Earth from harmful solar radiation and helps to keep the Earth's temperature stable. GHGs are necessary components in the atmosphere to trap heat from the sun. Without GHGs, the planet would be cold and deserted like Mars. On the other hand, with too much GHGs, the planet would be hot and lifeless like Venus. Hence, a proper amount of GHGs is needed to have an optimum climate on the Earth. Factories, vehicles, power plants, and some agricultural activities burn fossil fuels (coal, oil, natural gas, etc.) and emit substantial amounts of GHGs, making the Earth warm. The GHGs are typically emitted in smaller quantities. However, some, mainly *water vapor (H_2O), carbon dioxide (CO_2), methane (CH_4), nitrous oxide (N_2O), and fluorinated gases,*[1] are effective in thickening the Earth's blanket and changing its climate. For example, CO_2 is responsible for 80% of global temperature rise, making up only 0.04% of our atmosphere. Thus, even a small variation in CO_2 concentration can have significant consequences for the Earth.

Three questions determine the effectiveness of a GHG. These questions are (1) How much GHG is in the atmosphere? (2) How long does it remain? (3) How strongly does it affect the atmosphere? To address these questions, scientists calculate the GWP index for each GHG. The GWP index reflects the average time the gas remains in the atmosphere and causes "radiative forcing" (or heating effect) relative to CO_2. The GWP of CO_2 is 1, thus, a higher GWP means that the gas contributes more to warming the Earth. If the emission of GHG is multiplied by its GWP for a hundred-year time horizon, then one can determine the CO_2-eq emission. CO_2-eq emission denotes the amount of CO_2 emission that causes identical integrated radiative forcing (or global warming) as an emitted amount of another GHG. The metric CO_2-eq can be calculated for a mixture of GHGs by adding the CO_2-eq emissions of each GHG. CO_2-eq is commonly expressed as MMTCDE (million metric tons of carbon dioxide equivalents).

Aerosols are atmospheric particles suspended in a gas, like air, with tiny diameters (a few nanometers/micrometers). Aerosols can be suspended for several days in the troposphere and for years in the stratosphere. Aerosols in the troposphere might be natural or anthropogenic, whereas aerosols in the troposphere are generally natural (volcanic eruptions). Aerosols can scatter and absorb radiation, causing an effective radiative forcing. They also act as cloud condensation nuclei and ice nuclei, affecting the cloud's droplets, radiation, and precipitation characteristics. There are different types of aerosols, including sulfate, organic and black carbon, sea salt, and dust.

Carbon budget is the global amount of GHG emitted for a given level of global warming above a reference period. The distribution of the carbon budget to the regional level is established according to equity, cost, or efficiency considerations. Knowing the carbon budget is helpful in determining whether climate change mitigation plans are sufficient to limit global warming below a specific temperature. To curb excessive emissions and drive investment of cleaner and more efficient alternatives, such as renewable energies, climate scientists suggest putting a price on carbon emissions (carbon tax). Carbon tax is a way to

---

1 Fluorinated gases are human-made gases like hydrofluorocarbons (HFCs), perfluorocarbons (PFCs), and sulfur hexafluoride (SF_6).

raise fossil fuel prices, motivating users to switch to cleaner (non-carbon) and economically rewarding energy. A key drawback of the carbon tax is the increase in the cost of energy-related services, such as electricity, heating, and cooling, which increases the economic strain on people who are already struggling. This might be solved by giving some of the carbon tax revenue back to households.

Recent Climate Change is Anthropogenic

Although the climate has permanently changed, the phenomenon of climate change is relatively new. Humans began noticing the warming of the Earth's land surface in the 1930s. However, it was confusing as to whether to attribute this warming to a long-term trend in natural fluctuation or to recent human activity. After 1979, satellites provided a wealth of climatic data at finer resolutions that produced particular patterns of change, showing that the planet has been warming fast during the last few decades. The global air temperature has grown by at least 1 °C since the industrial revolution[2]. Two-thirds of the latter increase has happened since 1975 (see Figure 1.1). Projections show a warming trend of 20 times faster during the twenty-first century. Moreover, the observation-based datasets show that the troposphere and oceans have warmed while the stratosphere has cooled, less heat has escaped to space, ice and glacier masses have declined, sea levels have risen, extreme weather patterns have changed, and ocean salinity and acidity have increased.

Thus, climatic changes during the recent decades are a result of a surge in the concentration of GHGs. Additionally, other evidence from the past, such as fossils and sediments, bridges the gap in the evolution of Earth's climate [2]. Taken together, the evidence is overwhelming, and science confirms that recent climate change is mainly anthropogenic, i.e., caused by human activity.

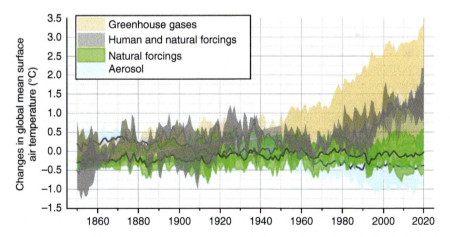

Figure 1.1 The changes in global mean surface air temperature in °C during 1850 and 2020, relative to the industrial period in HadCRUT4 observations, CMIP6 simulations of the response to greenhouse gases only (yellow band), natural forcings only (green band), aerosols only (blue band), and combined human and natural forcings (gray band). Solid colored lines present the CMIP6 multi-model mean.

---

2 The period from 1850 to 1900 is used as an approximate representation of the preindustrial period.

Figure 1.1 depicts the changes in global near-surface air temperature since 1850, compared with the industrial period, according to the sixth phase of the Coupled Model Intercomparison Project 6 (CMIP6). The CMIP6 models simulate changes due to natural forcings as well as human causes (GHG and aerosols). Natural forcings include internal climate variability related to global teleconnections, solar brightness variations, and volcanic emissions. The clearly observed global warming is simulated only when models include the human impact, particularly GHG emissions (orange band). The GHG emissions contribute substantially to such warming, partly offset by the cooling effect of increases in atmospheric aerosols (blue band). The models' simulations show that natural forcings cannot reproduce the observed global warming since these (natural) forcings simulate much smaller temperature trends (green band).

In short, only through the sustained reduction of GHG emissions can the world limit the globally averaged surface warming. This requires substantial efforts for reaching net-zero or net-negative CO_2 emissions and decreasing the net non-CO_2 forcings. Nonetheless, the resulting slowdown in warming would be masked by natural year-to-year variability, which means that mitigation benefits will take a few decades to be detected globally and regionally.

Spatial Distribution of Global Warming

Long-term global warming is not evenly distributed over the Earth [3]. Observational-based and model-based datasets substantiate this, and to prove this, we will depict the spatial pattern of near-surface air temperature at three specific levels of global warming. Figure 1.2 illustrates the spatial changes of surface warming under 1.5, 2, and 3 °C global warming levels. Generally, land areas are warmer than oceans. Very strong warming is observed over the Arctic. There is stronger warming in the Northern Hemisphere than in the Southern Hemisphere. High northern latitudes exhibit greater heat than tropical regions. Figure 1.2 c shows regional changes between global land areas under 3 °C global warming. Global hotspots for surface warming are the Russian Arctic, Central Asia, the Tibetan Plateau, the Middle East and North Africa, and North America.

The spatial patterns of other climatic parameters, e.g., precipitation, snow, and sea ice cover, may not be robust because changes in precipitation depend on several complex factors such as global mean temperature, aerosol emissions, and land use variations. Additionally, establishing a robust spatial pattern for snow and ice cover changes is difficult, as snow/ice vanishes completely if a certain temperature threshold is reached.

Modes of Climate Variability

They are also called "regimes" and "teleconnections." Teleconnections refer to modes of natural climate variability that link weather changes in widely separated points of the globe. Teleconnections are triggered by large changes in air movement around the atmosphere and occur over months and longer timescales. For additional details about the global modes of climate variability, the reader is directed to the NOAA website.

The two dominant annular modes that describe the total variation in the extratropical atmospheric flow are the Northern Hemisphere Annular Mode (NAM) and the Southern

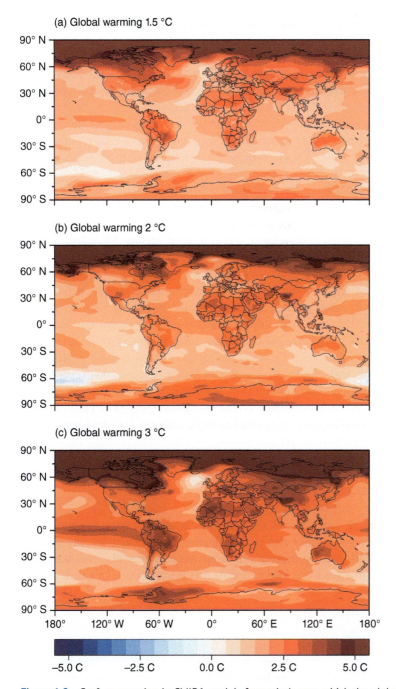

Figure 1.2 Surface warming in CMIP6 models for periods over which the global average near-surface warming is (a) 1.5 °C, (b) 2 °C, and (c) 3 °C, relative to the industrial period (1850–1900).

Hemisphere Annular Mode (SAM). The NAM, also known as the Arctic Oscillation (AO), is a winter fluctuation in the amplitude of a pattern described by low surface pressure in the Arctic and mid-latitude solid westerlies. Another mode of climate variability that has a strong correlation with the NAM is the North Atlantic Oscillation (NAO). The NAO is provided by the difference between sea-level pressure in Iceland and the Azores. The NAM and NAO are available from 1950 to the present. The SAM, also called the Antarctic Oscillation (AAO) or the High Latitude Mode, is the primary teleconnection of the Southern Hemisphere sea-level pressure and geopotential height and is related to changes in the latitude of the midlatitude jet. The SAM is a station-based index on monthly, seasonal, and annual time steps. It has been available since 1957.

Two key teleconnections that describe the climate variability of the tropical Atlantic on a year-to-year timescale are the Atlantic equatorial (or zonal) mode and the Atlantic meridional mode. The Atlantic zonal mode is associated with sea surface temperature anomalies near the equator. It peaks in the eastern basin. It is also called Atlantic Niño. The Atlantic meridional mode is characterized by an inter-hemispheric gradient of SST and wind anomalies.

The El Niño-Southern Oscillation (ENSO) is a coupled atmosphere–ocean phenomenon with time scales of two to seven years. It is often measured by the Southern Oscillation Index (SOI): the surface pressure anomaly difference between Tahiti and Darwin, Australia. The SOI has been available since 1951. The ENSO can also be measured by the sea surface temperatures in the equatorial Pacific Ocean (Niño regions). The long-lived El Niño-like pattern of Pacific climate variability is known as the Pacific Decadal Oscillation (PDO) index [4]. It has been available since January 1854. A broadened PDO index that occurs in the whole Pacific Basin is the Inter-decadal Pacific Oscillation (IPO). The PDO and IPO display temporal evolution. The PDO data go back to 1854, while the PDO data extend back to 1900.

Find, Read, and Process Climatic Data

Climatic data are available from several resources in different formats. The next lines show the primary sources of climate models, observations, and reanalysis data, respectively. The text then illustrates the development and characteristics of future scenarios and pathways and summarizes some essential tools that are necessary for handling and processing climatic data.

Climate Models (GCMs and RCMs)

General Circulation Models (GCMs) are time-dependent models representing atmosphere, ocean, and land interactions with discrete grid points distributed over the Earth. The GCMs use physics laws such as thermodynamics, fluid mechanics, momentum, continuity, and radiation. During the last half-century, humans have invented powerful computers that are able to run complex computer codes. These recent advances in technology have allowed GCM simulations to greatly improve. Scientists used advanced instruments to measure climatic parameters, e.g., temperature, precipitation, cloudiness, humidity, sea level, and wind. Climate scientists incorporated many more natural processes and provided much more data on the interaction between the atmosphere and land surfaces, ice, vegetation, and oceans. Thus, today we have evidence-based simulations on most aspects of the climate system. We can determine when the gigantic ice sheets of Greenland and Antarctica experience high melt rates, leading to major changes such as sea-level rise. Simulations are not restricted to oceans and land surfaces but include many other variables such as ice, snow, vegetation, and land use. Some climate models can simulate up to 50 km high in the atmosphere, enabling a better understanding of the processes between atmospheric layers. They can simulate eddies at a 100-km scale, improving our experience of heat transport in oceans. They can also determine the amount of CO_2 absorbed by plants and oceans under various climate and environmental changes.

The improvements in climate models become apparent when comparing most climatic variable simulations with observations. Consequently, climate models decrease spaces between grid points, producing higher (finer) resolutions. Nevertheless, the horizontal resolution in GCMs still cannot identify local weather conditions in many regions. For instance, urban development and the interactions between sea and land are main local forcings, but not easily captured through several GCMs [5, 6].

To overcome the issue of large horizontal resolution, two main downscaling types are used. The first is the statistical method, where empirical links are developed between atmospheric conditions at the grid points of the GCMs and the observed conditions of the specified domain. The second is the dynamic method, where outputs from GCMs force Regional Climate Models (RCMs) with higher resolutions to characterize the physical and dynamic features of a particular domain. Other types, like the weather regime method, where the outputs of the GCMs are categorized as a restricted number of weather regimes or analogs, are less common. Each method has its benefits and drawbacks [7–9]; however, the dynamic and weather regime methods consume time and money, as they require complex computations [8]. In contrast, the statistical approach mitigates systematic biases in the GCMs, improves spatial details, and generates variables that are not explicitly rendered by GCMs [8]. The results from the statistical method are reliable if an appropriate methodology with high-quality observations is applied [6, 8].

RCMs are climatic models forced by specific conditions from GCMs or observation-based reanalysis at finer resolutions over a specific area domain. The Coordinated Regional Climate Downscaling Experiment (CORDEX; www.cordex.org) involves the participation of 16 RCMs (according to the ESGF). These RCMs are COSMO-CLM, WRF, ALADIN, ALARO-0, CCAM, CRCM, ETa, HadGEM3-RA, HadRM3P, HIRHAM5, MAR36, RACMO, RCA, RegCM4, REMO, and RRCM. The COSMO-CLM and WRF models have become community models due to the interests, contributions, and long-term development of an international user base. Several climatic models participating in the CMIP5 were used as boundary conditions for different domains in CORDEX regional simulations. For instance, the Africa domain has one simulation available for the historical experiment, one simulation for the RCP4.5 experiment, and one simulation for the RCP8.5 experiment. The CanESM2-r0i0p0 was used as a boundary condition in the latter experiments. Similar simulations for these experiments are found under the boundary conditions of the CSIRO-Mk3-6-0-r0i0p0, EC-EARTH-r1i1p1, EC-EARTH-r3i1p1, GFDL-ESM2M-r0i0p0, IPSL-CM5A-MR-r0i0p0, and NorESM1-M-r1i1p1 models. The CORDEX has 14 domains with a prioritized horizontal grid resolution of 0.44° (~50 km). These domains are listed and described on the CORDEX website, https://cordex.org/domains/. The MENA domain has a higher resolution of 0.22° (~25 km). Two other domains (NAM and EURO) have higher resolutions of 0.11° (~12 km) and 0.22°.

Pathways and Scenarios

To better find, understand, and document climatic data, the CMIP has coordinated the design and distribution of climate model experiments from multiple international institutions and modeling groups. The CMIP began in 1995 as an initiative of the World Climate Research Programme (WCRP). It started with 10 climate centers carrying out global climate simulations. The number has increased to 17 climate centers in CMIP5 and 26 registered for CMIP6 [10]. Over time, experiments have been developed, and models run at higher resolutions, including integrations using idealized forcings. The latest phase of CMIP, i.e., CMIP6, has historical simulations (from 1850 to near-present), common experiments, the Diagnostic, Evaluation and Characterization of Klima (DECK), and an ensemble of CMIP-Endorsed Model Intercomparison Projects (MIPs).

Future simulations are represented in climate models in many forms. The CMIP3 multi-model dataset [11] includes future projections using the Special Report on Emissions Scenarios (SRES). The CMIP5 [12] uses the Representative Concentration Pathways (RCPs) to simulate the future climate [13]. Four RCPs are labeled by their projected radiative forcing values reached in the year 2100. These are RCP2.6, RCP4.5, RCP6.0, and RCP8.5. The radiative forcing in RCP2.6 peaks at ~3 Watts per square meter (Wm^{-2}) and then declines to 2.6 Wm^{-2} at the end of the twenty-first century. The RCP4.5 and RCP6.0 are intermediate pathways in which radiative forcing is limited at ~4.5 and ~6.0 Wm^{-2} in 2100, respectively. The RCP8.5 is the high pathway that leads to more than 8.5 Wm^{-2} radiative forcing in 2100. The RCPs in the CMIP5 models are purposefully separated from the socioeconomic drivers.

The latest CMIP simulations (CMIP6) draw on the Shared Socioeconomic Pathways (SSPs) [14] to complement the previous RCPs. Five SSPs establish a matrix of global forcing levels and socioeconomic storylines. Two abbreviations label these SSPs. The first abbreviation indicates five socioeconomic scenario families. These are SSP1 for sustainable pathways, SSP2 for middle-of-the-road, SSP3 for regional rivalry, and SSP5 for fossil-fuel-rich development. The second label (SSP1-1.9, SSP1-2.6, SSP2-4.5, SSP3-7.0, and SSP5-8.5) denotes the approximate level of global radiative forcing resulting from the scenario and reached by 2100. For example, the SSP2-4.5 comprises a radiative forcing level of 4.5 Wm^{-2} in 2100, consistent with the SSP2, where socioeconomic and technical trends do not shift their historical patterns significantly. The SSP5-8.5 comprises a radiative forcing level of 8.5 Wm^{-2} in 2100, consistent with the SSP5, where mitigating the unconstrained economic growth and energy use is very difficult.

Observations and Reanalysis

Observations and reanalysis are prevalent sources for monitoring historical climatic changes [15, 16]. Observational platforms include measurements of the land and ocean surfaces, upper-atmospheric observations such as aircraft, satellite-based retrievals, and paleoclimatic records. The observational datasets may not completely cover the global land because these observations are generated by gridding records at station locations onto an international grid. On the other hand, the reanalysis uses frozen data assimilation models to reanalyze archived observations and create a global dataset that describes the history of the atmospheric temperature and wind, land surface, and oceanographic temperature and currents. A wide range of observations are available with different spatial and temporal coverage. Table 1.1 lists information on some reanalysis datasets.

The reanalysis had some limitations. They may include several derived fields such as soil moisture over land and heating without direct observations. Reanalyses yield data that are as temporally homogeneous as possible. Additionally, the varying mix of observations as well as biases in observations and models may lead to false reanalysis output. Observational constraints may substantially differ according to the location, time period,

Table 1.1 Some of the reanalyses that have global coverage. All these reanalyses have sub-daily, daily, and monthly timesteps.

Name	Institution	Time coverage	Website
CERA-20C: Coupled Ocean-Atmosphere Reanalysis of the Twentieth Century	ECMWF	01/1901 to 12/2010	https://www.ecmwf.int/en/forecasts/datasets/reanalysis-datasets/cera-20c
Climate Forecast System Reanalysis (CFSR)	NOAA, NCEP	- CFSR is available from 1979/01 to 2011/03 - CFSv2 Operational Analysis is available from 2011/04 to the present	https://www.ncdc.noaa.gov/data-access/model-data/model-datasets/climate-forecast-system-version2-cfsv2
ERA5 atmospheric reanalysis	ECMWF	1979/01 to the present	https://www.ecmwf.int/en/forecasts/datasets/reanalysis-datasets/era5
JRA-55	Japanese Meteorological Agency	1957/12 to the present	https://jra.kishou.go.jp/JRA-55/index_en.html
NASA MERRA	National Aeronautics and Space Administration (NASA)	- MERRA1 is available from 1979/01 to 2016/02 - MERRA2 is available from 1980/01 to the present	https://gmao.gsfc.nasa.gov/reanalysis/
NCEP-DOE Reanalysis 2	NOAA, NCEP	1979/01 to the present	https://psl.noaa.gov/data/gridded/data.ncep.reanalysis2.html
NOAA-CIRES Twentieth Century Reanalysis (V2c)	NOAA/ESRL/PSL, CIRES CDC	1851/01 to 2014/12	https://psl.noaa.gov/data/gridded/data.20thC_ReanV2c.html

and variable considered. This variation affects the reanalysis' reliability. Therefore, global reanalysis still suffers from some biases. The reanalysis bias is apparent when simulating extreme climates.

Figure 1.3 compares maximum surface air temperature in 2010 in two reanalysis products (ERA5 [17] and JRA-55 [18]) with HadEX3 observations [19]. The figure also depicts the variation between both reanalysis datasets when generating the maximum surface air temperature in 2010. The ERA5 reanalysis was generated by the European Centre for Medium-Range Weather Forecasts (ECMWF). It is available on sub-daily, daily, and monthly timesteps from 1 January 1950 to the present on TL639(~31 km) spatial resolution. The JRA-55 reanalysis was generated by the Japan Meteorological Agency. It is available on sub-daily and monthly timesteps from 1958 to the present at TL319 (~55 km) resolution. The HadEX3 is an updated gridded observation generated by the Met Office Hadley Centre (UK). HadEX3 derived extreme indices from daily in situ records from 1901 to 2018 with a spatial resolution of 1.875°×1.25° longitude–latitude. The HadEX3 has no complete global land coverage.

Both reanalyses had similar simulations in 2010, as shown by Figure 1.3a,b. They tend to overestimate the maximum surface air temperature over the MENA, northern and eastern Australia, and South American Monsoon. At the same time, they underestimate the Tibetan Plateau and Iceland. The ERA5 and JRA55 reanalyses in other global regions are close to HadEX3 observations. Figure 1.3c depicts the difference between ERA5 and JRA55 datasets when generating the maximum surface air temperature in 2010. The ERA5 had larger values than JRA55 over West Siberia, central North America, northern South America, and East Australia. On the other hand, JRA55 had a larger reanalysis than ERA5 over other global land regions. According to Antarctica, the ERA5 clearly had a larger maximum near-surface air temperature than the JRA55.

Visualizing and Processing Climatic Data

Climatic data such as temperature, humidity, pressure, and wind are mainly produced in two formats: GRIB (GRIdded Binary or General Regularly distributed Information in Binary form) and netCDF (network Common Data Form). The GRIB format has been developed by the World Meteorological Organization (WMO) in three editions: 0, 1, and 2. GRIB2 is the latest format with some improvements in file compression and the inclusion of the parameter table that is required to unpack the data. Usually, one step is required to decode the GRIB file. The data, then, can be visualized and used as inputs for further applications. Data records of the GRIB file start with a header, followed by packed binary data. The header comprises unsigned 8-bit numbers and contains information about the field, level, time of production and forecasting, geographical location of the grid, and so on. In the netCDF format, data are stored in array form. A typical netCDF format should contain dimension names and sizes, the variables on the file (including the temporal/spatial coordinates), and the attributes (file contents). There are two versions of netCDF: netCDF-3 (also called the classic model) and netCDF-4. The netCDF-4 format supports compression, string variables, and parallel processing of large datasets. In comparison, the use of the netCDF-3 design is limited to smaller datasets with less complicated grids.

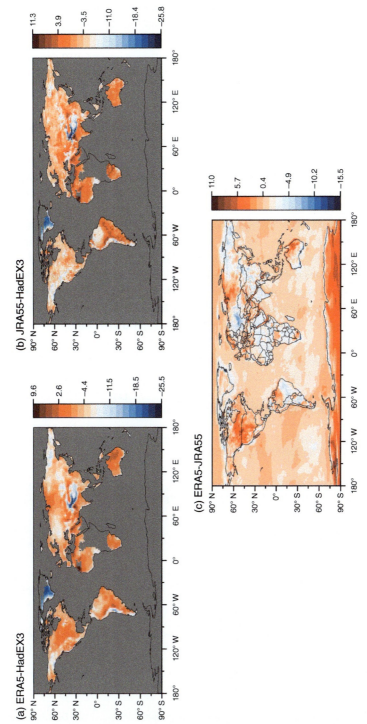

Figure 1.3 The differences in maximum surface air temperature for 2010 (a) between ERA5 and HadEX3, (b) between JRA55 and hadEX3, and (c) between ERA5 and JRA55 datasets. Gray colors in (a) and (b) indicate missing values in the hadEX3 dataset.

Several tools, freely available and commercially licensed, can visualize and process GRIB and netCDF files. Table 1.2 presents a list of free tools and provides helpful information on them. Some of these tools involve built-in statistical and arithmetic functions and support, producing high-quality figures. The ncl_convert2nc tool can convert GRIB and HDF files to netCDF format. Using the NASA tool (i.e., Panoply) is straightforward, but it requires installing a recent Java version. By implementing the CDO software, users can subsample data and spatially interpolate datasets. For further details on manipulating and displaying GRIB and netCDF, the University Corporation for Atmospheric Research (UCAR) summarized a list of software (Link: https://www.unidata.ucar.edu/software/netcdf/software.html).

The GCMs have different spatial resolutions, depending on the computational performance of each model and how it addresses pole singularities and physical constraints. Therefore, one may need to regrid (interpolate) these models to a typical resolution to evaluate models' data on different grids quantitively or get the multi-model ensemble (MME). There are several types of regridding[3]: bilinear (most common), nearest neighbor, spline, inverse distance, binning, spectral, and triangulation. Selecting the appropriate regridding method is challenging as it depends on the intended task; however, it should be noted that regridding does not provide any additional information from that of the original grid. Instead, using an inappropriate regridding method can lead to misleading outcomes. Climatic data are mostly georeferenced on a sphere. Regridding, therefore, should address pole singularities and the convergence of the longitude meridians issues.

Table 1.2 Freely available tools that are used to visualize and process climatic data.

Tool	File types supported	Institution name	Last version	Web page
NCAR Command Language (NCL)	GRIB, netCDF, HDF, HDF-EOS, shapefile, ASCII, binary	NCAR, Boulder, Colorado	6.6.2	http://www.ncl.ucar.edu/
Grid Analysis and Display System (GrADS)	GRIB, netCDF, HDF, Binary, and BUFR	George Mason University, Virginia	2.2.1	http://cola.gmu.edu/grads/grads.php
Panoply	GRIB, netCDF, and HDF	NASA	4.12.8	https://www.giss.nasa.gov/tools/panoply/
wgrib	GRIB	NCEP, NOAA	1.8	https://www.cpc.ncep.noaa.gov/products/wesley/wgrib.html
Climate Data Operators (CDO)	GRIB, netCDF, SERVICE, EXTRA and IEG	Max-Planck-Institut für Meteorologie, Germany	1.9.10	https://code.mpimet.mpg.de/projects/cdo

---

3 Here, we are interested in the rectilinear regridding where grids are described by 1-D latitude and longitude coordinates. Other regridding types, such as curvilinear and unstructured, are out of the scope of this chapter.

Conclusion

This chapter starts by defining some common terms in the climate change arena and concludes with the following points:

1) Human-induced climate change has catapulted our planet into a risky state that humans and ecosystems have never experienced before. Every day, people lose their lives due to extreme weather events, their houses due to sea-level rise and floods, and their crops due to drought. The implications of climate change include many irreversible actions of life. Therefore, the world should move toward a sustained reduction of GHG emissions. The benefits of such reductions will be evident in a few decades.
2) Climate change is not evenly distributed over the Earth. Land areas are warmed more than oceans, the Northern Hemisphere is heated more than the Southern Hemisphere, and extreme warming is observed over the Arctic.
3) Compared to previous generations, the latest advancements in climate models have improved their representations of physical and biogeochemical processes. Observations and reanalyses are powerful tools to monitor historical climatic changes; however, reanalyses have some limitations, especially when simulating extreme climates.

Although climate change data are readily available to the public, their explanation still needs to be more clearly communicated. To illustrate, previous studies documented a decrease of 0.1 in the pH of the surface ocean after the industrial era [1]. The reader may underestimate this decrease as it may seem like a slight change; however, it represents a 26% increase in ocean acidity.

References

1 Pachauri, R.K., Allen, M.R., Barros, V.R. et al. (2014). *Climate Change 2014: Synthesis Report. Contribution of Working Groups I, II and III to the Fifth Assessment Report of the Intergovernmental Panel on Climate Change*. [Core Writing Team (ed. R.K. Pachauri and L.A. Meyer), 151. Geneva, Switzerland: IPCC.
2 IPCC (2013). *Climate Change 2013: The Physical Science Basis: Contribution of Working Group I Contribution to the Fifth Assessment Report of the Intergovernmental Panel on Climate Change* (ed. T.F. Stocker, D. Qin, G.-K. Plattner, et al.), 1535. Cambridge, United Kingdom and New York, NY, USA: Cambridge University Press.
3 Ajjur, S.B. and Al-Ghamdi, S.G. (2021). Global hotspots for future absolute temperature extremes from CMIP6 models. *Earth Space Sci.* 8: e2021EA001817.
4 Zhang, Y., Wallace, J.M., and Battisti, D.S. (1997). ENSO-like interdecadal variability: 1900–93. *J. Climate* 10 (5): 1004–1020.
5 Giorgi, F. and Gutowski, W.J. (2015). Regional dynamical downscaling and the CORDEX initiative. *Ann. Rev. Environ. Resour.* 40 (1): 467–490.
6 Maraun, D. and Widmann, M. (2018). *Statistical Downscaling and Bias Correction for Climate Research*. Cambridge: Cambridge University Press.
7 Maraun, D., Widmann, M., Gutiérrez, J.M. et al. (2015). VALUE: A framework to validate downscaling approaches for climate change studies. *Earth's Future* 3 (1): 1–14.

8 Nath, M.J., Dixon, K.W., Lanzante, J.R. et al. (2018). Some pitfalls in statistical downscaling of future climate. *Bull. Am. Meteorol. Soc.* 99 (4): 791–803.

9 Warner, T.T. (2011). Quality assurance in atmospheric modeling. *Bull. Am. Meteorol. Soc.* 92 (12): 1601–1610.

10 Eyring, V., Bony, S., Meehl, G.A. et al. (2016). Overview of the coupled model intercomparison project phase 6 (CMIP6) experimental design and organization. *Geosci. Model Dev.* 9 (5): 1937–1958.

11 Meehl, G.A., Covey, C., Delworth, T. et al. (2007). The WCRP CMIP3 multimodel dataset: a new era in climate change research. *Bull. Am. Meteorol. Soc.* 88 (9): 1383–1394.

12 Taylor, K.E., Stouffer, R.J., and Meehl, G.A. (2012). An overview of CMIP5 and the experiment design. *Bull. Am. Meteorol. Soc.* 93 (4): 485–498.

13 Van Vuuren, D.P., Edmonds, J., Kainuma, M. et al. (2011). The representative concentration pathways: an overview. *Clim. Change* 109 (1–2): 5–31.

14 O'Neill, B.C., Tebaldi, C., van Vuuren, D.P. et al. (2016). The scenario model intercomparison project (ScenarioMIP) for CMIP6. *Geosci. Model Dev.* 9 (9): 3461–3482.

15 Ajjur, S.B. and Al-Ghamdi, S.G. (2021). Seventy-year disruption of seasons characteristics in the Arabian Peninsula. *Int. J. Climatol.* 41 (13): 1–18.

16 Ajjur, S.B. and Al-Ghamdi, S.G. (2021). Evapotranspiration and water availability response to climate change in the Middle East and North Africa. *Clim. Change* 166 (28): 1–19.

17 Hersbach, H., Bell, B., Berrisford, P. et al. (2020). The ERA5 global reanalysis. *Q. J. R. Meteorol. Soc.* 146 (730): 1999–2049.

18 Kobayashi, S., Ota, Y., Harada, Y. et al. (2015). The JRA-55 reanalysis: general specifications and basic characteristics. *J. Meteorol. Soc. Jpn. Ser. II* 93 (1): 5–48.

19 Dunn, R.J.H., Alexander, L.V., Donat, M.G. et al. (2020). Development of an updated global land in situ-based data set of temperature and precipitation extremes: HadEX3. *J. Geophy. Res. Atmos.* 125 (16): e2019JD032263.

2

Advancing Urban Resilience and Sustainability Through the WRF-Urban Model: Bridging Numerical Modeling and Real-World Applications

Safi Ullah[1,2] and Sami G. Al-Ghamdi[1,2]

[1] Environmental Science and Engineering Program, Biological and Environmental Science and Engineering Division, King Abdullah University of Science and Technology (KAUST), Thuwal, Saudi Arabia
[2] KAUST Climate and Livability Initiative, King Abdullah University of Science and Technology (KAUST), Thuwal, Saudi Arabia

Introduction

Urban sustainability refers to the ability of cities to meet the needs of current and future generations while maintaining the health and productivity of the urban environment [1, 2]. Urban resilience, on the other hand, is the ability of cities to cope with and recover from climate-related shocks and stresses [3–5]. Urban sustainability and resilience are essential aspects of creating livable, healthy, and prosperous cities. As climate change poses serious threats to the world, urban areas are becoming substantially vulnerable to extreme weather events such as floods, droughts, heat waves, storms, and sea level rise [6–8]. In this context, numerical modeling has become a critical tool for assessing and quantifying the potential impacts of climate change on urban systems and designing sustainable and resilient urban environments. Since more than half of the world's population lives in cities, it is highly critical to adopt sustainable measures to ensure the well-being of urban communities and reduce the adverse impacts of climate change on urban environments.

Numerical modeling refers to the use of mathematical models to simulate and analyze complex systems such as urban systems [9]. These high-resolution models can be used to understand the behavior of different components of the urban environment, including buildings, transportation systems, energy systems, and water and drainage systems [10]. In addition, numerical models can be used to assess and predict the impacts of climate change on urban systems, such as changes in temperature, precipitation, and sea levels [9, 11]. The need for numerical modeling in urban sustainability and resilience arises from the fact that cities are complex and dynamic that interact with their environment in multiple ways [12]. To design effective strategies for urban sustainability and resilience, it is essential to have a comprehensive understanding of the interactions between different

---

Sustainable Cities in a Changing Climate: Enhancing Urban Resilience. First Edition.
Edited by Sami G. Al-Ghamdi.
© 2024 John Wiley & Sons Ltd. Published 2024 by John Wiley & Sons Ltd.

components of the urban environment and their responses to different environmental conditions [13, 14]. The outcomes of numerical modeling allow urban planners and policy-makers to explore different scenarios and evaluate the effectiveness of different interventions before implementing them in the real world.

One of the most significant benefits of numerical modeling is that it can help identify potential risks and vulnerabilities of urban systems to climate change [15]. For example, numerical models can be used to simulate the impacts of climate extremes on urban areas and identify the most vulnerable areas that are likely to be affected by such extremes. The information obtained from such assessments can be used to devise effective strategies to reduce the potential risks of climate change, ensure sustainable urban planning, and design emergency response and public health interventions, which would eventually increase the resilience of urban areas to climate change. Numerical modeling can also be used to design sustainable and resilient green infrastructure in urban areas [16, 17]. For instance, numerical models can be used to optimize the design of buildings for energy efficiency and to identify the most appropriate renewable energy sources for different urban areas [18–21]. Numerical modeling can also be used to optimize the design of transportation systems to reduce emissions and improve air quality in urban areas [22, 23]. Furthermore, numerical modeling can predict the effects of climate change on the water sector by assessing the role of climatic factors in affecting the quantity and quality of water in urban areas [24]. This can inform decision-makers about sustainable water management, including the design of water supply and treatment systems. In addition, the high-resolution numerical models can simulate the spread of waterborne diseases in urban areas, considering local factors such as water quality, temperature, and population density [25]. The information can be used to adopt effective public health interventions, such as the distribution of clean water and sanitation infrastructure.

Overall, numerical modeling is an essential tool for urban sustainability and resilience in a changing climate. It provides a means to understand the complex interactions between different components of the urban environment and to design effective strategies for reducing risks and vulnerabilities to climate change. As cities continue to grow and face the impacts of climate change, numerical modeling will become an increasingly important tool for creating livable, sustainable, and resilient urban environments.

Nexus Between Urbanization and Climate Change

Urbanization and climate change are two significant global phenomena that are intertwined and interdependent. The nexus between urbanization and climate change is complex and has far-reaching implications for sustainable development and the well-being of people living in urban areas. The rapid growth of cities in recent years has led to an increase in greenhouse gas emissions, which are the main driver of climate change [26–28]. The transportation sector, which is a significant contributor to greenhouse gas emissions, is one of the main drivers of urbanization, as people migrate to cities in search of better economic opportunities [29]. The construction and maintenance of buildings and infrastructure in urban areas also contribute significantly to greenhouse gas emissions, as they require large amounts of energy to build and operate [30]. Urbanization also exacerbates the impacts of climate change on cities. The concentration of people and infrastructure in urban areas makes them more vulnerable to extreme weather events, such as floods, storms, and heat waves. For instance, the urban heat island (UHI) effect, in which urban areas are significantly warmer than surrounding rural areas due to the absorption and retention of heat by buildings and pavement, can lead to increased heat-related illnesses and mortality [31].

However, urbanization can also provide opportunities for addressing climate change. Urban areas can be designed to reduce greenhouse gas emissions and enhance resilience to climate change impacts. For example, urban planning can promote compact, walkable neighborhoods that reduce the need for automobile use and encourage the use of public transit, cycling, and walking [32]. Buildings can be designed to be more energy-efficient and incorporate renewable energy sources, such as solar and wind power [33, 34]. Urban green spaces, such as parks, green roofs, and green walls, can help mitigate the UHI effect and provide ecosystem services such as carbon sequestration and stormwater management [17].

To effectively address the nexus between urbanization and climate change, there is a need for integrated approaches that consider both the socio-economic and eco-environmental aspects of urbanization. These approaches should involve the adoption of sustainable and low-carbon urban development strategies, such as green infrastructure, low-carbon transport systems, and energy-efficient buildings. A multi-stakeholder, multi-disciplinary, and multi-sectoral approach is needed to address the complex challenges posed by urbanization and climate change and ensure livable, prosperous, sustainable, and resilient urban areas.

Urban Modeling Through WRF-Urban Model

The Weather Research and Forecasting Urban (WRF-Urban) model is a state-of-the-art numerical model used for predicting and simulating urban atmospheric processes [35]. It is a specialized version of the WRF model designed to simulate the interactions between the atmosphere and the urban environment. The WRF-Urban model was developed by a collaboration between the National Center for Atmospheric Research (NCAR), the Environmental Protection Agency (EPA), and other research organizations in 2007 [35]. Since then, it has been included with more advanced representations of urban features, such as urban land use, surface characteristics, street canyons, building morphology, and building effects. The model can provide more accurate and realistic forecasts of weather phenomena and atmospheric processes within cities, such as the UHI effect, urban boundary layer dynamics, air pollution, and the impact of buildings on local winds [36]. The real-time implementation of the WRF-Urban model is useful for metropolitan cities, which are characterized by complex urban environments and are highly vulnerable to extreme weather events [37].

The model is going through continuous developments and improvements. Recent advancements include incorporating machine learning techniques to improve model performance and incorporating land use changes to assess the impacts of urbanization on climate [38, 39]. The continued development of WRF-Urban model will be crucial for understanding and mitigating the impacts of urbanization on local and regional weather and air quality. By improving the accuracy of weather forecasts, the WRF-Urban model can help urban planners, environmental experts, and emergency responders make more informed decisions to protect public safety and reduce the risk of damage and loss.

Overview of the WRF-Urban Model

The WRF-Urban model is a three-dimensional model that simulates the atmospheric conditions in and around urban areas [35]. It takes into account the effects of buildings, streets, and other urban features on the flow of air, temperature, humidity, and other atmospheric variables [39]. The model is based on a grid system that divides the urban environment into smaller grid cells. Each grid cell represents a small portion of the urban environment, and the model simulates the atmospheric conditions within each cell. The size of the grid cells can be adjusted depending on the level of detail required for the simulation. The WRF-Urban model uses a variety of data sources to simulate atmospheric conditions, including surface observations, satellite data, reanalysis products, and other model outputs [40]. The model also incorporates detailed information about the urban environment, including the location and characteristics of buildings, streets, and other urban features [36].

Applications of the WRF-Urban Model

The WRF-Urban model has a wide range of applications in atmospheric science, urban planning, and environmental management. Some of the key applications of *the model* include:

Urban weather forecasting: The WRF-Urban model can be used to provide detailed and accurate weather forecasts specifically tailored for urban areas. It considers the influence of urban surfaces, buildings, and anthropogenic heat sources on weather patterns, allowing for improved predictions of temperature, wind, precipitation, and other meteorological variables within cities.

Urban air quality forecasting: The WRF-Urban model can be used to predict the concentration and distribution of air pollutants in urban areas. This information can be used to issue air quality warnings and guide efforts to reduce air pollution.

Urban heat island (UHI) effect studies: The WRF-Urban model can be used to study the effects of UHI, which are areas of higher temperatures within urban areas. By simulating the atmospheric conditions in and around urban areas, the model can help researchers understand the causes and effects of UHI and develop strategies to mitigate their effects.

Urban climate change studies: The WRF-Urban model can be used to study the effects of climate change on urban environments. By simulating future atmospheric conditions, the model can help researchers understand how cities will be affected by climate change and extremes and how they can prepare for these changes and extremes.

Urban hydrological modeling: The WRF-Urban model can be employed to simulate urban hydrological processes, including rainfall–runoff interactions, drainage systems, and flooding in urban areas. It assesses the impact of land use changes, stormwater management strategies, and climate change on urban water resources, allowing for effective urban water management and urban flood risk mitigation.

- *Urban energy modeling:* The WRF-Urban model allows for the evaluation of energy consumption and energy demand within urban areas. It takes into account the effects of building materials, urban geometry, and vegetation on energy fluxes, allowing for the accurate assessment of urban energy-efficient measures, such as green roofs/walls, cool pavements, and building design modifications.
- *Urban land use planning:* The WRF-Urban model can be used to evaluate the impact of new buildings, roads, and other infrastructure on the urban environment. By simulating the atmospheric conditions before and after the construction of new developments, the model can help planners make informed decisions about the design and location of new infrastructure.
- *Urban emergency response planning and management:* The WRF-Urban model provides valuable information for emergency response planning in urban areas. It can assist in predicting and understanding the dispersion of hazardous substances in the event of accidental or intentional releases, helping emergency management authorities make informed decisions regarding safe evacuation routes, sheltering, and resource allocation for an effective emergency response.

Relevant Case Studies

The WRF-Urban model has been widely used for numerical modeling in urban environments around the world. Here, we listed three case studies of the WRF-Urban model to understand its efficacy in representing the interactions between urban features and meteorology.

Case Study 1: Urban Climate Modeling in Singapore Using WRF-Urban

In a recent study, Singh et al. [41] used the WRF-Urban model to simulate the urban climate in Singapore and evaluate the impacts of different urbanization scenarios on the UHI effect and local climate. The simulation was run for the period from January to December 2019, and the model was calibrated and validated using observations from several weather stations in the study area. The simulation results showed that the WRF-Urban model could capture the spatial and temporal variations of the UHI effect in Singapore, with the strongest effect observed in the downtown area during the nighttime. Increasing vegetation cover and water bodies were found to be the most effective strategies for mitigating the UHI effect and improving the local climate. The study provides useful findings and implications for urban planners and policymakers in designing sustainable and resilient cities.

Case Study 2: Summertime Air Conditioning Electric Loads Modeling in Beijing, China, Using WRF-Urban

Xu et al. [42] conducted a study on the impacts of summertime air conditioning (AC) electric loads on urban weather in Beijing using the WRF-Urban model. The results revealed that the AC electric loads contribute significantly to the UHI effect and the formation of atmospheric pollutants. The simulations showed that the AC electric loads increased the near-surface temperature by up to 1.5 °C, which resulted in a higher surface energy balance and increased atmospheric instability. The increased atmospheric instability led to stronger convective mixing and higher concentrations of pollutants near the surface. Furthermore, the study also demonstrated that reducing AC electric loads by 20% could lead to a decrease in near-surface temperature by up to 0.3 °C and a decrease in the concentration of pollutants by up to 10%. Overall, this study highlights the importance of considering the impacts of AC electric loads on urban weather and the potential benefits of reducing these loads to mitigate the UHI effect and air pollution. The reported results could be useful for policymakers and city planners in developing strategies to reduce energy consumption from ACs and improve urban air quality and livability.

Case Study 3: Coastal-Urban Meteorology Study in the Metropolitan Region of Vitória, Brazil, Using the WRF-Urban Model

Kitagawa et al. [43] conducted a sensitivity study to evaluate the performance of the WRF-Urban model in a tropical coastal-urban area of the Metropolitan Region of Vitória, Brazil, using high-resolution numerical experiments over one year. Despite some biases, the model realistically reproduced the daily patterns of near-surface parameters. However, it overestimated air temperature and wind speed, underestimated relative humidity, and had difficulties in simulating wind speed over urban areas. To address these issues, the study explored the use of urban canopy schemes, different model heights, and land use and land cover datasets. The building effect parameterization (BEP) scheme improved wind speed over urban areas, whereas lowering the model height showed slight differences, particularly for wind speed. The land use and land cover data had insignificant impacts on urban areas but affected temperature and precipitation patterns over natural surfaces. The study demonstrated the WRF-Urban model's performance over a long time frame and its sensitivity to various configurations in the tropical coastal-urban context. Additionally, modifying land use and cover data impacted temperature and precipitation fields, providing valuable insights for air pollution and climate assessments.

Limitations of the WRF-Urban Model

While the WRF-Urban model offers significant advantages for urban weather forecasting and simulations, it also has some limitations/disadvantages and potential sources of uncertainty that should be considered prior to its application:

Computational capabilities: The WRF-Urban model requires substantial computational resources and capabilities due to its high-resolution simulations and complex parameterizations. Running the model at a fine spatiotemporal scale may require powerful computing resources and long-run simulations.

Data requirements: The WRF-Urban model often relies on accurate and detailed input data, including urban morphology, land-use and land cover information, building characteristics, emission inventories, and meteorological observations. The acquisition and preparation of such datasets for the model can be challenging, especially in data-sparse regions or when data quality is poor.

User expertise: Utilizing the WRF-Urban model effectively requires a certain level of expertise in atmospheric modeling, data processing, and model configuration. Training and experience are necessary to optimize model settings, interpret results, and address potential issues or errors.

Urban complexity representation: The model's representation of, and sensitivity to, urban complexity, such as the interactions between buildings, roads, and vegetation, is still an active research area. While the WRF-Urban model incorporates urban canopy schemes, there are integral generalizations and uncertainties in capturing the complex urban processes at a microscale.

Parameterization uncertainties: The selection and calibration of model parameterizations, such as those related to turbulent mixing, radiation, and land surface processes, induct a certain level of uncertainty. These parameterizations may not accurately represent the unique characteristics of different urban areas and complex urban microscale processes, leading to potential biases in the model output.

Sensitivity to initial and boundary conditions: The accuracy of WRF-Urban simulations is highly dependent on the quality of initial and boundary conditions. Errors or uncertainties in the provided meteorological data may affect the model's boundaries that can propagate and influence the model's outcomes, particularly for long-term simulations and regional-scale studies.

Limited chemical mechanisms: Since the WRF-Urban model includes a chemical transport module, the representation of complex atmospheric chemical processes and the full spectrum of air pollutants may be limited. It is important to assess the model's suitability for specific chemical types and reactions of interest.

Model evaluation and validation: The validation of WRF-Urban simulations can be challenging due to the lack of high-resolution observational or reference datasets specific to urban areas. Evaluating the model's performance and validating its results against limited and sparse urban measurements may lead to uncertainties and errors in fully assessing its accuracy.

Ways Forward for Improvement

There are several areas where improvements can be made in the WRF-Urban model in the future:

1) *Improved representation of urban morphology:* The current WRF-Urban Model incorporates urban morphology through the use of building data, but there is room for improvement in terms of the resolution and accuracy of the input data. The model could benefit from more detailed information on building height, layout, and materials.

2) *Better representation of urban features:* The model could be improved by incorporating more detailed information on urban features such as land cover, vegetation, green spaces, buildings, roads, and pavements. This would improve the accuracy of the model's simulations of UHI effects and other urban climate processes. In addition, various socioeconomic factors (i.e., gross domestic product [GDP], urbanization, and population) of the urban environment can also be integrated into the model to accurately

represent the degree and spatial extent of urban exposure to various environmental and climatic hazards.

3) *Enhanced representation of anthropogenic emissions and physical parameterizations:* The anthropogenic emissions data in the WRF-Urban model need improvement in terms of resolution and accuracy. More detailed information on emission sources and their temporal and spatial variability could improve the accuracy of the model's simulations of air pollution. In addition, the model's parameterizations for urban surfaces and processes need to be improved to better represent the complex interactions between urban surfaces and the atmosphere. This includes the development of more accurate and advanced physical schemes for simulating boundary layer processes, land surface processes, energy fluxes, radiation, and turbulent processes in urban environments.

4) *Increased computational efficiency and high-resolution modeling:* The computational efficiency of the WRF-Urban Model could be optimized to run faster and more efficiently, allowing for larger and more detailed simulations. The high-resolution modeling can benefit from better capturing the fine-scale features of the urban environment.

5) *Improved data assimilation:* The ability to assimilate data from various sources, such as satellite, radar, reanalysis, and ground-based observations, can significantly improve the accuracy of the WRF-Urban model's predictions. Future developments could focus on improving the data assimilation techniques to better incorporate these data sources and reduce the underlying uncertainties in the prediction of urban features.

6) *Coupling with other models:* Future developments could focus on developing coupling frameworks between the WRF-Urban model and other models to improve its capabilities in simulating urban environmental conditions. For instance, the WRF-Urban model can be coupled with air quality models to better understand the interactions between urban meteorology and air quality. Similarly, the WRF-Urban model can be integrated with building energy model (BEM) to accurately simulate and predict building energy consumption and requirement based on the surrounding microclimate.

Conclusions

Urban sustainability and resilience play a crucial role in establishing livable, healthy, and prosperous cities. With the increasing challenges posed by climate change, population growth, and urbanization, the use of numerical modeling has emerged as a fundamental tool for designing sustainable and resilient urban environments. Numerical modeling helps identify potential risks and vulnerabilities of urban systems to climate change, enabling urban planners and policymakers to assess various scenarios and evaluate the efficacy of different interventions before implementing them in the real world. Among the advanced numerical models, the WRF-Urban model stands out as a state-of-the-art tool specifically designed for predicting and simulating atmospheric processes within urban areas. By providing accurate and realistic forecasts of weather phenomena, the WRF-Urban model equips urban planners, hydrometeorologists, environmental experts, emergency responders, and other relevant stakeholders with valuable insights to make informed and timely decisions. These decisions can range from protecting public safety during extreme weather events to minimizing damages and losses by taking appropriate preventive measures. Ultimately, the utilization of the WRF-Urban model contributes to the creation of cities that are better prepared, resilient, and adaptable in the face of a changing climate, thereby fostering urban sustainability and improving the overall quality of urban life.

References

1 Elmqvist, T., Andersson, E., Frantzeskaki, N. et al. (2019). Sustainability and resilience for transformation in the urban century. *Nat. Sustain.* 2 (4): 267–273.
2 Acuto, M., Parnell, S., and Seto, K.C. (2018). Building a global urban science. *Nat. Sustain.* 1 (1): 2–4.
3 Meerow, S., Newell, J.P., and Stults, M. (2016). Defining urban resilience: a review. *Landscape Urban Plann.* 147: 38–49.
4 Leichenko, R. (2011). Climate change and urban resilience. *Curr. Opin. Environ. Sustain.* 3 (3): 164–168.
5 Tayyab, M., Zhang, J., Hussain, M. et al. (2021). GIS-based urban flood resilience assessment using urban flood resilience model: a case study of Peshawar city, Khyber Pakhtunkhwa, Pakistan. *Remote Sens.* 13 (10): 1864.
6 Cobbinah, P.B. (2021). Urban resilience in climate change hotspot. *Land Use Policy* 100: 104948.
7 Ullah, S., You, Q., Chen, D. et al. (2022). Future population exposure to daytime and nighttime heat waves in South Asia. *Earth's Future* 10 (6): 1–16.
8 Ajjur, S.B. and Al-Ghamdi, S.G. (2021). Global hotspots for future absolute temperature extremes from CMIP6 models. *Earth Space Sci.* 8 (9): e2021EA001817.
9 Krishnamupti, T.N. and Bounoua, L. (2018). *An Introduction to Numerical Weather Prediction Techniques*. CRC Press.
10 Bauer, P., Thorpe, A., and Brunet, G. (2015). The quiet revolution of numerical weather prediction. *Nature* 525 (7567): 47–55.
11 Ali, G., Bao, Y., Asmerom, B. et al. (2021). Assessment of the simulated aerosol optical properties and regional meteorology using WRF-Chem model. *Arabian J. Geosci.* 14 (18): 1871.
12 Salimi, M. and Al-Ghamdi, S.G. (2020). Climate change impacts on critical urban infrastructure and urban resiliency strategies for the Middle East. *Sustain. Cities Soc.* 54: 101948.
13 Hassler, U. and Kohler, N. (2014). Resilience in the built environment. *Build. Res. Inf.* 42 (2): 119–129.
14 Romero-Lankao, P., Gnatz, D., Wilhelmi, O., and Hayden, M. (2016). Urban sustainability and resilience: from theory to practice. *Sustainability* 8 (12): 1224.
15 Baklanov, A., Grimmond, C.S.B., Carlson, D. et al. (2018). From urban meteorology, climate and environment research to integrated city services. *Urban Clim.* 23: 330–341.
16 Razzaghmanesh, M., Beecham, S., and Salemi, T. (2016). The role of green roofs in mitigating Urban Heat Island effects in the metropolitan area of Adelaide, South Australia. *Urban For. Urban Greening* 15: 89–102.

17 Sun, S., Xu, X., Lao, Z. et al. (2017). Evaluating the impact of urban green space and landscape design parameters on thermal comfort in hot summer by numerical simulation. *Build. Environ.* 123: 277–288.
18 Jayasooriya, V.M., Ng, A.W.M., Muthukumaran, S., and Perera, B.J.C. (2017). Green infrastructure practices for improvement of urban air quality. *Urban For. Urban Greening* 21: 34–47.
19 Tahir, F. and Al-Ghamdi, S.G. (2023). Climatic change impacts on the energy requirements for the built environment sector. *Energy Rep.* 9: 670–676.
20 Waqas, H., Shang, J., Munir, I. et al. (2023). Enhancement of the energy performance of an existing building using a parametric approach. *J. Energy Eng.* 149 (1): 04022057.
21 Al Huneidi, D.I., Tahir, F., and Al-Ghamdi, S.G. (2022). Energy modeling and photovoltaics integration as a mitigation measure for climate change impacts on energy demand. *Energy Rep.* 8: 166–171.
22 Taniguchi, E., Thompson, R.G., and Yamada, T. (2014). Recent trends and innovations in modelling city logistics. *Procedia Soc. Behav. Sci.* 125: 4–14.
23 Miranda, A., Silveira, C., Ferreira, J. et al. (2015). Current air quality plans in Europe designed to support air quality management policies. *Atmos. Pollut. Res.* 6 (3): 434–443.
24 Arnell, N.W., Halliday, S.J., Battarbee, R.W. et al. (2015). The implications of climate change for the water environment in England. *Prog. Phys. Geog. Earth Environ.* 39 (1): 93–120.
25 Tahir, F., Bansal, D., Rehman, A.U. et al. (2023). Assessing the impact of climate conditions on the distribution of mosquito species in Qatar. *Front. Public Health* 10: 970694.
26 An, N., Mustafa, F., Bu, L. et al. (2022). Monitoring of atmospheric carbon dioxide over Pakistan using satellite dataset. *Remote Sens.* 14 (22): 5882.
27 Ali, G., Bao, Y., Ullah, W. et al. (2020). Spatiotemporal trends of aerosols over urban regions in Pakistan and their possible links to meteorological parameters. *Atmosphere* 11 (3): 306.
28 Khan, S., Zeb, B., Ullah, S. et al. (2023). Assessment and characterization of particulate matter during the winter season in the urban environment of Lahore, Pakistan. *Int. J. Environ. Sci. Technol.* 20: 1–12.
29 Huang, S.Z., Sadiq, M., and Chien, F. (2021). Dynamic nexus between transportation, urbanization, economic growth and environmental pollution in ASEAN countries: does environmental regulations matter? *Environ. Sci. Pollut. Res.* 30 (15): 42813–42828.
30 Li, Y., Han, M.Y., Liu, S.Y., and Chen, G.Q. (2019). Energy consumption and greenhouse gas emissions by buildings: a multi-scale perspective. *Build. Environ.* 151: 240–250.
31 Sachindra, D.A., Ullah, S., Zaborski, P. et al. (2023). Temperature and urban heat island effect in Lublin city in Poland under changing climate. *Theor. Appl. Climatol.* 151 (1–2): 667–690.
32 Sallis, J.F., Bull, F., Burdett, R. et al. (2016). Use of science to guide city planning policy and practice: how to achieve healthy and sustainable future cities. *Lancet* 388 (10062): 2936–2947.
33 Kylili, A. and Fokaides, P.A. (2015). European smart cities: the role of zero energy buildings. *Sustain. Cities Soc.* 15: 86–95.

34 Zhang, J., You, Q., and Ullah, S. (2022). Changes in photovoltaic potential over China in a warmer future. *Environ. Res. Lett.* 17 (11): 114032.

35 Chen, F., Kusaka, H., Bornstein, R. et al. (2011). The integrated WRF/urban modelling system: development, evaluation, and applications to urban environmental problems. *Int. J. Climatol.* 31 (2): 273–288.

36 Salamanca, F., Zhang, Y., Barlage, M. et al. (2018). Evaluation of the WRF-urban modeling system coupled to Noah and Noah-MP land surface models over a semiarid urban environment. *J. Geophys. Res. Atmos.* 123 (5): 2387–2408.

37 Li, H., Yuan, F., Shen, L. et al. (2022). Improving the WRF/urban modeling system in China by developing a national urban dataset. *Geosci. Front.* 13 (4): 101385.

38 He, X., Li, Y., Wang, X. et al. (2019). High-resolution dataset of urban canopy parameters for Beijing and its application to the integrated WRF/urban modelling system. *J. Clean. Prod.* 208: 373–383.

39 Wong, N.H., He, Y., Nguyen, N.S. et al. (2021). An integrated multiscale urban microclimate model for the urban thermal environment. *Urban Clim.* 35: 100730.

40 Vahmani, P. and Ban-Weiss, G.A. (2016). Impact of remotely sensed albedo and vegetation fraction on simulation of urban climate in WRF-urban canopy model: a case study of the urban heat island in Los Angeles. *J. Geophys. Res. Atmos.* 121 (4): 1511–1531.

41 Singh, V.K., Mughal, M.O., Martilli, A. et al. (2022). Numerical analysis of the impact of anthropogenic emissions on the urban environment of Singapore. *Sci. Total Environ.* 806: 150534.

42 Xu, X., Chen, F., Shen, S. et al. (2018). Using WRF-urban to assess summertime air conditioning electric loads and their impacts on urban weather in Beijing. *J. Geophys. Res. Atmos.* 123 (5): 2475–2490.

43 Kitagawa, Y.K.L., de Almeida Albuquerque, T.T., Kumar, P. et al. (2022). Coastal-urban meteorology: a sensitivity study using the WRF-urban model. *Urban Clim.* 44: 101185.

3

Assessing and Projecting Climatic Changes in the Middle East and North Africa (MENA) Region: Insights from Regional Climate Model (RCM) Simulations and Future Projections

Salah B. Ajjur[1,2] and Sami G. Al-Ghamdi[1,3,4]

[1] Division of Sustainable Development, College of Science and Engineering, Hamad Bin Khalifa University, Qatar Foundation, Doha, Qatar
[2] Department of Earth, Environmental and Planetary Sciences, Brown University, Providence, RI, USA
[3] Environmental Science and Engineering Program, Biological and Environmental Science and Engineering Division, King Abdullah University of Science and Technology (KAUST), Thuwal, Saudi Arabia
[4] KAUST Climate and Livability Initiative, King Abdullah University of Science and Technology (KAUST), Thuwal, Saudi Arabia

Introduction

The impact of climate change is not homogeneous, neither in space nor in time [1]. This makes providing practical and actionable climatic information at regional scales vital for climate change mitigation and adaptation plans. It is the only way to build urban resilience systems, protect lives, and limit damage to properties. Two primary sources of climatic information are observational and experimental datasets. Observational data, including in situ records, reanalyses, and paleo-observations, are essential to understand the present-day climate and evaluate whether climate models realistically simulate the required regional climatic aspects. If climate models can reproduce the current climate, then they can project future climatic changes. Experimental data include general circulation models (GCMs), regional climate models (RCMs), statistically downscaled data, bias-corrected data, and idealized scenarios. This chapter introduces GCMs and RCMs and illustrates a step-by-step methodology for evaluating and projecting climatic changes at a regional scale.

GCMs are fundamental tools for understanding climate. They help develop our understanding of how climate systems respond to different types of large-scale forcings such as greenhouse gases (GHGs), variation in solar radiation, and volcanic aerosols. The GCMs have undergone significant improvements in computational capabilities and represent atmospheric, land, and ocean interactions through time. These improvements have made GCM outputs at global to subcontinental scales valuable; however, there is still a gap in providing regional climatic information at finer spatial resolutions. Representing the regional climate is a challenging task because it is not only controlled by large-scale

Sustainable Cities in a Changing Climate: Enhancing Urban Resilience. First Edition.
Edited by Sami G. Al-Ghamdi.
© 2024 John Wiley & Sons Ltd. Published 2024 by John Wiley & Sons Ltd.

forcings; local-scale forcings and modes of climate variability (teleconnections) are also major regional climatic features. Local-scale forcings, including land use and urban development, topography, sea–land interactions, mesoscale circulations, and natural aerosols, cannot be described entirely by GCMs. Such descriptions are not yet economically feasible, as their computational costs are very high. Additionally, GCMs cannot accurately describe climate extremes at regional scales; this information is vital to avoiding climate risk disasters. Therefore, there is a dire need to describe local-scale forcings, not resolved in GCMs, while simultaneously bridging the spatial resolution gap. To this end, regional climate downscaling (RCD) is a great tool.

The RCD is a complementary technique to increase the reliability and effectiveness of local and regional climatic data, as it adds further details (e.g., finer resolutions and longer model runs) to GCM simulations [2]. The RCD outputs (i.e., RCMs) should not be seen as in competition with each other or with GCMs [2, 3]. The RCD has two methods with both strengths and weaknesses: dynamical and statistical downscaling. The primary assumption in RCD methods is to use boundary conditions from large-scale models (such as GCMs or reanalyses) to derive a higher resolution RCM for a specific region. Accurate management of boundary conditions is a vital issue in RCD [3]. It provides a way to present how the studied region has been affected by the rest of the world. Managing 100% boundary conditions is impossible since, typically, GCMs and RCMs have different resolutions and process descriptions. Therefore, there might be several solutions (i.e., representations). Over the past three decades, improvements in RCD techniques have produced high-resolution downscaled climatic data, which have narrowed (but not wholly resolved) the uncertainty in climate change impact studies.

Some uncertainties are inevitably inherited in RCMs. First, since RCMs are generally driven by boundary conditions from GCMs, the quality of GCMs is crucial. GCMs have, and still, suffer from systematic biases, especially at regional scales. In some GCMs, the bias is due to nonlocal processes, which complicates the model improvement. Hall [4] referred to this issue as garbage in, garbage out. He explained that the jet stream, which brings winter storms to the Northern Hemisphere's mid-latitudes, is simulated by all GCMs, but with different positions. Only a few GCMs had properly positioned the jet stream. As a result, the bias in jet stream positioning in GCMs leads to misleading downscaled signals at regional scales. Additionally, RCMs can be strongly affected by interpolation methods over complex orography and data-scarce regions. RCMs generally have atmospheric and land components (i.e., they exclude some critical processes in the Earth's system, such as air–sea coupling and the chemistry of cloud–aerosol interactions). In fact, some recent initiatives in RCMs involve a high-resolution high-frequency coupling of atmosphere–land–river–ocean components over the Mediterranean region. The advanced RCMs produced from earlier initiatives are called regional climate system models (RCSMs) [5]. Still, much must be done to lengthen the experiments and provide a higher temporal frequency (six hours or higher).

The Coordinated Regional Climate Downscaling Experiment (CORDEX), [6] was motivated by improving, fostering, and popularizing the international efforts in RCD. It has a diverse range of RCMs controlled by standard worldwide regional settings. Since its beginning in 2013, CORDEX developed a two-stream simulation protocol. In the first stream (evaluation), the participating RCMs performed equivalent domain downscaling of

ERA-Interim reanalysis from 1989 and 2008, and the RCMs were evaluated against observed data during the simulation period. In the second stream (projection), the models ran with GCM boundary conditions from January 1950 to December 2005, the fifth phase of the Coupled Model Intercomparison Project (CMIP) [7] for the historical period, and from January 2006 to December 2100 for the future period [6]. Future projections were run under two Representative Concentration Pathways (RCP4.5 and RCP8.5). A typical RCM resolution was initially set to 0.44° (~50 km) to facilitate participation from a broad community. These actions enabled scientists all over the world to engage in state-of-the-art research without requiring large infrastructures that are typically needed to run high-quality GCMs. Currently, the CORDEX is an accessible, solid base for providing robust climatic data for vulnerability, impacts, and adaptation research at regional (spatial scale <10,000 km^2) to local scales [8]. Some new RCMs in CORDEX have a grid spacing of 10 km.

Methodology

This section projects and evaluates the changes in surface air temperature and precipitation parameters, through the twenty-first century, over the Middle East and North Africa (MENA)-CORDEX domain (6.88°S–45.13°N and 26.63°W–75.63°E). Temperature and precipitation simulations were obtained based on a monthly timestep from two RCMs (RegCM4-3 and RCA4) participating in the CORDEX. The r1i1p1 ensemble was selected for each RCM. The CORDEX data can be accessed through the Earth System Grid Federation (ESGF) nodes, for instance, esgf-data.dkrz.de. There are four ESGF nodes (in Sweden, Germany, United Kingdom, and France), but they all provide access to the same data.

The National Center for Atmospheric Research (NCAR) generated the RegCM in 1989. The most recent development, RegCM4 (used in this study), was built on successive predecessor improvements in 1993 (RegCM2), 1999 (RegCM2.5), and 2006 (RegCM3). The RegCM4 runs at 0.44° grid spacing (~50 km) and is driven by the GCM MPI-ESM-MR, while the MPI-ESM-MR runs at an atmospheric resolution of 1.875° (~200 km). The RCA RCM was generated by the Swedish Meteorological and Hydrological Institute (SMHI). This study used its latest development (RCA4) since it includes significant physical and technical improvements. The RCA4 runs at 0.22° (~25 km) and 0.44° grid spacings and is driven by the GFDL-ESM2M GCM. The GFDL-ESM2M runs at an atmospheric resolution of 2° latitude × 2.5° longitude. This study used RCA4 simulations at 0.22° resolution.

First, a comparison between GCMs and RCMs in simulating MENA temperature and precipitation was made. Then, the NCEP/NCAR derived data were used to assess the RCMs' ability to identify MENA weather conditions. The NCEP/NCAR reanalysis is provided at a spatial resolution of 1.875 longitude × 1.915 latitude from the National Oceanic and Atmospheric Administration (NOAA) website. To ensure spatial resolution consistency, the GCMs and RCMs were bilinearly interpolated to a spatial grid resolution like the NCEP/NCAR. To investigate the changes in temperature and precipitation parameters through 2100, we used RegCM4-3 projections. The period 1976–2005 was selected as a reference period to compare future changes relative to the recent past. Two Representative Concentration Pathways (RCP4.5 and RCP8.5) represented the future periods through 2100.

The near-term future was represented by 2031–2050, while 2081–2100 represented the long-term future. We averaged months across years for reference and future periods. Then, we selected July surface air temperature and calculated the difference between reference and near/long-term periods. Regarding precipitation, we calculated the monthly averages through each period. After that, we aggregated monthly records to get the yearly sum. Similar to temperature, the difference between reference and near/long-term periods was calculated. All calculations were replicated under RCP4.5 and RCP8.5. Python and ArcGIS software were used to implement the abovementioned steps.

GCMs vs. RCMs in Simulating MENA Temperature and Precipitation

To better comprehend the differences between GCMs and RCMs in simulating regional climate, we compared RCM simulations with their corresponding GCM during 1976–2005 over the MENA region. Figure 3.1 compares the MENA mean surface air temperature for July in RegCM4-3 and SMHI-RCA4 RCMs, and MPI-ESM-MR and GFDL-CM4 GCMs. It is hard to get spatially detailed local information from GCMs. On the other side, The RCMs help deepen our understanding of climate change in the regions. Using Figure 3.1b,d, one can identify areas with the highest or lowest air temperatures. The RCMs show that the MENA regions most affected by high temperatures are eastern North Africa and the Arabian Gulf. The MENA has complex coastlines such as the Red Sea and the Arabian Gulf and complex topography in south Saudi Arabia and east Yemen. It is challenging to get informative climatic data from GCMs in these regions.

The RCMs also provide more climatic information than GCMs, especially in coastal areas. The minimum surface air temperature in the MPI-ESM-MR simulation was 2 °C, whereas RegCM4-3 showed a minimum value of −7 °C. In other words, the RegCM4-3

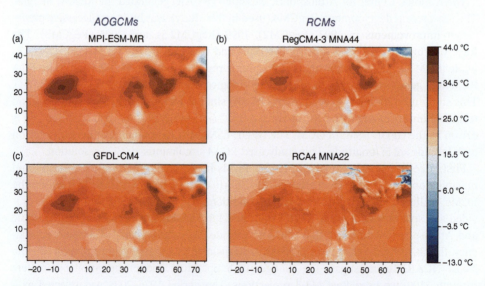

Figure 3.1 The mean surface air temperature for July in RCMs (right column) and the corresponding GCMs (left column) during 1976–2005 over the MENA-CORDEX domain; (a) MPI-ESM-MR, (b) RegCM4-3 MNA44, (c) GFDL-CM4 GCMs, and (d) RCA4 MNA22.

RCM showed some areas in the eastern north MENA-CORDEX domain cooled by a further 9 °C than that projected by the MPI-ESM-MR GCM. Similarly, the minimum surface air temperature in the GFDL-CM4 simulation was −6 °C, whereas RCA4 showed a minimum value of −13 °C. Thus, the RCA4 RCM showed some areas in the eastern north MENA-CORDEX domain cooled by a further 7 °C than that projected by the GCM. These added values of RCMs were made possible because the GCMs did not resolve the well-understood temperature impact over the whole MENA region. These additional findings from RCMs carry additional implications on aspects of water deficit, power supply, and livelihood. Since most MENA citizens live in coastal cities, these details help increase the ability of these communities to set up climate change resilience plans.

Figure 3.2 depicts the MENA mean annual precipitation in the RegCM4-3 and RCA4 and the corresponding driving GCMs: MPI-ESM-MR and GFDL-CM4. Generally, the annual precipitation over land in MENA areas did not exceed 101 mm during 1976–2005. Only minimal parts had higher precipitation in Sudan and Iran. Similar to surface temperature, getting spatially detailed precipitation information from GCMs was challenging. Annual precipitation ranged up to 121 mm in MPI-ESM-MR and up to 113 mm in GFDL-CM4. On the other side, the RCM simulations provided extra information. In both RegCM4-3 and RCA4, the annual precipitation varied by up to 202 mm. In some southern parts of the MENA-CORDEX domain, the annual precipitation exceeded 202 mm (see Figure 3.2b). The RCA4 at 0.22° resolution (Figure 3.2d) even showed higher precipitation values over eastern Saudi Arabia and Yemen, east Senegal, and eastern Cameron. The GCMs did not capture such precipitation extremes in these regions. It can be concluded that the RCMs have strengthened our understanding of the spatial patterns of precipitation over the MENA.

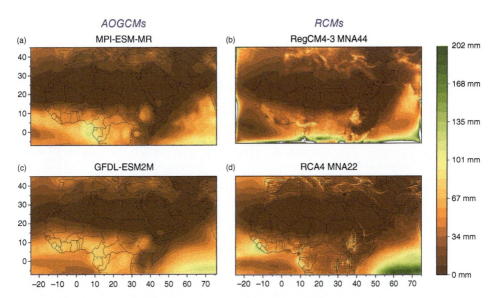

Figure 3.2 The mean annual precipitation (mm) in RCMs (right column) and the corresponding GCMs (left column) during 1976–2005 over the MENA-CORDEX domain; (a) MPI-ESM-MR, (b) RegCM4-3 MNA44, (c) GFDL-CM4 GCMs, and (d) RCA4 MNA22.

RCMs Performance in Simulating MENA Climatic Changes

Although RCMs offer more climatic details than GCMs, the downscaled changes from RCMs should be examined to determine their credibility. Otherwise, outputs of the original GCMs, though in coarser resolution, probably better represent the current limits of scientific understanding [4]. Typically, the RCMs are compared with observations and reanalyses. Although, observations are more reliable with more rigorous quality control, they may not completely cover the studied area, especially if they were in situ measurements taken at stations' locations. Several reanalyses had complete global spatial coverage, however (see Chapter 2). Figure 3.3 shows the performance of both RCMs in simulating the historical climate (1976–2005) by comparing their simulations of surface air temperature and annual precipitation with the NCEP/NCAR reanalysis. The differences between RCA4 and NCEP/NCAR simulations are shown in the left column, while the right column shows the differences between RegCM4-3 and NCEP/NCAR. A standard color scale is used for each climatic parameter with a white middle color assigned to zero value (i.e., no difference between RCMs and reanalysis).

The RCMs can generally capture the MENA climate, especially for July surface air temperature, during the reference period. In general, the RCA4 had better performance than the RegCM4-3. However, some regions in both RCMs showed bias values compared with the NCEP/NCAR. Both RCMs projected higher temperatures over southern parts of the MENA-CORDEX (near 10 North latitude) and northern Iran. The RCA4 presented higher values, up to 17 °C more, than NCEP/NCAR, as Figure 3.3a shows. The RegCM4-3 presented higher values, up to 13 °C more, than NCEP/NCAR (Figure 3.3b). In contrast, some northern areas of Iran exhibited cooler surface air temperatures in both RCMs compared to NCEP/NCAR. The discrepancy with the NCEP/NCAR projection hit −26 °C in RCA4 and −33 °C in RegCM4-3.

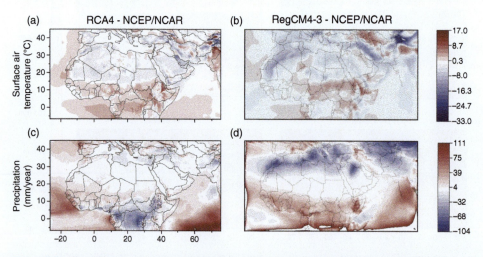

Figure 3.3 The difference between July mean surface air temperature (°C) (a, b) and annual precipitation (mm) (c, d) in RCA4 (first row) and RegCM4-3 (second row) and NCEP/NCAR reanalysis during 1976–2005.

Figure 3.3c,d depicts the variations in average annual precipitation during 1976–2005 between RCA4 (RegCM4-3) and NCEP/NCAR. Compared with temperature, precipitation variations are more considerable between RCMs and reanalysis. In general, RCA4 can better describe annual precipitation over the MENA than RegCM4-3. Compared with NCEP/NCAR, the RCA4 had some biases in southern African parts. The RCA4 underestimated annual precipitation values for up to 80 mm. On the other hand, the RegCM4-3 overestimated annual precipitation in these regions. The RegCM4-3 underestimates annual precipitation around the Mediterranean and over North Iran by up to 104 mm. The biased values in temperature and precipitation were mainly located over the MENA mountainous and coastal regions, which shows an urgent need for more relevant simulations.

Projected Future Changes Over MENA-CORDEX

This section projects the future climatic changes through 2100 over the MENA-CORDEX domain, relative to the recent past (1976–2005). Although RCA4 had finer resolution than RegCM4-3 and showed better performance in simulating the MENA temperature and precipitation, future projections were obtained from RegCM4-3. The RCA4 analysis showed only one available future scenario (i.e., RCP8.5). Alternatively, RegCM4-3 showed two future projections (RCP4.5 and RCP8.5). Therefore, we aimed to show the reader how climatic parameters vary under different future possibilities. Figure 3.4 depicts the evolution of July surface air temperature and annual precipitation during the near term (2031–2050) and long term (2081–2100) under RCP4.5 and RCP8.5. The first and second columns in Figure 3.4 describe the future changes during the near-term future under RCP4.5 and RCP8.5, respectively. At the same time, the third and fourth columns display the future changes during the long-term period. Change anomalies are presented in °C for temperature and mm for annual precipitation, relative to the average for 1976–2005.

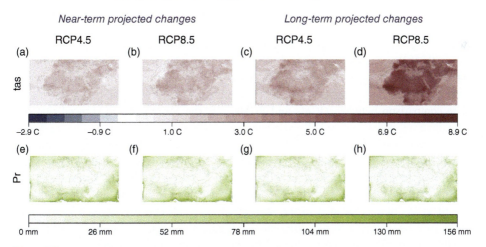

Figure 3.4 The near-term (2031–2050) and long-term (2081–2100) projected future changes in July surface air temperature (a–d) and annual precipitation (e–h) under RCP4.5 (first and third columns) and RCP8.5 (second and fourth columns), relative to the recent past (1976–2005). Projections were obtained from RegCM4-3. "tas" and "Pr" denote surface air temperature and precipitation, respectively.

Overall, the surface air temperature is expected to increase over the MENA-CORDEX domain during the twenty-first century under both RCPs. Only a tiny area in the northeastern parts of the MENA is expected to cool down to −2.9 °C. Most of the MENA areas exhibit temperature rises of up to 8.9 °C. Rises in temperature were more considerable under RCP8.5. During 2031–2050, the MENA is projected to warm up to 4.6 °C under both RCPs. These changes are more pronounced during 2081–2010, when the MENA is projected to warm up to 5.2 and 8.9 °C under the RCP4.5 and RCP8.5, respectively. The analysis shows that the MENA regions most affected by temperature increases are the western Sahara countries, particularly north Libya, north Algeria, Tunisia, and southern Mauritania. This does not essentially mean that these areas had the highest temperature values in the recent past. It means that these regions are warming much more, compared with their historical experience, than other MENA areas.

The future trend in annual precipitation was not clear over the MENA area compared with the temperature trend. No significant changes in annual precipitation projections were noticed among the two future periods and under both RCPs (see Figure 3.4e–h). A typical small growth (<26 mm) was observed over the MENA land regions under four possibilities. Considering the minimal precipitation in the region, this small growth might be substantial, presenting more than a 100% change [9]. Other areas, such as the southern parts of the MENA-CORDEX domain and north Iran, may exhibit precipitation increases of up to 156 mm per year.

Conclusion

This chapter complements the previous one, which focused on climate change basics and summarized the latest improvements in climatic models and data. It provides a step-by-step methodology for comparing GCMs and RCMs in simulating two climatic parameters (temperature and precipitation) over the MENA-CORDEX domain. This chapter then assesses the performance of two RCMs (RCA4 and RegCM4-3) by comparing their simulations with NCEP/NCAR reanalysis during 1976–2005. Using RegCM4-3 simulations at 0.44° (~50 km) grid resolution, the MENA surface air temperature and precipitation evolution through the twenty-first century was documented.

The comparison between GCMs and RCMs over the MENA region showed that our understanding of the regional climatic changes is improved when the GCMs are downscaled. Nevertheless, the RCMs' projections should be validated before being treated as credible data for practical purposes. A typical way to achieve this is to use observations and reanalyses. Our analysis showed an inconsistency between the RCMs' simulations and NCEP/NCAR over the MENA-CORDEX domain, which is a major concern. The inconsistency in RCMs may be related to the scarcity of observational records in mountainous regions (see the spatial pattern of differences in Figure 3.3). Even if the measurements were available, mountains are highly affected by the gauge location, networking, and set-up, especially for precipitation. If the gauges are located without allowing/considering the influence of topography, major alterations to the actual value ensue. Additionally, since boundary conditions are mainly obtained from GCMs, reducing the error in GCMs is crucial to improve the performance of RCMs. Alternatively, RCMs may use observed reanalysis as boundary conditions. A possible improvement for the MENA regional modeling includes involving all relevant forcings in RCMs. These forcings include air–sea coupling, the chemistry of cloud–aerosol interactions, and land-use changes.

The RegCM4-3 simulations projected that surface air temperatures would generally increase over the MENA region, with notable changes (up to 8.9 °C) observed under RCP8.5 during 2081–2100. The most affected MENA areas by the surge in temperature are western Sahara countries (north Libya, north Algeria, Tunisia, and south Mauritania). The projected surface warming trend over the MENA is 4.5 times greater than the global average under RCP8.5 (the mean value was obtained from Figure 3.4). These findings agree, to some extent, with Cook and Vizy [10], who observed that the warming of the Arabian Peninsula is 2.3–3.1 greater than the global mean. And, according to CMIP6 multi-model ensemble median, the MENA surface air temperature will increase by up to 6 °C through the twenty-first century, compared with 1981–2010. Although reference periods between our study and Ajjur and Al-Ghamdi [9] differ, the discrepancy between both projections is high (2.9 °C in 2100). This discrepancy might be related to the difference between the projections of the GCMs and RCMs. Regarding precipitation, a minimal growth (<26 mm in most land regions) was found in annual precipitation through the twenty-first century, compared with the 1976–2005 average. These quantities should not be underestimated, considering the scarcity of rainfall over the MENA. Ajjur and Al-Ghamdi [9] evaluated the future rainfall changes in 10 GCMs participating in the CMIP6. They found that the MENA annual precipitation increased, through 2100, up to 105% under SSP2-4.5 and up to 250% under SSP5-8.5, relative to 1981–2010. The highest values of annual precipitation were found over southern Saudi Arabia, Yemen, and Oman.

The warming trend in surface air temperature over the MENA can be linked primarily to human influences, such as GHG emissions and land use actions [11]. Kim and Min [12] used observations and CMIP5 models during 1951–2010 to emphasize the human contribution to the increasing extreme temperatures over the Northern Hemisphere. Similar attribution is valid for future trends in the MENA surface air temperature. According to precipitation changes, the attribution on a regional scale is not yet clear. The model errors (see Figure 3.3c,d) and the large uncertainty in precipitation observations hinder the scientific attribution of precipitation changes to anthropogenic forcings. Therefore, further examination is recommended.

This study showed a pressing need to obtain ground measurements and high-resolution projections for all climatic parameters over the MENA region. Besides temperature and precipitation parameters, parameters are needed for wind speed and direction, humidity, soil moisture, radiation, and so on. Such measurements and predictions would allow for practical and applicable decisions to be made to help increase the resilience of the MENA fragile environments against climate change [13]. The MENA is predominantly arid to semi-arid areas with minimal natural water resources [9]. Therefore, these findings should be followed by investigating how the projected changes in the MENA climate influence the water cycle, land cover, and soil at a regional scale.

References

1 Pachauri, R.K., Allen, M.R., Barros, V.R. et al. (2014). *Climate Change 2014: Synthesis Report. Contribution of Working Groups I, II and III to the Fifth Assessment Report of the Intergovernmental Panel on Climate Change.* Core Writing Team (ed. R.K. Pachauri and L.A. Meyer), 151. Geneva, Switzerland: IPCC.

2 Giorgi, F. and Gutowski, W.J. (2015). Regional dynamical downscaling and the CORDEX initiative. *Annu. Rev. Environ. Resour.* 40 (1): 467–490.

3 Rummukainen, M. (2010). State-of-the-art with regional climate models. *WIREs Clim. Change* 1 (1): 82–96.

4 Hall, A. (2014). Projecting regional change. *Science* 346 (6216): 1461–1462.

5 Somot, S., Ruti, P., Ahrens, B. et al. (2018). Editorial for the Med-CORDEX special issue. *Climate Dyn.* 51 (3): 771–777.

6 Gutowski, W.J. Jr., Giorgi, F., Timbal, B. et al. (2016). WCRP COordinated Regional Downscaling EXperiment (CORDEX): a diagnostic MIP for CMIP6. *Geosci. Model Dev.* 9 (11): 4087–4095.

7 Taylor, K.E., Stouffer, R.J., and Meehl, G.A. (2012). An overview of CMIP5 and the experiment design. *Bull. Am. Meteorol. Soc.* 93 (4): 485–498.

8 Giorgi, F., Jones, C., and Asrar, G.R. (2009). Addressing climate information needs at the regional level: the CORDEX framework. World Meteorological Organization Bulletin, p. 175. https://public.wmo.int/en/bulletin/addressing-climate-information-needs-regional-level-cordex-framework (accessed 11 July 2021). Report No.: 0042-9767 Contract No.: 3.

9 Ajjur, S.B. and Al-Ghamdi, S.G. (2021). Evapotranspiration and water availability response to climate change in the Middle East and North Africa. *Clim. Change* 166 (3): 28.

10 Cook, K.H., Vizy, E.K., Liu, Y., and Liu, W. (2021). Greenhouse-gas induced warming amplification over the Arabian Peninsula with implications for Ethiopian rainfall. *Clim. Dyn.* 57: 3113–3133.

11 Ajjur, S.B. and Al-Ghamdi, S.G. (2021). Global hotspots for future absolute temperature extremes from CMIP6 models. *Earth Space Sci.* 8: e2021EA001817.

12 Kim, Y.-H., Min, S.-K., Zhang, X. et al. (2016). Attribution of extreme temperature changes during 1951–2010. *Clim. Dyn.* 46 (5): 1769–1782.

13 Ajjur, S.B. and Al-Ghamdi, S.G. (2021). Seventy-year disruption of seasons characteristics in the Arabian Peninsula. *Int. J. Climatol.* 41: 1–18.

4

Building for Climate Change: Examining the Environmental Impacts of the Built Environment

Mehzabeen Mannan[1] and Sami G. Al-Ghamdi[1,2,3]

[1] *Division of Sustainable Development, College of Science and Engineering, Hamad Bin Khalifa University, Qatar Foundation, Doha, Qatar*
[2] *Environmental Science and Engineering Program, Biological and Environmental Science and Engineering Division, King Abdullah University of Science and Technology (KAUST), Thuwal, Saudi Arabia*
[3] *KAUST Climate and Livability Initiative, King Abdullah University of Science and Technology (KAUST), Thuwal, Saudi Arabia*

Introduction

As the global population continues to grow, urban regions across the world are witnessing a surge in development and construction activities. The magnitude of this expansion is staggering, with an estimated need for an additional 2.4 trillion ft^2 (230 billion m^2) of new floor space per month for the next four decades, equivalent to constructing an entire New York City [1]. However, this construction boom is not without consequences. Buildings and their construction are responsible for a staggering 36% of worldwide energy use and 39% of energy-related CO_2 emissions per year, making them key contributors to the climate crisis [2, 3].

The environmental impact of buildings is multifaceted. Building-related emissions are commonly measured as a combination of two factors: operational carbon emissions and embodied carbon (EC) emissions. Operational carbon emissions refer to the daily energy use of a building, including lighting, heating, and cooling. Building operations alone are responsible for approximately 28% of global emissions annually [4]. On the other hand, EC emissions refer to the carbon generated by the manufacturing of building materials, the transportation of supplies to construction sites, and the construction process itself. They account for around a quarter of a building's total lifetime carbon emissions and represent approximately 11% of global emissions. If left unchecked, building emissions are projected to double by 2050 [5]. Between 2020 and 2060, construction material-related emissions, that is, the embodied emissions in residential and commercial buildings, are expected to increase by 3.5 to 4.6 Gt CO_2eq/yr [6].

Sustainable Cities in a Changing Climate: Enhancing Urban Resilience. First Edition.
Edited by Sami G. Al-Ghamdi.
© 2024 John Wiley & Sons Ltd. Published 2024 by John Wiley & Sons Ltd.

In contrast to operational carbon emissions, which can be gradually reduced through building energy upgrades and the adoption of renewable energy sources, embedded carbon emissions remain fixed once a building is constructed. While there has been significant research and focus from governmental institutions on improving operational energy efficiency, the measurement and mitigation of EC in the building environment have received comparatively less attention [7]. Understanding the magnitude and identifying the most promising areas for mitigation and reduction of EC in buildings is of paramount importance. It is crucial to urgently address the issue of EC at a large scale to effectively mitigate the impacts of the built environment on climate change. This chapter provides a comprehensive examination of the influence of the building environment on climate change, with a specific emphasis on the potential for mitigating EC emissions as well as operational carbon emissions.

Embodied Carbon Emission in Building Environment

EC encompasses all the greenhouse gas (GHG) emissions associated with the construction of buildings, which include emissions from material extraction, transportation, manufacturing, and on-site installation, as well as the operational and end-of-life emissions linked to those materials [8]. However, measuring and tracking EC proves to be more challenging compared to operational carbon, which can be relatively straightforwardly inferred from occupants' energy consumption data. Assessing the EC of a building material cannot be determined solely by examining the final product; instead, it requires manufacturers to engage in self-assessment and disclose their production processes transparently.

It is essential to recognize that two seemingly identical materials, with the same cost and performance standards, may possess distinct EC characteristics [9]. For instance, consider a scenario where a recycled-steel beam, manufactured using renewable energy, bears an indistinguishable appearance from a virgin-steel beam produced using a coal-fired furnace. However, their levels of EC can significantly differ. Factors such as the origin of each steel beam and the distance it was transported introduce additional complexity in evaluating EC.

To gain a comprehensive understanding of EC, it is necessary to delve into the intricacies of material sourcing, manufacturing methods, and transportation logistics. Only through self-assessment and transparent reporting by manufacturers can accurate assessments of EC be achieved. This underscores the importance of collaboration and knowledge-sharing across the building industry to drive toward greater transparency and enable informed decision-making for minimizing EC in construction projects.

Embodied Carbon Emission for Selected Building Materials

The construction process includes numerous phases, beginning with the manufacture of materials (nonmetallic minerals, oil, cement mortar, iron, steel, concrete, etc.) and material transportation, which generates 82–96% of total embodied CO_2 emissions during the production and construction phases (Figure 4.1) [11]. In the following paragraphs, we will discuss the carbon footprint of several important building materials.

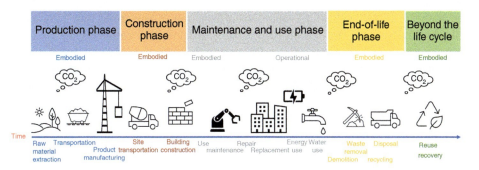

Figure 4.1 Life cycle stages of buildings and associated embodied carbon emissions. Source: Adapted from [10].

Embodied Carbon Emission of Limestone Quarrying

Limestone is one of the most widely manufactured crushed rocks, and it is a fundamental component of construction materials such as aggregate, lime, cement, and building stones. The energy needed for lime quarrying is linked to the machine fuel, diesel, and electricity necessary for limestone processing. Figure 4.2 shows the machines used, together with their energy consumption and CO_2 emissions, for limestone quarrying [12].

In Thailand, the effects of limestone quarrying on ecology and climate change were studied in depth [13]. The findings emphasize the need to comprehend and mitigate the resource depletion and GHG emissions associated with limestone quarrying. Using the method of life cycle assessment (LCA), the production processes were divided into three phases: extraction of primary materials, transportation, and comminution. The analysis incorporated both the IMPACT 2002+ and Greenhouse Gas Protocol methodologies.

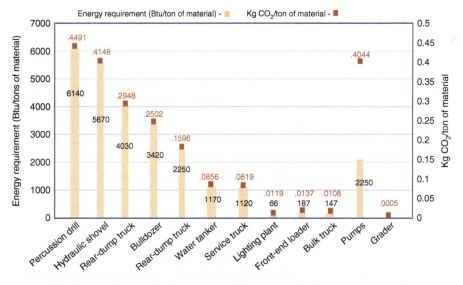

Figure 4.2 The machines used, together with their energy requirements and CO_2 emissions, in limestone processing. Source: Adapted from Choate (2003) [12].

According to the IMPACT 2002+ analysis, for every ton of pulverized limestone rock produced, there was a total resource depletion of 79.6 MJ and GHG emissions of 2.76 kg of CO_2. Among the four categories of damage examined, resource extraction and climate change emerged as the most significant environmental effects of limestone production. These effects were predominantly attributable to diesel fuel and electricity consumption during mining operations.

Concerning climate change, the study expressed emissions in terms of CO_2 equivalents, with an estimated value of 3.13 kg CO_2 per ton of limestone rock produced. The results from the IMPACT 2002+ analysis and the Greenhouse Gas Protocol method were comparable. Notably, electrical energy consumption emerged as the leading contributor to GHG emissions, accounting for approximately 46.8% of total fossil fuel CO_2 emissions. The identification of diesel fuel and electricity consumption as major contributors to these impacts underscores the need for targeted strategies and sustainable practices in the limestone industry to reduce its environmental footprint.

Embodied Carbon Emission from Cement and Concrete Manufacturing

Concrete, a widely used material in infrastructure and construction projects, is the second-most consumed product on Earth after water [14]. In the cement production process, limestone, clay, and various additives are subjected to high temperatures in a kiln, resulting in the release of CO_2 through both the energy-intensive firing process and the chemical reactions that occur within the mixture under heat. Considering their significant GHG emissions and essential role in society, cement and concrete stand out as two crucial focus areas for mitigating EC emissions [15].

Without substantial reductions in global emissions, the concrete industry will persist in annually releasing over 4 billion tons of CO_2, perpetuating environmental pollution [16]. It is estimated that the production of 1 kg of Portland clinker emits almost an equal amount of CO_2 into the atmosphere [11]. CO_2 emissions can be categorized into two primary sources: combustion, accounting for 40% of emissions, and calcination, responsible for 60% of emissions. The calcination process in cement kilns is responsible for approximately 0.55 kg of CO_2 emissions per kilogram of cement clinker produced [17].

It is important to note that the quantity of CO_2 released during the cement manufacturing process is influenced by several factors, including the specific manufacturing process employed (such as the type of kiln utilized), the type of fuel utilized (such as petroleum, natural gas, and coal), and the ratio of clinker to cement (referring to the percentage of additives present) (Table 4.1). Therefore, these three aspects present significant opportunities for enhancing and innovating the production process to effectively manage energy consumption and mitigate embodied CO_2 emissions.

Concrete is constructed using cement mixed with an aggregate, a grainy blend of materials such as stone and sand. The process of manufacturing concrete releases nearly no carbon emissions; cement is the actual reason when it comes to carbon footprint. The cement process is the sole reason why the concrete industry makes up 8% of overall global emissions (Figure 4.3). In most cases, the rate of CO_2 emission that occurs during the manufacture of concrete falls somewhere in the range of 347 and 351 kg CO_2-e/m^3. Specific examples further highlight the emissions associated with cement and concrete. In an insightful study

Table 4.1 CO_2 emissions per kilogram of cement produced for a range of fuel types and clinker-to-cement ratios, for both dry and wet cement manufacturing methods, in kilograms.

Clinker/cement ratio (%)	Process emissions Clinker	Process and fuel-related emissions (CO_2)							
		Dry process				Wet process			
		Coal	Fuel oil	Natural gas	Waste	Coal	Fuel oil	Natural gas	Waste
55	0.28	0.55	0.5	0.47	0.36	0.67	0.59	0.53	0.36
75	0.38	0.72	0.66	0.61	0.47	0.88	0.77	0.69	0.47
(Portland) 95	0.49	0.89	0.81	0.75	0.57	1.09	0.95	0.9	0.57

Source: Fayomi et al. [18]/MDPI/Public Domain CC BY 3.0.

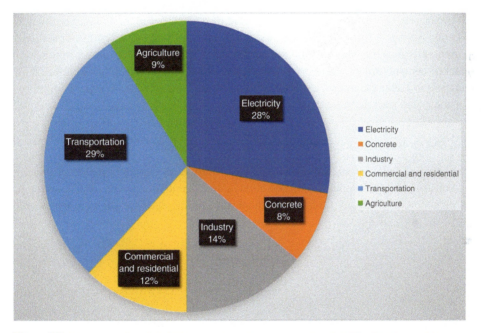

Figure 4.3 Category-wise global CO_2 emissions. *Source:* Ramsden [16]/The Trustees of Princeton University.

conducted by Solís-Guzmán, the use of Cement II/AL 32.5 N in the construction of two four-story residential blocks, comprising a total of 107 dwellings with a combined area of 10,243.69 m², resulted in an annual emission of 148,180 kg CO_2 equivalents [19]. Additionally, the utilization of concrete with the composition of HA25/B/40 during the construction process contributed to a substantial emission of 312,596.55 kg CO_2 equivalents within a single year. These findings emphasize the critical need to carefully evaluate and select cement types and concrete mixes to effectively address and mitigate the environmental impact associated with EC emissions. Focusing on a ready-mixed concrete facility in Cuenca, Ecuador, a comprehensive assessment of the environmental impact associated with concrete production was conducted in another study [20]. Following the rigorous methodologies specified in ISO 14040, ISO 14044, and the IPCC Guidelines, this study analyzed the embodied energy and CO_2 emissions throughout the entire production life cycle of concrete. Using a "gate-to-gate" methodology, the analysis included both the transport of raw materials to the concrete facility and the transport of the final product to the construction site. The study revealed that the production of 1 cubic meter of ready-mixed concrete required an energy input of 568.69 MJ and resulted in the emission of 42.83 kg of CO_2. Notably, indirect transport activities emerged as the primary contributor to environmental impact, with "transport of raw materials" accounting for about 80% of embodied energy and 79% of CO_2 emissions. These results illuminate the significant role that transportation plays in the overall environmental footprint of concrete production. It highlights the significance of optimizing logistics and investigating more sustainable transportation methods in order to reduce EC and GHG emissions.

The year 2019 witnessed significant emissions of CO_2 equivalents from cement plants, with a staggering 67 million metric tons reported by 92 facilities to the US Environmental Protection Agency (EPA) [21]. These emissions constitute approximately 10% of the direct reported emissions attributed to the industrial sector. Understanding the environmental impact of cement and concrete production is crucial to promoting sustainable building practices and seeking alternative solutions that mitigate emissions and reduce the EC footprint associated with these materials (Figure 4.4).

Embodied Carbon from Asphalt Production and Construction

In the building industry, asphalt pavements have emerged as the predominant choice for road construction due to their multitude of advantages compared to concrete pavements. The economic viability of asphalt pavement stems from its ease of installation and

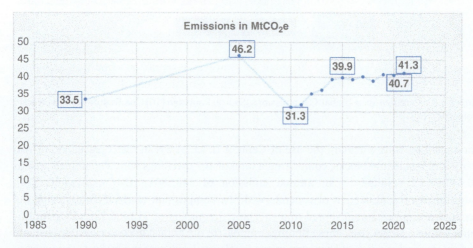

Figure 4.4 Greenhouse gas emissions from cement production in the United States between 1990 and 2021. *Source:* Adapted from [22]. Statista, 2023.

maintenance, resulting in reduced costs and efficient use of resources. Notably, the inherent recyclability of asphalt allows for sustainable practices and minimizes the required landfill area. The flexibility exhibited by asphalt pavements further contributes to their widespread adoption, as they can accommodate varying loads and traffic conditions while maintaining structural integrity [23].

The asphalt industry is currently confronted with the necessity of sustainable operations and reducing climate-damaging emissions, in line with global efforts to combat climate change. The 2015 Paris Climate Accords, which seek to reduce GHG emissions by 43% by 2030 and eliminate them entirely by 2050, have set ambitious goals for mitigating the effects of climate change [24]. Despite its relatively smaller size compared to other industries, the asphalt sector has the potential to make a significant contribution toward these goals, as asphalt production presents promising opportunities for reducing emissions, particularly CO_2 and total carbon (C_{tot}).

It is important to recognize that the production of asphalt binder involves several energy-intensive processes, including crude oil extraction, transportation, and refining. These processes collectively consume significant amounts of energy. For every ton of asphalt binder produced, approximately 4900 MJ of energy is required, leading to the emission of around 285 kg of CO_2 [25]. In terms of energy consumption, the production of bitumen for asphalt pavement is less demanding than the production of cement. It is essential to observe, however, that bitumen is the only energy-intensive component in asphalt pavement [23]. Due to the need for elevated temperatures, the process of mixing asphalt concrete requires a considerable amount of energy. In addition, the essential action of preheating aggregate during the mixing process contributes to the overall energy intensity of asphalt pavement production.

A case study based in China proposed a comprehensive method for assessing GHG emissions from asphalt pavement construction using a process-based LCA technique [26]. The study sheds light on the GHG emissions associated with various phases of bitumen pavement construction. The results demonstrate that the phase of mixture blending accounts for approximately 54% of the total GHG emissions. In the case of cement-stabilized base/subbase, where it accounts for approximately 98% of emissions, the fabrication of raw materials also plays a significant role. When implementing measures to reduce GHG emissions in asphalt pavement construction, these findings highlight the significance of focusing on the raw material manufacturing phase. However, it should be noted that, excluding the production of basic materials, efforts to reduce emissions should be focused on the mixture mixing phase. This study offers insights that can inform strategies and interventions for reducing GHG emissions in the asphalt pavement construction industry (Figure 4.5).

Embodied Carbon Emission of Steel Production

The production of steel is a complex process that begins with the reaction between iron ore and a reducing agent, typically coking coal, in blast furnaces. This reaction results in the production of molten iron, which is further processed to obtain steel. It is important to note that the primary source of CO_2 emissions in steel production is the reaction of iron ore with carbon, contributing to approximately 70–80% of the total CO_2 emissions [13].

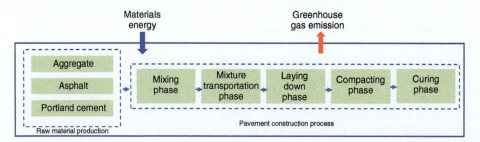

Figure 4.5 Evaluation system boundary for GHG emissions during asphalt pavement construction. *Source:* Ma [26]/MDPI / Public Domain CC BY 4.0.

Utilizing input–output analysis, a comparative study examined the energy consumption and CO_2 emissions associated with building materials during the construction stage [27]. Two types of public buildings, one with a reinforced concrete structure and the other with a steel skeleton structure, are analyzed. The research reveals that the energy consumption units for the steel structure are 1126 and 1283 Mcal/m^2, which are higher compared to the reinforced concrete structure with values of 848.8 and 809.7 Mcal/m^2. These variations in energy consumption are attributed to the different building materials required for each structure. Similarly, the findings demonstrate a similar trend for CO_2 emissions, with shaped steel and cold strip iron contributing significantly to the energy consumption and CO_2 emissions in the steel structure. In the reinforced concrete building, steel bars and ready-mixed concrete were identified as having higher energy consumption and CO_2 emissions compared to other building materials.

As professionals in the building industry, it is essential for us to consider the environmental impact of steel production and explore strategies to mitigate its carbon footprint. This can involve adopting sustainable practices, such as using recycled steel or implementing energy-efficient technologies in steel production processes. By making informed choices regarding steel materials and production methods, we can contribute to reducing the overall carbon emissions associated with building construction and promote a more sustainable built environment.

Embodied Carbon Mitigation Strategies

The role and importance of lowering EC emissions from the building sector are growing in the face of the impending climate crisis. This chapter yielded five major mitigation strategies (MSs), each of which is presented and addressed separately below (Figure 4.6).

MS1: Using Materials with a Lower Embodied Carbon

Alternative materials with low EC is a popular strategy for reducing the built environment's impact to climate change. Increasing the use of wood in construction to reduce the EC emissions of buildings has been considered as one of the potential strategies. Wood, being a renewable resource that is grown and harvested, possesses the unique ability to

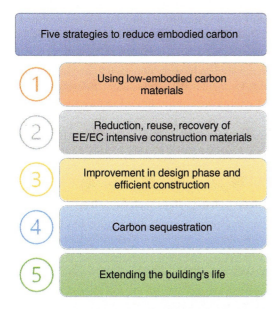

Figure 4.6 Five strategies to tackle embodied carbon emissions and contribute to decarbonizing the built environment. *Source:* Adapted from RPS, 2023 [28].

sequester CO_2 from the atmosphere through sustainable forest management practices. Moreover, specific wood products such as cross-laminated timber (CLT) and glued laminated timber (glulam) exhibit remarkable carbon storage capabilities, extending over several decades. This characteristic plays a pivotal role in offsetting the EC associated with a building, consequently reducing its overall carbon footprint.

It is crucial to highlight the significance of sourcing wood from environmentally responsible origins. The provenance of timber becomes a pivotal factor, as the EC content can vary based on whether the wood is obtained from deforested areas or sustainably managed forests. To ensure effective carbon reduction, it is imperative to have a comprehensive understanding of the timber's origin and its compliance with sustainable forestry practices. By embracing wood as a construction material and promoting responsible sourcing practices, stakeholders in the building industry can maximize the potential for EC reduction. This not only contributes to the immediate goal of mitigating carbon emissions but also fosters the establishment of sustainable, climate-resilient structures. As the adoption of wood gains momentum, it becomes essential for industry professionals to make informed choices that prioritize sustainable wood procurement, effectively harnessing its carbon sequestration capabilities to build a greener and more sustainable future. In this scenario, one study conducted a systematic literature review, analyzing 79 papers and 226 scenarios, revealing that wooden buildings exhibit significantly lower embodied emissions on average compared to buildings in general [29]. However, the study pointed out that addressing methodological variability and improving transparency in biogenic carbon accounting are crucial for accurate comparisons and reliable embodied GHG emission results.

In cement manufacturing, substituting a portion of the clinker content with industrial by-products is recommended to reduce energy requirements. It has been found that substituting up to 30% of the clinker content (by weight of total binder) does not compromise performance or strength [30, 31]. High-energy milling techniques can also be employed to enhance the reactivity and surface area of constituents, thereby improving compressive strength development [32]. Recent studies have demonstrated the potential of alkali-activated slag mortars as a replacement for regular Portland cement, offering significant environmental benefits through lower energy consumption and reduced CO_2 emissions [33]. The use of admixtures such as Peramin SRA 40, polyethylene glycol (PEG), and polypropylene glycol (PPG) further contributes to CO_2 emission reduction [34]. Alternative clinker chemistries and more energy-efficient production methods, such as fluidized bed kilns and oxy-fuel technologies, have also shown promise in reducing energy consumption and CO_2 emissions in cement manufacturing processes [35]. Garcia-Segura et al. came up with a novel way where they calculated the reduction in GHG emissions owing to the usage of blended cement, which contains more fly ash (FA) and blast furnace slag (BFS), rather than Portland cement. As a result of this strategy, emissions are expected to drop by 7–20% [36].

Below are some other examples of innovative low-carbon building materials based on the study by Chan et al. [37].

Precast Hollow-Core Slabs
Precast hollow-core slabs are widely favored in innovative construction projects, particularly for concrete flooring systems. Renowned for their self-supporting nature, superior surface finish, and efficient on-site construction, these slabs offer a host of advantages and added value. From office buildings and educational facilities to warehouses, factories, shopping centers, and residential apartments, the versatility of precast hollow-core slabs makes them a preferred choice for various structures, including wall panels, due to their exceptional quality and performance.

Steel Framework System
Steel framework is a commonly employed construction technique, employing panels constructed from thin steel plates reinforced by small steel angles along the edges. These panel units can be seamlessly interconnected using suitable clamps, allowing for flexible customization in terms of dimensions and shapes. Primarily utilized in the construction of high-rise buildings, steel formwork demands meticulous handling due to its substantial initial and maintenance costs, emphasizing the need for proper care to ensure its longevity and durability in construction projects.

Use of Unfired Brick
Unfired brick, also known as earth masonry, is a construction material that undergoes a manufacturing process involving the addition of additives to enhance its resilience. Widely employed in structural construction, the preference for reduced wall thickness, approximately 100 mm, akin to standard concrete blockwork and fired clay bricks, serves to optimize construction costs and maximize usable space within buildings by minimizing structural loading. Notably, the hygroscopic nature of this brick offers environmental regulation benefits, while its usage as a construction material boasts a commendably low carbon footprint, aligning with sustainable building practices [38].

Ethylene Tetrafluoroethylene

Ethylene tetrafluoroethylene (ETFE), renowned for its remarkable flexibility and impressive light transmission capacity of up to 95%, has emerged as a compelling choice for building cladding applications. With embodied energy values ranging from 26.5 to 210 MJ/kg, ETFE presents itself as a durable and high-strength fluorine-based plastic, providing a robust alternative to polytetrafluoroethylene (PTFE) for various architectural purposes. Its exceptional properties make it an attractive solution for architects and designers seeking innovative and sustainable building envelope solutions.

MS2: Reducing, Reusing, and Recovering–Heavy Building Materials

Implementing effective strategies for resource conservation and waste reduction in the construction industry entails adopting practices such as material reclamation from deconstructed structures, procurement of recycled or reconstituted construction and demolition (C&D) materials, and the incorporation of prefabricated building components designed for easy reusability. These sustainable approaches not only minimize environmental impact but also offer significant benefits in terms of cost-effectiveness, project efficiency, and the promotion of circular economy principles within the built environment. For example, the carbon footprint of high recycled content steel is significantly lower compared to virgin steel, with potential reductions of up to five times [39]. Prioritizing the use of recycled steel in construction projects offers substantial environmental benefits and promotes sustainability.

A comprehensive analysis comparing manufacturer-reported environmental product declarations and estimated energy inputs for deconstruction reveals that reusing materials can result in CO_2e emissions reductions of up to 99% [40]. For instance, the production of a standard commercial door system emits 155 kg CO_2e, whereas salvaging the same door system emits less than 1 kg CO_2e. This highlights the substantial environmental benefits and carbon emission savings achievable through material reuse practices in the construction industry. In order to assess the energy implications during the decommissioning phase of a concrete high-rise commercial structure, another study performed a comprehensive life cycle energy assessment [41]. The study aimed to determine the most effective waste management strategies with significant potential for embodied energy savings. The results indicated that recycling offered the highest energy-saving potential, with a significant reduction of 53%, compared to 6.2% for reuse and 0.4% for incineration. Structures containing substantial amounts of concrete, such as upper-floor construction, should prioritize recycling strategies, while high-aluminum-content building materials, like windows, are better suited for repurposing rather than recycling, ensuring optimal energy utilization and environmental benefits (Figure 4.7).

MS3: Improvement in Design Phase and Efficient Construction

In the pursuit of reducing EC in the construction industry, the implementation of design for deconstruction methodologies and sound design practices during the early stages of a project have been recognized as essential. Acquaye and Duffy conducted an insightful input–output analysis of the Irish construction industry, revealing that improved design

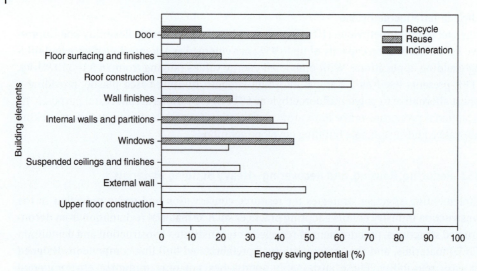

Figure 4.7 Potential energy savings from incorporating alternative approaches into various construction components. *Source:* Adapted from Ng and Chau [41]/with permission of Elsevier.

decisions could potentially yield a remarkable 20% reduction in indirect emissions and a 1.6% reduction in direct emissions. This significant improvement translates into a substantial total reduction of 3.43 MtCO$_2$e, highlighting the pivotal role of proactive design strategies in achieving EC reduction and mitigation objectives [42]. In another study conducted by Basbagill et al. the application of LCA was explored as a valuable tool for designers to enhance their understanding of the environmental impacts. Through the optimization of critical characteristics, including pile and footing thickness, as well as external and interior wall thickness, significant reductions in the building's maximum total embodied effect ranging from 63% to 75% were achievable. This research underscores the importance of leveraging LCA to inform design decisions and drive substantial improvements in the environmental performance of buildings [19].

On the other hand, the strategy of "designing for disassembly" plays a crucial role in reducing the EC of buildings [43]. By adopting this approach, designers and construction professionals prioritize methods that enable buildings to be easily renovated or dismantled at the end of their life cycle. This involves careful consideration of the selection and arrangement of building systems, materials, and components with the aim of maximizing their potential for reuse or recycling as reconstituted building materials. By facilitating the efficient disassembly and reuse of building elements, we can significantly minimize the environmental impact associated with new material extraction and production, thereby reducing the overall EC footprint of buildings. Through innovative design and construction practices, we can embrace the principles of circularity and create a more sustainable built environment that fosters resource conservation and minimizes waste generation. By carefully considering disassembly and dismantling during the planning stage, designers can facilitate the efficient modification of buildings, enabling them to evolve alongside advancements in technology. This forward-thinking approach recognizes the dynamic nature of our technological environment and emphasizes the importance of designing buildings that can seamlessly accommodate future changes. By embracing design strategies that prioritize adaptability, the built environment can remain responsive and resilient, enabling occupants to harness the benefits of emerging technologies without the need for extensive renovations or demolitions. This approach not only reduces the EC associated with frequent building alterations but also supports sustainability by prolonging the lifespan and relevance of structures in a rapidly evolving world.

MS4: Carbon Sequestration

Incorporating agricultural products with carbon sequestration potential presents a compelling opportunity for reducing the EC of construction projects. While wood stands as a well-known choice, it is crucial to explore alternative options such as fiber or hemp insulation [39]. These innovative materials offer distinct advantages, being annually renewable resources that contribute to sustainable building practices. Moreover, the integration of plant-based materials in structural insulated panels (SIPs) and other exterior panels presents a promising avenue [44]. These materials have the potential to contribute to net carbon sequestration or significantly reduce EC when compared to conventional petroleum- or gypsum-based alternatives. While straw bale SIPs and prefabricated straw bale wall panels have demonstrated successful implementation in selective residential projects, their adoption in the commercial market is not yet widespread. Nevertheless, these innovative applications showcase the growing potential of plant-based solutions in the construction industry, offering environmentally friendly alternatives that align with the principles of sustainable building practices.

MS5: Extending the Building's Life

An essential approach to mitigating EC is to prioritize the design and construction of buildings that exhibit durability, adaptability, and ease of maintenance, with the ultimate objective of extending their lifespan. By adopting a "building to last" mindset, we not only minimize the EC emissions associated with end-of-life demolition and disposal but also effectively reduce the need for frequent renovations and maintenance. This holistic approach not only ensures the longevity of the building but also contributes significantly to reducing the overall EC footprint throughout its entire life cycle. By implementing sustainable materials, robust structural systems, and efficient maintenance practices, we can optimize the building's performance, reduce resource consumption, and enhance its resilience to meet the challenges of the future.

Operation Carbon Emissions in Building Environment

The building industry is a significant consumer of nonrenewable energy resources and a substantial emitter of CO_2 during the operation phase, contributing to environmental challenges and climate change [45]. Residential and commercial structures consume considerable amounts of energy for heating, ventilation, lighting, and other purposes. The subsequent paragraph illustrates a case study of operational energy utilization and the corresponding carbon emissions (Figure 4.8).

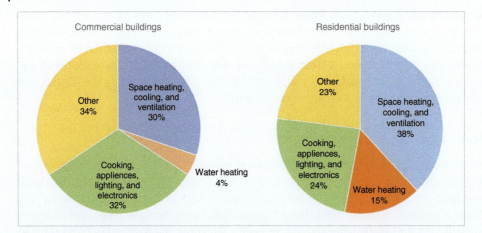

Figure 4.8 Carbon dioxide emissions from US commercial and residential structures for the year 2016. *Source:* Adapted from Conti [46].

Evaluation of CO_2 emissions associated with electricity use, heating, ventilation, and air conditioning (HVAC) system usage, and hot water usage has been performed in a US university [47]. The research begins by assessing the overall on-campus consumption of electricity, natural gas, and oil for a complete year. Based on this data, the carbon emissions per unit of energy consumption were calculated for each category. By extrapolating these figures, the total CO_2 emissions for the entire university were estimated. The results indicated that the university's activities contribute approximately 4 tons of CO_2 emissions per student annually, with a total of nearly 38,000 tons emitted during the 2007 fiscal year. Furthermore, it is identified that on-campus steam production accounts for approximately 57% of the total CO_2 emissions, and the implementation of two proposed cogeneration facilities is suggested to improve these emissions.

Operation Carbon Mitigation Strategies

Within the building sector, there is a common belief that the emissions generated during the operational phase of a building exceed its embodied emissions. As a result, significant emphasis has been placed on reducing energy consumption during this phase. Encouragingly, advancements in renewable energy technologies, energy efficiency measures, and strategies to promote behavioral change have shown great potential in achieving substantial reductions in operational emissions for buildings.

Enhanced building codes and efficiency standards for equipment and appliances have played a crucial role in achieving notable advancements in energy efficiency within the building sector. Analysis of data provided by the US Energy Information Administration reveals a substantial reduction in energy intensity between 2007 and 2017, with a decrease of 19% in residential buildings (measured in million Btu per household) and 15% in commercial buildings (measured in thousand Btu per square foot) [48].

Efficient HVAC Systems in Buildings

In the context of addressing climate change, the reduction of carbon emissions from HVAC systems becomes paramount. Thus, it is crucial to prioritize the seamless integration of highly efficient HVAC systems within buildings. By implementing efficient operational practices and adhering to regular maintenance protocols, buildings can significantly reduce their carbon footprint and contribute to climate change mitigation efforts.

One effective strategy is to adopt scheduled heating and cooling cycles during predetermined hours, adjusting temperature settings based on seasonal variations. This optimized approach minimizes energy usage, thereby mitigating the impact of HVAC systems on climate change. Similarly, leveraging off-peak hours to adjust temperature settings intelligently contributes to energy conservation and emission reduction.

To further curb carbon emissions, attention should be directed toward the continuous operation of ventilation systems, even when outside air is unnecessary. By implementing smart controls and sensors to monitor indoor air quality, buildings can tailor ventilation rates to actual needs. This not only improves energy efficiency but also reduces reliance on electricity and natural gas for HVAC systems, resulting in tangible reductions in both energy bills and carbon footprint.

Additionally, replacing conventional electric steam humidifiers with low-energy alternatives represents an effective measure for reducing the carbon footprint of buildings. This simple yet impactful step can lead to substantial energy savings, contributing to climate change mitigation.

By adopting these comprehensive strategies, buildings can actively promote sustainable practices and play a pivotal role in addressing climate change. The integration of highly efficient HVAC systems, coupled with efficient operational practices and the implementation of innovative technologies, sets a solid foundation for achieving climate change mitigation targets and establishing environmentally conscious building standards.

Renewable Resources Integration

Integrating renewable solutions into the project, such as solar panels, water recycling systems, bioenergy, and efficient waste management systems, presents viable options for reducing operational carbon emissions and mitigating the impacts of climate change. By embracing these sustainable technologies, we can not only achieve significant cost savings but also contribute to a greener future. Solar panels harness the power of the sun to generate clean electricity, reducing our reliance on fossil fuels and minimizing carbon emissions. Water recycling systems help conserve this precious resource while reducing the energy needed for water treatment. Implementing bioenergy solutions allows us to tap into renewable sources for heating and power generation, further reducing our carbon footprint. Lastly, efficient waste management systems promote recycling and waste-to-energy conversion, diverting waste from landfills and minimizing methane emissions. By embracing these renewable options, we can make substantial strides toward achieving our sustainability goals, reducing operational costs, and combating climate change.

Strategy for Water Use

Water management plays a significant role in the carbon footprint of buildings, as the energy required for water supply, treatment, and usage contributes to GHG emissions. Implementing water-oriented strategies offers a promising avenue for reducing energy consumption and mitigating the impacts of climate change [49]. By prioritizing water conservation, efficiency, and reuse in building design, we can unlock substantial energy and carbon reduction opportunities that are both effective and cost-efficient [50].

One key aspect is addressing water losses due to leaks, known as "non-revenue water," which significantly impact water efficiency. By specifying plumbing equipment that prevents leaks, we can minimize water waste and associated energy consumption, thereby reducing our carbon footprint. Additionally, incorporating high-efficiency fixtures and appliances can significantly decrease water usage. For example, opting for toilets with reduced flush volumes compared to conventional ones can effectively conserve water.

Embracing rainwater as a valuable water source presents another sustainable strategy. Designing buildings to harvest rainfall and creating outdoor spaces that primarily rely on precipitation for irrigation can lead to substantial water savings. Rainwater harvesting, coupled with xeriscaping techniques that promote water-efficient landscaping, can result in up to a 50% reduction in outdoor water use. Moreover, rainwater harvesting boasts a low environmental footprint since it eliminates the need for long-distance pumping. By utilizing rainwater directly from a building's roof for activities such as flushing toilets, irrigation, and operating washing machines, we can further minimize our carbon footprint.

These water-oriented strategies not only contribute to efficient resource management but also align with the urgent need to address climate change. By reducing energy consumption and GHG emissions associated with water use in buildings, we actively mitigate the environmental impacts and promote sustainable practices [51]. Integrating these approaches into building design and operations enables us to make substantial progress toward our climate change mitigation goals while ensuring a more resilient and environmentally conscious future.

Use of Lighting

Optimizing lighting systems and incorporating advanced daylighting techniques are vital to reducing the carbon footprint of buildings. Lighting alone accounts for approximately 40% of energy consumption in typical commercial buildings, making it a significant area for energy savings and climate change mitigation [52]. Balancing solar thermal gain is crucial, as excessive gain in summer can lead to overheating and increased cooling demands, while insufficient gain in winter requires additional heating.

To address these challenges, the utilization of solar control window films emerges as an effective solution, offering potential energy savings of up to 30% by reducing overall energy expenditures [53]. By implementing these films, we can achieve a balance between natural lighting and solar heat gain, optimizing energy efficiency and reducing carbon emissions.

Promoting daylighting strategies is another valuable approach, as endorsed by Steve Fronek, a recognized expert in sustainable building design. Open-plan offices and light-colored interior finishes help distribute daylight more evenly throughout the building, while high-performance curtain wall systems with integrated sunshades and light shelves enhance natural light penetration. Effective daylighting strategies not only enhance the visual environment but also contribute to significant reductions in HVAC peak loads. This, in turn, results in decreased mechanical equipment capacity requirements and a reduced carbon footprint.

When implementing daylighting strategies, it is crucial to consider site-specific and building-specific conditions. Evaluating window properties such as U-factor, solar heat gain coefficient, and condensation resistance provides valuable insights to inform a sound daylighting strategy. By leveraging the right combination of these properties, we can optimize energy performance and enhance occupant comfort while effectively addressing climate change challenges.

Incorporating state-of-the-art lighting systems, leveraging daylighting techniques, and deploying solar control window films all play a pivotal role in reducing energy consumption, mitigating climate change, and fostering sustainable building practices. By embracing these strategies, we can achieve substantial energy savings, lower carbon emissions, and create healthier, more environmentally responsible built environments.

Conclusion

Buildings are a major contributor to climate change and one of the primary sectors where a lot of innovation is required if we are to avert a climate catastrophe. Building emission reduction is more than a technological challenge. Government and corporate policies can also be quite beneficial. Also, building codes must be strengthened to ensure that buildings are not only energy efficient, but also constructed using low-carbon materials. Policymakers have at their disposal two key approaches to address energy consumption in the building sector: constructing new energy-efficient structures and renovating existing ones [54]. The effectiveness of energy efficiency policies in reducing energy consumption can be evaluated using two methods: real-world experimentation and computer simulations that replicate real-world dynamics. These simulations serve as valuable tools for assessing the impact of energy efficiency policies in a controlled environment, enabling policymakers to make informed decisions based on comprehensive data analysis. By employing a combination of real-world trials and advanced computer simulations, policymakers can refine energy efficiency policies, optimize energy consumption in the building sector, and contribute to sustainable and environmentally responsible development.

The decarbonization challenge for buildings is influenced by the various phases a building goes through and the different stakeholders involved, including the construction team, owners, and occupants. One significant aspect to consider is the EC of building materials, which refers to the emissions associated with the production, transportation, and construction of these materials. It is crucial to address both operational and EC emissions to achieve comprehensive decarbonization.

A notable issue arises when the entity responsible for constructing or owning a building may not prioritize the installation of energy-efficient appliances or equipment, thereby neglecting the potential reduction in operational carbon emissions. This split incentive tends to favor lower upfront costs, overlooking the long-term savings that could be achieved through enhanced energy efficiency and reduced operational carbon footprint.

Furthermore, the transition of existing buildings to alternative fuels or technologies, such as heat pumps, can be hindered by upfront financial barriers. The EC of building materials also comes into play here, as the replacement or retrofitting of certain components may be required. In both the commercial and residential sectors, there are potential financial incentives available to encourage such investments while considering the EC implications. These incentives encompass streamlined loan processes, rebates, favorable loan terms, assistance programs for weatherization in low-income households, funding options like Property Assessed Clean Energy (PACE), as well as tax credits for the installation of on-site renewable energy systems and the pursuit of specific green building certifications. These measures aim to address the economic challenges associated with decarbonizing buildings, both in terms of operational and EC, and contribute to mitigating climate change by promoting the adoption of sustainable and energy-efficient practices.

References

1 International Energy Agency (2022). Global buildings sector CO_2 emissions and floor area in the Net Zero Scenario, 2020-2050. https://www.iea.org/data-and-statistics/charts/global-buildings-sector-co2-emissions-and-floor-area-in-the-net-zero-scenario-2020-2050.
2 United Nations Environment Programme (2021). Global status report 2021 for buildings and construction: towards a zero-emission, efficient and resilient buildings and construction sector. Nairobi. Onu. https://globalabc.org/sites/default/files/2021-10/GABC_Buildings-GSR-2021_BOOK.pdf.
3 Röck, M., Saade, M.R.M., Balouktsi, M. et al. (2020). Embodied GHG emissions of buildings – The hidden challenge for effective climate change mitigation. *Appl. Energy* 258.
4 World Green Building Council (2019). Bringing embodied carbon upfront. World Green Building Council. https://www.worldgbc.org/sites/default/files/WorldGBC_Bringing_Embodied_Carbon_Upfront.pdf.
5 IEA (2014). Technology roadmap: energy efficient building envelopes. https://www.iea.org/publications/freepublications/publication/technology-roadmap-energy-efficient-building-envelopes.html.
6 Zhong, X., Hu, M., Deetman, S. et al. (2021). Global greenhouse gas emissions from residential and commercial building materials and mitigation strategies to 2060. *Nat. Commun.* 12 (1), 6126.
7 Akbarnezhad, A. and Xiao, J. (2017). Estimation and minimization of embodied carbon of buildings: a review. *Buildings* 7 (1): 5.
8 Mannan, M. and Al-Ghamdi, S.G. (2020). Life cycle embodied energy analysis of indoor active living wall system. *Energy Rep.* 6: 391–395.
9 Cameron, L. (2020). Data to the rescue: Embodied carbon in buildings and the urgency of now. , McKinsey & Company. https://www.mckinsey.com/business-functions/operations/our-insights/data-to-the-rescue-embodied-carbon-in-buildings-and-the-urgency-of-now#.

10 British Standards Institution (2011). Sustainability of construction works – Assessment of environmental performance of buildings – Calculation method. British Standards Institution.

11 Sizirici, B., Fseha, Y., Cho, C.S. et al. (2021). A review of carbon footprint reduction in construction industry, from design to operation. *Materials* 14 (20): 6094.

12 Choate, W.T. (2003). Energy and emission reduction opportunities for the cement industry. Energy Efficiency and Renewable Energy, pp. 1–41.

13 Kittipongvises, S. (2017). Assessment of environmental impacts of limestone quarrying operations in Thailand. *Environ. Clim. Technol.* 20 (1): 67–83.

14 Mohammad, M., Masad, E., and Al-Ghamdi, S.G. (2020). 3d concrete printing sustainability: a comparative life cycle assessment of four construction method scenarios. *Buildings* 10 (12): 1–20.

15 Khan, S.A., Koç, M., and Al-Ghamdi, S.G. (2021). Sustainability assessment, potentials and challenges of 3D printed concrete structures: a systematic review for built environmental applications. *J. Clean. Prod.* 303: 127027.

16 Ramsden, K. (2020). *Cement and Concrete: The Environmental Impact*. Princeton Student Climate Initiative.

17 Nielsen, C.V. (2008). Carbon footprint of concrete buildings seen in the life cycle perspective. In: *Proceedings NRMCA 2008 Concrete Technology Forum*, 1–14.

18 Fayomi, G.U., Mini, S.E., Fayomi, O.S.I., and Ayoola, A.A. (2019). Perspectives on environmental CO_2 emission and energy factor in cement industry. *IOP Conf. Ser.: Earth Environ. Sci.* 331 (1): 012035.

19 Basbagill, J., Flager, F., Lepech, M., and Fischer, M. (2013). Application of life-cycle assessment to early stage building design for reduced embodied environmental impacts. *Build. Environ.* 60: 81–92.

20 Vázquez-Calle, K., Guillén-Mena, V., and Quesada-Molina, F. (2022). Analysis of the embodied energy and CO_2 emissions of ready-mixed concrete: a case study in Cuenca, Ecuador. *Materials* 15 (14): 4896.

21 EPA (2021). U.S. Cement Industry Carbon Intensities (2019). https://www.ecfr.gov/cgi-bin/text-.

22 Statista 2023. Greenhouse gas emissions from cement production in the United States from 1990 to 2021(in million metric tons of CO_2 equivalent). https://www.statista.com/statistics/451804/green-house-gas-emissions-in-united-states-from-cement-production/.

23 Abey, S.T. and Kolathayar, S. (2020). Embodied energy and carbon emissions of pavements: a review. *Lect. Notes Civ. Eng.* 36: 167–173.

24 The United Nations (2022). *Sustainable Development Goals Report Infographics*, 1–18.

25 Chehovits, J. and Galehouse, L. (2010). Energy usage and greenhouse gas emissions of pavement preservation processes for asphalt concrete pavements. In: *First International Conference on Pavement Preservation*, 27–42.

26 Ma, F., Sha, A., Lin, R. et al. (2016). Greenhouse gas emissions from asphalt pavement construction: a case study in China. *Int. J. Environ. Res. Public Health* 13 (3): 351.

27 Chae, C.-U., Lee, K.H., and Jung, C.S. (2005). Comparative study on the amount of Co_2 emission of building materials between reinforced concrete and steel structure buildings using the input-output analysis. In: *The 2005 World Sustainable Building Conference*.

28 rps 2022. Embodied carbon: what it is and how to tackle it | RPS. https://www.rpsgroup.com/services/environment/sustainability-and-climate-resilience/expertise/what-is-embodied-carbon/.
29 Andersen, C.E., Rasmussen, F.N., Habert, G., and Birgisdóttir, H. (2021). Embodied GHG emissions of wooden buildings—challenges of biogenic carbon accounting in current LCA methods. *Front. Built Environ.* 7: 729096.
30 Naqi, A. and Jang, J.G. (2019). Recent progress in green cement technology utilizing low-carbon emission fuels and raw materials: A review. *Sustainability* 11 (2): 537.
31 Abdul-Wahab, S.A., Hassan, E.M., Al-Jabri, K.S., and Yetilmezsoy, K. (2019). Application of zeolite/kaolin combination for replacement of partial cement clinker to manufacture environmentally sustainable cement in Oman. *Environ. Eng. Res.* 24 (2): 246–253.
32 Kumar, R., Kumar, S., and Mehrotra, S.P. (2007). Towards sustainable solutions for fly ash through mechanical activation. *Resour. Conserv. Recycl.* 52 (2): 157–179.
33 Mikhailova, O., Šimonová, H., Topolár, L., and Rovnaník, P. (2018). Influence of polymer additives on mechanical fracture properties and on shrinkage of alkali activated slag mortars. *Key Eng. Mater.* 761: 39–44.
34 Keches, C. (2007). *Reducing Greenhouse Gas Emissions from Asphalt Materials*, vol. 66.
35 United State Environmental Protection Agency (EPA) (2010). Available and emerging technologies for reducing greenhouse gas emissions from the Portland cement industry. In: *Office of Air and Radiation*, 1–43.
36 García-Segura, T., Yepes, V., and Alcalá, J. (2014). Life cycle greenhouse gas emissions of blended cement concrete including carbonation and durability. *Int. J. Life Cycle Assess.* 19 (1): 3–12.
37 Chan, M., Masrom, M.A.N., and Yasin, S.S. (2022). Selection of low-carbon building materials in construction projects: construction professionals' perspectives. *Buildings* 12 (4): 486.
38 Sutton, A., Black, D., and Walker, P. (2011). Unfired clay masonry: an introduction to low-impact building materials. In: *Information Paper*, 1–6. BRE Publisher https://files.bregroup.com/bre-co-uk-file-library-copy/filelibrary/pdf/projects/low_impact_materials/IP16_11.pdf.
39 Strain, L. (2017). 10 Steps to reducing embodied carbon. American Institute of Architects. https://www.aia.org/articles/70446-ten-steps-to-reducing-embodied-carbon.
40 Ellsworth, A. 2018. Building material reuse: the overlooked solution to carbon reduction. Research to Action: The Science of Drawdown. https://drawdown.psu.edu/poster/retaining-embodied-carbon-energy-through-commercial-building-material-reuse.
41 Ng, W.Y. and Chau, C.K. (2015). New life of the building materials-recycle, reuse and recovery. *Energy Procedia* 75: 2884–2891.
42 Acquaye, A.A. and Duffy, A.P. (2010). Input-output analysis of Irish construction sector greenhouse gas emissions. *Build. Environ.* 45 (3): 784–791.
43 Vestian (2018). Designing for disassembly: sustainable building construction. https://www.vestian.com/blog/designing-disassembly-sustainable-building-construction/.
44 Jungclaus, M., Esau, R., Olgyay, V., and Rempher, A. (2021). Reducing embodied carbon in buildings: low-cost, high-value opportunities. https://rmi.org/insight/reducing-embodied-carbon-in-buildings?submitted=ecrpfgerbh.

45 Al-Ghamdi, S.G. and Bilec, M.M. (2015). Life-cycle thinking and the LEED rating system: Global perspective on building energy use and environmental impacts. *Environ. Sci. Technol.* 49 (7): 4048–4056.

46 Conti, J.J. (2018). Annual Energy Outlook 2018. https://www.eia.gov/pressroom/presentations/conti_020132018.pdf.

47 Riddell, W., Bhatia, K.K., Parisi, M. et al. (2009). Assessing carbon dioxide emissions from energy use at a university. *Int. J. Sustain. High. Educ.* 10 (3): 266–278.

48 Leung, J. (2018). Decarbonizing U.S. buildings. In: *Climate Innovation*, 1–6. https://www.c2es.org/wp-content/uploads/2018/06/innovation-buildings-background-brief-07-18.pdf.

49 Mannan, M. and Al-Ghamdi, S.G. (2020). Environmental impact of water-use in buildings: latest developments from a life-cycle assessment perspective. *J. Environ. Manag.* 261: 110198.

50 Yoonus, H., Mannan, M., and Al-Ghamdi, S.G. (2020). Environmental performance of building integrated grey water reuse systems: life cycle assessment perspective. In: *World Environmental and Water Resources Congress 2020*, 1–7.

51 Mannan, M. and Al-Ghamdi, S.G. (2022). Water consumption and environmental impact of multifamily residential buildings: a life cycle assessment study. *Buildings* 12 (1): 48.

52 Office of Energy Efficiency & Renewable Energy (2022). About the commercial buildings integration program. https://www.energy.gov/eere/buildings/about-commercial-buildings-integration-program.

53 Pereira, J., Teixeira, H., Gomes, M.d.G., and Rodrigues, A.M. (2022). Performance of solar control films on building glazing: a literature review. *Appl. Sci.* 12 (12): 5923.

54 Kamal, A., Al-Ghamdi, S.G., and Koç, M. (2019). Role of energy efficiency policies on energy consumption and CO_2 emissions for building stock in Qatar. *J. Clean. Prod.* 235: 1409–1424.

5

Unveiling the Nexus: Human Developments and Their Influence on Climate Change

Mehzabeen Mannan[1] and Sami G. Al-Ghamdi[1,2,3]

[1] Division of Sustainable Development, College of Science and Engineering, Hamad Bin Khalifa University, Qatar Foundation, Doha, Qatar
[2] Environmental Science and Engineering Program, Biological and Environmental Science and Engineering Division, King Abdullah University of Science and Technology (KAUST), Thuwal, Saudi Arabia
[3] KAUST Climate and Livability Initiative, King Abdullah University of Science and Technology (KAUST), Thuwal, Saudi Arabia

Introduction

For more than a century, scientists have recognized and investigated the possibility that human actions may alter Earth's climate through greenhouse gas (GHG) emissions. There is now little room for question, according to comprehensive scientific evidence across many fields, that human impact is a major cause of climate change. The Intergovernmental Panel on Climate Change's (IPCC) Fourth Assessment Report in 2013 unequivocally stated that human influence on the climate system is clear, as evident from the increasing concentrations of GHGs in the atmosphere, positive radiative forcing, observed warming, and our understanding of the climate system [1].

Today, the use of fossil fuels, coupled with excessive water consumption, is accelerating the rise in Earth's temperature at an unprecedented rate due to excessive carbon dioxide (CO_2) emissions, surpassing anything experienced since the dawn of civilization (Figure 5.1). This rapid warming has far-reaching consequences, impacting our planet's ability to provide food and freshwater, as well as jeopardizing human health and overall well-being. The intensification of climate-related events, such as extreme weather phenomena and wildfires, is occurring with greater frequency and severity than scientists had predicted even just a decade ago. As we look ahead, the projected substantial growth of the global population by the end of this century adds to the urgency of addressing our impact on the environment. Without a transformation in our energy and water-use patterns, the degradation of ecosystems and the persistent threat of climate change will continue to undermine human prospects. The need to limit the impacts of climate change demands a collective global effort to shift toward renewable energy sources, adopt sustainable water production practices, and embrace responsible consumption patterns.

Sustainable Cities in a Changing Climate: Enhancing Urban Resilience. First Edition.
Edited by Sami G. Al-Ghamdi.
© 2024 John Wiley & Sons Ltd. Published 2024 by John Wiley & Sons Ltd.

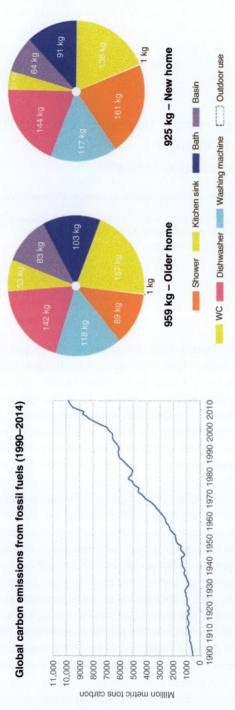

Figure 5.1 Trends in global carbon emissions from fossil fuels and carbon emissions from daily water use.
Source: Boden et al. [2]; Timperly [3].

Transitioning away from fossil fuels to renewable energy sources holds immense potential for mitigating climate change. Harnessing the power of solar, wind, hydro, and geothermal energy can reduce GHG emissions and promote a sustainable energy future. Alongside this shift, sustainable water production practices should prioritize efficient water usage, reuse, and conservation to alleviate pressure on freshwater resources and prevent further ecological imbalances. However, addressing climate change goes beyond energy and water. It necessitates transforming our consumption patterns and embracing a circular economy, where resources are utilized efficiently, waste is minimized, and materials are recycled or repurposed. Sustainable land-use practices, including reforestation, land restoration, and the preservation of natural habitats, can help sequester CO_2 and restore ecological balance.

Overall, human activities have undeniably played a significant role in driving climate change. The scientific consensus is clear, and the evidence is overwhelming. We must act swiftly and decisively to limit the impacts of human development on climate change and safeguard our planet for future generations. This requires a global commitment to transitioning to renewable energy sources, implementing sustainable water production practices, and adopting responsible consumption patterns.

Life Cycle Assessment for Environmental Impact

Life cycle analysis is a holistic analytical approach that integrates the principles of life cycle assessment (LCA), life cycle cost analysis (LCCA), and other related methodologies [4]. Throughout its evolution, LCA has progressively expanded its focus to encompass the environmental consequences of energy technologies as well as water systems [5–8]. This comprehensive methodology enables the evaluation of the environmental, economic, and social dimensions of energy systems, encompassing the entire life cycle of a product or process, from the extraction of raw materials to its utilization in performing work. By adopting a system thinking approach, LCA examines the interconnectedness and interdependencies of these stages, providing a comprehensive understanding of the environmental footprint associated with energy and water systems [9].

By utilizing LCA, stakeholders can gain a deeper understanding of the potential environmental impacts, resource consumption, GHG emissions, and other relevant factors associated with different energy and water systems. This knowledge can inform policy development, technological advancements, and investments in new infrastructure, facilitating the transition toward more sustainable and environmentally friendly energy and water solutions.

One of the key phases in LCA is the life cycle impact assessment (LCIA). During this phase, the evaluation is conducted to assess the potential environmental impacts that arise from the elementary flows, which are the environmental resources and releases identified in the life cycle inventory phase [10]. The LCIA in LCA plays a crucial role in translating the material and energy inputs of a system into meaningful impact indicators. These indicators serve as a means to quantify the severity of various impact categories on the environmental burden. The process of deriving these indicators follows a set of recommended steps outlined by ISO 14040 and ISO 14044 standards (ISO 2006a, 2006b).

Goedkoop et al. (2009) introduced a method known as ReCiPe for conducting LCIA [11]. This innovative approach offers a standardized set of characterization factors at both midpoint and endpoint levels, ensuring harmonization and consistency in the assessment process. Midpoint indicators capture the impacts at an intermediate level, such as resource depletion, global warming potential (GWP), and acidification. Endpoint indicators, on the other hand, provide a broader assessment of the overall environmental impacts, including human health effects, ecosystem damage, and resource scarcity.

ReCiPe Impact Category: Climate Change

Within the ReCiPe methodology of LCIA, the "Climate Change" impact category focuses on quantifying and evaluating the environmental consequences associated with changes in the Earth's climate system due to GHG emissions and their effects on temperature rise, sea-level rise, and other related factors. It considers the emissions of GHGs throughout the life cycle of a product, process, or system and quantifies their impact on radiative forcing, temperature change, and subsequent impacts on ecosystems and human health (Figure 5.2).

Within the impact category of climate change, the damage modeling process consists of several distinct steps. Initially, the release of a GHG (expressed in kilograms) contributes to an elevated atmospheric concentration of these gases (measured in parts per billion, ppb). Subsequently, this increase in concentration enhances the radiative forcing capacity (measured in watts per square meter, W/m^2), thereby influencing the global mean temperature (measured in degrees Celsius, °C). The rising global mean temperature resulting from increased radiative forcing has far-reaching consequences for both human health and

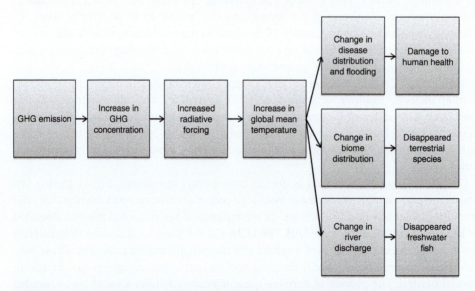

Figure 5.2 The link between greenhouse gas emissions and their impacts on human health and biodiversity loss in terrestrial and freshwater ecosystems. *Source:* National Institute for Public Health and the Environment [11]; De Schryver et al. [12].

ecosystems. These temperature elevations can lead to adverse impacts such as heat stress, changes in disease patterns, and disruptions to agricultural productivity, thereby affecting human well-being. Additionally, ecosystems face significant challenges, including altered habitats, loss of biodiversity, and increased vulnerability to extreme weather events as a result of temperature increases.

The climate change impact category is commonly characterized using the GWP. The GWP quantifies the additional radiative forcing, integrated over a specific time period (such as 20, 100, or 1000 years), resulting from the emission of 1 kg of GHGs compared to the radiative forcing caused by the release of 1 kg of CO_2 over the same time horizon. The measure of radiative forcing integrated over time resulting from the emission of 1 kg of GHGs is known as the absolute global warming potential (AGWP) and is expressed in the unit of $W/m^2 \cdot yr \cdot kg$. This allows for a standardized assessment of the warming potential of different GHGs relative to CO_2. Given a greenhouse gas (x) and a temporal horizon (TH), we may get their midpoint characterization factor as follows:

$$GWPx, TH = \frac{AGWPx, TH}{AGWPCO_2, TH}$$

Energy Sector Impact on Climate Change

The year 2021 witnessed a disheartening record in the realm of climate change. Global CO_2 emissions from energy combustion and industrial processes surged to an all-time high, marking a significant setback in our efforts to address the climate crisis. According to the International Energy Agency (IEA), these emissions rose by 6% compared to 2020, reaching a staggering 36.3 gigatonnes (Gt) [13]. The Covid-19 pandemic had initially brought about a glimmer of hope, as it led to a 5.2% reduction in global CO_2 emissions in 2020. However, the subsequent year witnessed a rapid economic recovery driven by unprecedented fiscal and monetary stimulus, coupled with the uneven rollout of vaccines. As a result, energy demand rebounded significantly in 2021, compounding the adverse effects of extreme weather events and volatile energy market conditions. Despite witnessing the largest-ever annual growth in renewable power generation, there was a surge in coal consumption. In the year 2022, the electric power sector of the United States accounted for approximately 1539 million metric tons (MMmt) of CO_2 emissions [14]. This figure represents approximately 31% of the total energy-related CO_2 emissions in the country, which amounted to approximately 4964 MMmt (Table 5.1).

This section describes the environmental impacts of energy production as well as electricity generation technologies, in different countries, in terms of GWP based on LCA.

Case Study 1: Electricity Generation in Turkey

Turkey's vibrant economy has spurred a remarkable surge in electricity demand, reflecting the nation's growing energy needs. Comparing the figures from 1990 to 2010, there has been a staggering fourfold increase in electricity generation. The total installed capacity of 49,524 MW facilitated the production of 211,208 GWh of electricity. While Turkey has harnessed the power of renewable sources such as hydropower, wind power, and geothermal

Table 5.1 US electric power sector CO_2 emissions by fuel type in 2022.

Source	Million metric tons	Share of total (%)
Coal	847	55
Natural gas	661	43
Petroleum	20	1
Other[1]	11	<1

[1] Includes carbon dioxide released by fossil fuel waste incineration and certain geothermal power facilities.
Source: EIA [14]/U.S. Energy Information Administration/Public Domain.

energy, the predominant share of electricity generation, approximately 73%, still relies on coal and natural gas.

Recognizing the diverse technological landscape in Turkey's electricity sector, a study conducted by Atilgan and Azapagic delved into the long-term sustainability of the country's energy infrastructure using LCA methodology. By employing LCA, the researchers aimed to comprehensively evaluate the environmental impacts associated with various electricity generation technologies employed in Turkey [15]. The functional unit of this LCA study was 1 kWh of electricity. The scope of the study incorporates the entire life cycle of electricity generation, including extraction, processing, and transportation of raw materials, as well as construction, operation, and decommissioning of power plants. The research considers all 516 power plants in Turkey, but excludes transmission, distribution, and consumption of electricity (Figure 5.3).

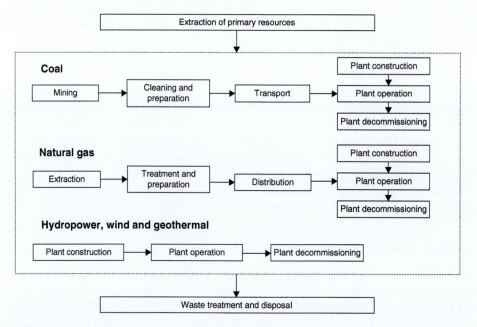

Figure 5.3 Life cycle of current Turkish electricity options. *Source:* Atilgan and Azapagic [15]/with permission of Elsevier.

To evaluate the viability of various electricity options and the Turkish electricity industry as a whole, copious amounts of data from a variety of sources have been compiled. The background life cycle inventory data are primarily derived from Ecoinvent v2.2, which has been modified to account for Turkey's particular circumstances.

Among the renewable energy alternatives, small reservoir and run-of-river hydropower exhibit the lowest GWP, estimated at 4.2 and 4.1 g CO_2-eq./kWh, respectively. Wind power is also considered relatively favorable with a GWP of 7.3 g CO_2-eq./kWh. Conversely, large reservoir hydropower shows a higher GWP of 8.3 g CO_2-eq./kWh, primarily due to emissions of CO_2 and CH_4 resulting from the degradation of submerged biomass. Geothermal power, despite being renewable, has a higher estimated GWP of 63 g CO_2-eq./kWh, making it the least favorable option among the renewable alternatives.

In contrast, fossil fuel options demonstrate significantly higher emissions. Hard coal is identified as the most environmentally damaging option, with an estimated GWP of 1126 g CO_2-eq./kWh, followed by lignite at 1062 g CO_2-eq./kWh, and gas with less than half of that, at 499 g CO_2-eq./kWh. Fuel combustion accounts for the majority of GWP associated with all three fossil fuel alternatives. Despite the higher efficiency per unit of electricity generated, hard coal exhibits a higher GWP than lignite due to additional GHG emissions from long-range transport. Moreover, hard coal plants have lower recycling credits as they require less construction material per unit of electricity generated.

Turkey's current energy policy is predominantly focused on enhancing energy security and lowering GHG emissions. However, in designing a sustainable strategy for the energy sector, the government must take into account wider environmental, economic, and social implications to avoid tackling one problem at the expense of another. By integrating these factors into the decision-making process, more sustainable decisions can be made, thereby ensuring a sustainable future. This case study highlights the importance of exhaustive evaluations and the incorporation of multiple dimensions when assessing the sustainability of electricity alternatives. It provides valuable information to policymakers and stakeholders, enabling them to make informed choices and develop a comprehensive strategy for achieving sustainable growth in the electricity sector while also contributing to the attainment of climate change goals.

Case Study 2: Coal Power Plant with Carbon Capture Technology in Czech Republic

Carbon capture, utilization, and storage is a compelling strategy for reducing CO_2 emissions in the fossil fuel-dependent power and industrial sectors. It functions as a bridge to a sustainable transition to a future energy system with zero net emissions [16]. In coal-based energy industries, the Czech Republic is proposing innovative technological approaches for the reduction of GHG emissions. One of these technologies is the post-combustion capture of CO_2 from flue gases using activated carbon adsorption in power plants. Zakuciová et al. (2020) investigated a reference power plant and then compared it with a reference power unit with a CO_2 adsorption process [17].

In this case study, LCA was conducted to evaluate the environmental impacts of temperature-swing adsorption technology for capturing CO_2 from power plant flue gases. The focus of the study was on the conditions specific to the Czech Republic. The aim was to develop a model and analyze the potential environmental effects of CO_2 adsorption on activated carbon integrated with a 250 MW brown coal power unit. Two scenarios were considered: scenario 1 assessed electricity production solely from the 250 MW coal power unit, while scenario 2 assessed electricity production from the integrated coal power unit with the CO_2 adsorption unit. The functional unit for both scenarios was defined as the power capacity (250 MW) of the coal power unit. The system boundaries encompassed the operational aspects of the power plant, including activated carbon production, emission treatment, CO_2 capture process, and waste generation. However, CO_2 compression, transport, and final storage were excluded due to limited information. The LCA analysis employed the ReCiPe 1.08 method to evaluate the environmental impacts.

The study revealed significant changes in the categories of climate change potential, terrestrial acidification, and particulate matter formation. Scenario 1 exhibited the highest contribution to the overall impacts, with fossil depletion accounting for 46.81% and climate change potential contributing 29.27%. Fossil depletion was primarily influenced by brown coal mining, while the combustion of brown coal and thermal energy production for the power unit operation had a significant impact on the climate change category. Comparatively, scenario 2, which incorporated the CO_2 adsorption process, showed a notable reduction in CO_2 levels, resulting in a substantial decrease in climate change potential (Figure 5.4).

The findings emphasize the importance of CO_2 capture technologies in mitigating climate change impacts. The integration of temperature-swing adsorption technology with a brown coal power unit demonstrated significant potential for reducing CO_2 emissions. The results highlight the significance of considering the entire life cycle of energy production and adopting sustainable practices to minimize the impacts of climate change.

Case Study 3: Solar Power with Energy Storage

Despite the widespread deployment of renewable energy technologies, the evaluation of their environmental impact remains an open question. LCA of a concentrated solar power (CSP) tower plant, with the storage of molten salts for baseload configuration, was evaluated by Gasa et al. and compared to a CSP plant without storage [18] (Figure 5.5).

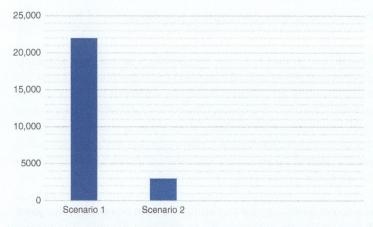

Figure 5.4 Climate change scenario comparison. *Source:* Zakuciová et al. [17]/MDPI/Public Domain CC BY 4.0.

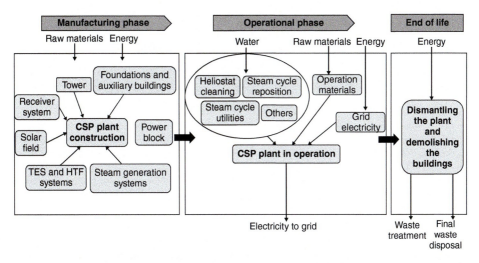

Figure 5.5 Concentrated solar power plant's system boundary. TES, thermal energy storage; HTF, heat transfer fluid. *Source:* Gasa et al. [18]/MDPI/Public Domain CC BY 4.0.

This case study focuses on the evaluation of environmental impacts using the ReCiPe and IPCC2013 GWP impact evaluation methods. The study compares two CSP tower plants with a net production of 110 MWel in a tower configuration. Both CSP plants have a projected lifespan of 30 years. The analysis considers two time horizons, GWP20a and GWP100a, to account for the different atmospheric lifetimes of gases.

One of the CSP plants in the comparison includes thermal energy storage with a capacity of 17.5 hours. The data for the LCA of the CSP plant without energy storage were derived by scaling the data of the plant with energy storage. The functional unit chosen for this study, in accordance with ISO 14040 and 14044 standards, is "1 kWh of net electricity fed to the grid." All results of the LCA are reported based on this functional unit.

To assess the environmental impacts associated with materials and energy, the Ecoinvent v3.6 database was utilized, specifically focusing on the data from the geographical area RoW (rest of the world). The ReCiPe2016 and IPCC2013 GWP were selected as the quantitative indicators for this study. The findings indicate that for the CSP plant without storage, the operational phase contributes the most to climate change impacts (Figure 5.6). In contrast, for the CSP plant with storage, the manufacturing phase is responsible for the majority of impacts, with minimal impacts observed during the operational phase.

Emissions Savings from Energy Sector

The surge in emissions underscores the pressing challenges we face in addressing climate change. It calls for bold and transformative actions from governments, businesses, and individuals to curtail GHG emissions. Transitioning to clean and renewable energy sources, promoting energy efficiency, and adopting sustainable practices across all sectors are critical steps. Prioritizing the development and deployment of innovative technologies and policies that promote decarbonization and enhance resilience against climate change impacts is essential.

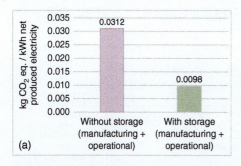 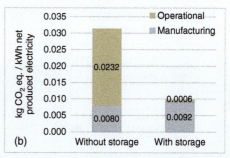

Figure 5.6 GW20a indicator per kWh of net produced electricity (IPCC technique). (a) Total impact and (b) production stage and operational stage impacts. *Source:* Gasa et al. [18]/MDPI/Public Domain CC BY 4.0.

The window of opportunity to mitigate the worst effects of climate change is rapidly closing. Seizing this moment requires substantial changes to our energy systems, industrial processes, and consumption patterns. By doing so, we can pave the way for a sustainable and prosperous future that safeguards the well-being of current and future generations. The challenge is immense, but failure is not an option if we are to protect our planet and secure a livable climate for all. A few potential ways have been discussed below to reduce the emissions from the energy sector [19].

Energy Efficiency Increase

Energy efficiency is a paramount factor in addressing global emissions reduction, as emphasized by the IEA in 2017. It holds immense potential and should be an integral component of the global energy strategy for several compelling reasons.

One of the key advantages of energy efficiency lies in its ability to substantially reduce global electricity demand while maintaining high levels of service, such as heating and lighting. Current rates in the US power sector indicate that energy efficiency measures could potentially cut global electric demand by more than 20%. This reduction can be achieved without compromising the quality of services, ensuring a sustainable and energy-efficient future. Furthermore, energy efficiency offers an economically viable pathway for large-scale decarbonization. By avoiding the need for increased energy production, capacity, and grid services, it emerges as the most cost-effective means of decarbonizing our energy systems. Notably, it can lead to the reduction of billions of tons of CO_2 emissions, making a substantial contribution toward mitigating climate change.

The significance of energy efficiency becomes evident when examining specific examples. Take lighting, for instance. In 2017, the United States achieved a significant milestone by installing over one billion LED and CFL lights. This resulted in a remarkable avoidance of 142 million tons of CO_2 emissions annually, at an estimated cost of approximately $7 per ton. In comparison, the cumulative capacity of rooftop solar reached 8000 MW in the same year, saving eight million tons of CO_2 annually but at a significantly higher cost of $360 per ton. The immense impact of lighting efficiency is evident, yielding nearly 18 times the CO_2 emissions reduction compared to rooftop solar while costing only around 2% per ton. Google exemplifies the power of energy efficiency through its data centers. The company has made impressive strides in reducing energy consumption, with its data centers now utilizing 50% less energy than the average center. Moreover, Google's development of the Tensor Processing Unit (TPU), a highly efficient chip, has revolutionized computing capabilities. The TPU outperforms standard CPUs by 83 times, enabling Google to decrease power consumption in its data centers by a staggering factor of 30. This achievement surpasses the impact of all the company's renewable energy purchases on its carbon footprint.

These examples highlight the transformative potential of energy efficiency. Its ability to achieve significant emissions reductions at a fraction of the cost compared to alternative measures makes it a compelling choice for policymakers, businesses, and individuals alike.

Wind and Solar Plant Installation

The landscape of renewable energy has witnessed remarkable advancements, with wind and solar emerging as highly cost-effective sources of power. Over the past decade, the price of solar energy has witnessed a staggering 85% reduction since 2009, while wind power costs have dropped by 50%. These substantial cost reductions have made wind and solar energy increasingly affordable and accessible to a wide range of users. Furthermore, portfolio models, backed by studies conducted by experts such as Frew, highlight the enormous potential of solar and wind energy [20]. These models indicate that, when integrated strategically with dispatchable assets such as hydroelectric power, natural gas, and electrical storage, solar and wind power can contribute up to 80% of zero-carbon energy in many grids. This integration is crucial for ensuring grid reliability and stability, as it allows for intelligent management and optimization of diverse energy sources.

The combination of solar and wind power with dispatchable assets creates a comprehensive and reliable energy portfolio. While solar and wind are renewable and abundant sources of energy, their intermittent nature requires complementary resources that can provide consistent power when sunlight or wind availability fluctuates. Dispatchable assets, such as hydroelectric power, natural gas, and electrical storage systems, play a vital role in filling the gaps and ensuring a constant and reliable supply of electricity.

The intelligent integration of these diverse energy sources paves the way for a decarbonized grid. By harnessing the abundant and renewable energy generated by solar and wind and combining it with dispatchable assets, we can establish a sustainable and low-carbon energy system. This holistic approach not only reduces dependence on fossil fuels but also contributes significantly to mitigating climate change and achieving our decarbonization goals.

As we embrace the full potential of solar and wind energy, it is essential to recognize their role as critical building blocks for a decarbonized grid. Their declining costs, coupled with strategic integration, make them powerful tools for transitioning to a sustainable energy future. By maximizing the utilization of renewable resources and intelligently managing our energy portfolio, we can create a resilient and environmentally friendly grid that serves the needs of communities while minimizing our carbon footprint.

Keep Running the Nuclear Plants

Nuclear power is a significant source of zero-carbon energy in the United States. If these nuclear plants were to close, it would require constructing three times the capacity of solar and wind plants, integrated with hydropower, natural gas, or electrical storage. Replacing existing nuclear power could take over a decade, impeding urgent climate action. The Union of Concerned Scientists recommends maintaining nuclear plants to diversify the energy mix. Advances in technology have reduced costs and improved safety, as demonstrated in countries like France, Sweden, the United Kingdom, and China. Despite varying views, nuclear power is recognized in climate models, including the IPCC's, for its role in climate stabilization. Addressing public concerns, waste management, and costs is vital for considering nuclear power as part of the solution to combat climate change.

Freshwater Sector Impact on Climate Change

The water sector's energy demand plays a substantial role in the generation of GHG emissions [21, 22]. In the United States, water and wastewater utilities' energy consumption is estimated to contribute between 1% and 4% to the nation's total electricity generation. Currently, tracking water-related GHG emissions primarily relies on accounting for electrical consumption, natural gas demand, and fuel oil usage. However, it is important to note that the carbon footprint associated with our water usage is likely increasing for several reasons. As water demand continues to rise and local, low-energy water sources become depleted, water providers are compelled to seek more distant or alternative sources that often come with higher energy and carbon costs compared to existing supplies. Additionally, the adoption of stricter water treatment standards at the state and federal levels will further increase the energy and carbon expenditures associated with treating water and wastewater. The following are a few case studies demonstrating the effects of water production using LCA.

Case Study 1: Water Supply in Singapore

Despite the country's water shortage, Singapore has developed a complex urban water system that includes various sources, innovative treatment technologies, and recycling for both potable and non-potable reuse. Singapore's public water supply, which includes tap water and recycled water, was studied using LCA in order to inform industry as well as policymakers on the environmental impact of these sources [23].

Singapore's water system incorporates diverse water sources, advanced treatment technologies, and wastewater recycling, resulting in two types of public water supply: NEWater (recycled purified water) and tap water (potable). This study utilizes a LCA approach to evaluate the environmental impacts of supplying these water sources. The LCA considers treatment plants, wastewater collection, and water distribution networks, following ISO 14040/44 guidelines. The functional unit is defined as $1\,m^3$ of delivered water. The findings indicate that $1\,m^3$ of NEWater contributes $2.19\,kg\ CO_2$-eq., while $1\,m^3$ of tap water contributes $1.30\,kg\ CO_2$-eq. to climate change potential. This LCA approach can guide improvements in the water system and serve as a reference for other cities developing their urban water systems.

Case Study 2: Seawater Desalination in South Africa

The growing urban population in South Africa is placing an increasing strain on the country's water resources, which has prompted several initiatives aimed at alleviating this problem. Both desalination of saltwater and the use of reverse osmosis (RO) membranes are currently under consideration. Both methods have come under scrutiny because of their high energy usage and environmental implications; hence, an LCA was conducted for both approaches [24].

The first case study examined a proposed RO desalination plant in the Southern eThekwini region, while the second case study focused on a mine-water reclamation process in Mpumalanga, utilizing ultrafiltration (UF) and RO technologies. The functional unit was defined as 1 kiloliter (kl) of water meeting potable water standards produced over the lifespan of each process unit. Data collection and compilation proved to be the most labor-intensive and time-consuming tasks for this study. The total energy consumption for the desalination plant was determined to be 3.69 kWh/kl, while the mine-water reclamation plant consumed 2.16 kWh/kl. Consequently, the desalination plant emitted 4.17 kg CO_2-eq./kl, whereas the mine-affected plant released 2.44 kg CO_2-eq./kl. The results indicate that the operational phase primarily drives the majority of environmental impacts associated with both systems, with energy consumption being the largest contributor, followed by chemical usage. Detailed analysis revealed that the desalination process has a greater overall environmental impact compared to the mine-water reuse process, primarily due to higher energy requirements. To mitigate these impacts, exploration into substituting fossil fuel-based energy with renewable sources, such as solar or wind energy, was undertaken. It was found that such a transition could significantly reduce the climate change effect, bringing GHG emissions down to levels comparable to conventional water treatment processes used in the eThekwini Municipality. Additionally, other technological advancements should be considered to minimize energy and chemical usage, particularly in the planning process of the desalination plant. This includes investigating pre-treatment stages, optimizing overall chemical usage, and replacing environmentally burdensome chemicals. For the long-term development of RO processes in potable water production, the promotion of alternative energy sources such as solar and wind power is recommended.

Case Study 3: Multistage Flash Desalination in Qatar

The comprehensive understanding of the environmental impact of multistage flash (MSF) desalination, particularly in the Middle Eastern and North African region, is currently limited. In Qatar, MSF desalination accounts for approximately 75% of the municipal water supply due to its process reliability and other advantages. However, this method is highly energy-intensive and imposes a significant environmental burden. Therefore, this study aimed to develop a holistic life-cycle-based framework to assess the overall environmental and human health impacts of MSF desalination in Qatar [25]. Three different MSF systems were examined by varying the gain ratio (GOR) through a LCA. Various environmental indicators, including climate change, freshwater eutrophication, fossil fuel depletion, ozone depletion, and human toxicity, were evaluated. The findings revealed that the modified MSF configuration with a higher GOR emitted 7.32 kg CO_2 per 1 m^3 of water production, while the

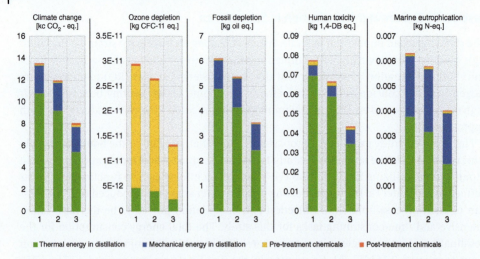

Figure 5.7 Impact results in five categories for every MSF desalination facility. The numbers 1, 2, and 3 in the illustration above represent the three MSF plants: plant 1, plant 2, and plant 3. The following is a description of the plants: the gain ratio of plant 1 is 8.21, the gain ratio of plant 2 is 9.73, and the gain ratio of plant 3 is 16.07. Source: Mannan et al. [25]/with permission of Elsevier.

plant with the lowest GOR released 12.6 kg (see Figure 5.7). This quantitative analysis of the environmental degradation associated with desalination reflects the water consumption reality in Qatar and can serve as a motivation for users to reduce their water consumption in alignment with Qatar's national vision for 2030. The implications of this study are particularly significant at a regional level, providing a preliminary baseline for developing a more efficient water strategy to mitigate the impacts on climate.

Emission Savings from Water Sector

Groundwater Management

The management of groundwater resources is crucial for addressing the challenges posed by climate change and the increasing demands of a growing population. In various regions, groundwater is being excessively extracted and contaminated, while in some areas, its availability and characteristics remain uncertain. A comprehensive understanding of groundwater systems, along with their exploration, protection, and sustainable utilization, is essential for effective adaptation to climate change and meeting the water requirements of a growing population. As a water LCA expert, it is imperative to consider the impacts of groundwater extraction and its sustainable management on the overall water life cycle, including aspects such as resource depletion, water quality, and ecological implications. By incorporating a life cycle perspective, we can assess the environmental and social implications associated with groundwater use and develop strategies to ensure its efficient and responsible utilization for long-term water security.

Energy Management in Water System

In recent years, there has been an increasing adoption of energy management practices among water utilities. These strategies encompass various approaches such as water conservation, water loss control, pump optimization, prioritizing water sources, upgrading equipment, and implementing real-time control systems. Water utilities are also becoming more involved in demand response events, leveraging their storage capacity and redundancy features, which aligns with the efforts of power providers to balance supply and demand in the grid, especially with the growing integration of intermittent renewable energy sources such as wind and solar [26].

While the primary objective of energy management programs is typically to achieve cost savings, it is important to recognize that these initiatives also have a significant impact on reducing the energy intensity of water utilities [27]. As a result, they contribute to the reduction of carbon emissions associated with water operations. It is crucial to advocate for the implementation of energy management strategies in water systems, not only for financial benefits but also to enhance overall operational efficiency, reduce environmental impacts, and promote sustainability in the water sector.

Smart Wastewater Treatment Technology

In recent years, wastewater treatment plants have shifted their focus from simply minimizing energy consumption to maximizing the net energy surplus. This shift has led to the implementation of various optimization measures to improve energy efficiency and increase the amount of carbon retained within the system [28].

One such measure is the optimization of the nitrogen removal process using online sensor control. By utilizing frequency converters, the level of aeration can be adjusted precisely according to the actual demand, resulting in reduced energy consumption while increasing carbon retention. Another effective approach is upgrading blower technology to high-speed turbo blowers. This upgrade achieves further energy reduction in the aeration process, contributing to overall energy efficiency improvements. Aerobic sludge age control, based on temperature and plant load, is another strategy employed to reduce energy consumption and enhance carbon retention. This control is achieved through the use of frequency converters, which regulate the operation of return sludge pumps. Moreover, wastewater treatment plants have seen benefits from upgrading their combined heat and power (CHP) processes for energy production. By improving the efficiency of CHP systems with up to 90% energy efficiency, plants can generate energy more effectively while utilizing the by-products for various heating and power needs.

Concluding Remarks

In conclusion, the human development activities associated with water and energy production have a significant impact on GHG emissions, thereby contributing to climate change. The demand for water and energy continues to grow, driving the need for sustainable solutions that prioritize energy efficiency, renewable resources, and carbon reduction strategies. It is crucial for policymakers, industries, and individuals to recognize the interdependency between water and energy systems and adopt integrated approaches that minimize GHG emissions throughout the entire life cycle of these essential resources. By promoting innovation, adopting sustainable practices, and investing in renewable technologies, we can mitigate the environmental consequences of human development and pave the way for a more sustainable and resilient future. Only through collective action can we create a sustainable and resilient future for ourselves and the planet we call home.

References

1 The Intergovernmental Panel on Climate Change. Climate Change 2013 - The Physical Science Basis. 2014. https://www.ipcc.ch/report/ar5/wg1/.
2 Boden, T.A., Marland, G., and Andres, R.J. (2017). *Global, Regional, and National Fossil-Fuel CO2 Emissions*. United States, Department of Energy.
3 Timperly, J. (2021). The hidden impact of your daily water use. BBC Future. https://www.bbc.com/future/article/20200326-the-hidden-impact-of-your-daily-water-use.
4 Mannan, M. and Al-Ghamdi, S.G. (2021). Complementing circular economy with life cycle assessment: deeper understanding of economic, social, and environmental sustainability. In: *Circular Economy and Sustainability, Vol. 1 Management and Policy* (ed. A. Stefanakis and I. Nikolaou), 145–160. Amsterdam, Netherlands: Elsevier.
5 Bergerson, J.A. and Lave, L.B. (2005). Should we transport coal, gas, or electricity: cost, efficiency, and environmental implications. *Environ. Sci. Technol.* 39 (16): 5905–5910.
6 Weber, C.L., Jaramillo, P., Marriott, J., and Samaras, C. (2010). Life cycle assessment and grid electricity: what do we know and what can we know? *Environ. Sci. Technol.* 44 (6): 1895–1901.
7 Mannan, M. and Al-Ghamdi, S.G. (2022). Water consumption and environmental impact of multifamily residential buildings: a life cycle assessment study. *Buildings.* 12 (1): 48.
8 Yoonus, H., Mannan, M., Al-Ghamdi, S.G. (2020). Environmental performance of building integrated grey water reuse systems: life cycle assessment perspective. In: *World Environmental and Water Resources Congress 2020: Water, Wastewater, and Stormwater and Water Desalination and Reuse - Selected Papers from the Proceedings of the World Environmental and Water Resources Congress 2020*, Henderson, Nevada, (17–21 May 2020), pp. 1–7.
9 Mannan, M. and Al-Ghamdi, S.G. (2021). Life cycle thinking and environmental assessment of energy systems from supply and demand perspectives. *Green Energy Technol.* 1: 107–126.
10 Nieuwlaar, E. (2004). Life cycle assessment and energy systems. *Encycl. Energy* 3: 647–654.
11 M. A. J. Huijbregts et al., "ReCiPe2016: a harmonised life cycle impact assessment method at midpoint and endpoint level," *Int. J. Life Cycle Assess.*, vol. 22, no. 2, pp. 138–147, Feb. 2017.
12 De Schryver, A.M., Brakkee, K.W., Goedkoop, M.J., and Huijbregts, M.A.J. (2009). Characterization factors for global warming in life cycle assessment based on damages to humans and ecosystems. *Environ. Sci. Technol.* 43 (6): 1689–1695.

13 IEA (2022). Global energy review: CO_2 emissions in 2021. IEA. https://www.iea.org/articles/global-energy-review-co2-emissions-in-2020.

14 EIA (2021). How much of U.S. carbon dioxide emissions are associated with electricity generation?, pp. 1–1. https://www.eia.gov/tools/faqs/faq.php?id=77&t=11.

15 Atilgan, B. and Azapagic, A. (2016). An integrated life cycle sustainability assessment of electricity generation in Turkey. *Energy Policy.* 93: 168–186.

16 Hasan, M.M.F., Zantye, M.S., and Kazi, M.K. (2022). Challenges and opportunities in carbon capture, utilization and storage: a process systems engineering perspective. *Comput. Chem. Eng.* 166: 107925.

17 Zakuciová, K., Štefanica, J., Carvalho, A., and Kočí, V. (2020). Environmental assessment of a coal power plant with carbon dioxide capture system based on the activated carbon adsorption process: a case study of the Czech Republic. *Energies.* 13 (9): 2251.

18 Gasa, G., Lopez-roman, A., Prieto, C., and Cabeza, L.F. (2021). Life cycle assessment (LCA) of a concentrating solar power (CSP) plant in tower configuration with and without thermal energy storage (TES). *Sustain.* 13 (7): 3672.

19 Woolard, J. (2019). Beyond renewables: how to reduce energy-related emissions by measuring what matters. World Resources Institute: Insights – Energy, pp. 2–14. https://www.wri.org/insights/beyond-renewables-how-reduce-energy-related-emissions-measuring-what-matters (accessed 20 November 2020).

20 Frew, B.A., Becker, S., Dvorak, M.J. et al. (2016). Flexibility mechanisms and pathways to a highly renewable US electricity future. *Energy.* 101: 65–78.

21 Mannan, M. and Al-Ghamdi, S.G. (2020). Environmental impact of water-use in buildings: Latest developments from a life-cycle assessment perspective. *J. Env. Manage.* 261 (1): 110198.

22 Mannan, M. and Al-Ghamdi, S.G. (2019). Life-cycle assessment of thermal desalination: environmental perspective on a vital option for some countries. In: *World Environmental and Water Resources Congress 2019 Groundwater, Sustainability Hydro-Climate/Climate Change, and Environmental Engineering - Selected Papers from World Environmental and Water Resources Congress 2019*, Pittsburgh, Pennsylvania (19–23 May 2019), pp. 449–460.

23 Hsien, C., Choong Low, J.S., Chan Fuchen, S., and Han, T.W. (2019). Life cycle assessment of water supply in Singapore – a water-scarce urban city with multiple water sources. *Resour. Conserv. Recycl.* 151: 104476.

24 Goga, T., Friedrich, E., and Buckley, C.A. (2019). Environmental life cycle assessment for potable water production – a case study of seawater desalination and mine-water reclamation in South Africa. *Water SA.* 45 (4): 700–709.

25 Mannan, M., Alhaj, M., Mabrouk, A.N., and Al-Ghamdi, S.G. (2019). Examining the life-cycle environmental impacts of desalination: a case study in the State of Qatar. *Desalination.* 452: 238–246.

26 Sowby, R.B. (2021). Increasing water and wastewater utility participation in demand response programs. *Energy Nexus.* 1: 100001.

27 Sowby, R.B. and Capener, A. (2022). Reducing carbon emissions through water conservation: an analysis of 10 major U.S. cities. *Energy Nexus.* 7: 100094.

28 Warming, M. (2014). Generating surplus power from wastewater treatment, pp. 30–33. http://www.danfoss.com/

Part II

Quantifying Resilience and Its Qualities

Part 2 of the book delves into the assessment and enhancement of resilience in urban environments and critical infrastructures. As cities continue to grow and face the challenges of climate change, it becomes imperative to understand and strengthen the resilience of interdependent infrastructures. Chapter 6 emphasizes the importance of assessing the resilience of transportation, water, and electricity systems, highlighting the need to consider their interdependencies. Chapter 7 expands on this topic by exploring various approaches for quantifying infrastructure resilience, taking into account the specific characteristics and goals of the assessment.

In Chapter 8, the focus shifts to the structural resilience of buildings, particularly in earthquake-prone areas. The chapter provides insights into the different systems within buildings and presents a representation of structural resilience, enabling the quantification and comparison of resilience across different development options. Chapter 9 broadens the perspective to encompass the resilience of the built environment in the face of climate change stresses and shocks. It discusses the essential qualities of a resilient built environment, such as reflectivity, robustness, flexibility, and inclusivity, while considering the projected impacts of climate change.

Chapter 10 explores the definitions and interrelations of urban systems' resilience qualities. It synthesizes various resilience qualities and their interconnections, drawing on evidence from current research on urban systems' resilience. Lastly, Chapter 11 focuses on the assessment of urban resilience, covering both qualitative and quantitative methods and approaches. It discusses the basis for recognizing resilience, describes different assessment methods, and presents frameworks, indices, and indicators used to evaluate the resilience of physical infrastructures.

Overall, Part 2 provides a comprehensive understanding of assessing and enhancing resilience in urban environments, critical infrastructures, and buildings. It emphasizes the need for interdependent approaches, considering multiple dimensions and qualities of resilience, and offers valuable insights and tools for researchers, practitioners, and policymakers involved in urban planning and climate change adaptation.

6

Assessing Resilience in Urban Critical Infrastructures: Interdependencies and Considerations

Mohammad Zaher Serdar[1] and Sami G. Al-Ghamdi[1,2,3]

[1] Division of Sustainable Development, College of Science and Engineering, Hamad Bin Khalifa University, Qatar Foundation, Doha, Qatar
[2] Environmental Science and Engineering Program, Biological and Environmental Science and Engineering Division, King Abdullah University of Science and Technology (KAUST), Thuwal, Saudi Arabia
[3] KAUST Climate and Livability Initiative, King Abdullah University of Science and Technology (KAUST), Thuwal, Saudi Arabia

Introduction

Over the past decades, the world's population has grown in an accelerated pattern, and this growth has been accompanied by an increased concentration in urban areas. The improvements in living standards, availability of better-paid jobs, and accessibility to medical services in urban areas are all factors that have contributed to the increased urban population over the past several decades [1, 2]. These features are all powered and supported by a substantial network of interconnected and interdependent infrastructures that provide many essential services and ensure far better living standards than those previously available in the countryside [2, 3]. These infrastructures include transportation networks, such as roads, metros, and waterways; electrical networks; water distribution; sewer systems; and communication networks [4, 5]. These networks are exposed to several sources of hazards, and their vulnerability to each type of disturbance depends on various factors such as location, magnitude, alternative availability, and recoverability, among other characteristics that describe their resilience [6, 7]. Furthermore, these threats, or hazards, have increased in both magnitude and intensity due to increased political and economic instability and the advent of climate change impacts [6, 8, 9].

Resilience thinking is considered the future approach to designing and developing new infrastructures that resolve increasing challenges and unprecedented disturbances such as climate change. The concept of resilience is relatively new and was initially introduced in ecological science by Holling in 1973 to describe an ecological system's ability to limit degradation and rebalance after a disturbance [10]; eventually, this definition has taken many forms in engineering fields, as presented in [11, 12]. However, the emergence of new

---

Sustainable Cities in a Changing Climate: Enhancing Urban Resilience. First Edition.
Edited by Sami G. Al-Ghamdi.
© 2024 John Wiley & Sons Ltd. Published 2024 by John Wiley & Sons Ltd.

challenges facing engineered systems, especially in the last two decades, and rapid urbanization have highlighted the limited efficiency of conventional design approaches and have promoted the development of resilience-based thinking in urban infrastructure design [5, 6, 13]. Furthermore, the diverse types of urban infrastructures limit the ability to provide a single and holistic resilience approach. Thus, a wide array of resilience assessment approaches are presented in the literature with varying accuracy, depending on the application and the investigated infrastructure [4–6]. This variation in assessment approaches also complicates the development of holistic assessment approaches for interdependent infrastructures, a critical component for the development of future efficient infrastructures as we increasingly move toward more propped, monitored, controlled, and automated "smart" infrastructures [4, 5, 14].

Urban areas, such as cities, can be described as giant cells having various economic and community components with the infrastructure networks facilitating their interaction. One can imagine a city as a giant cell with input, internal reaction, and output processes that are all determined by its internal structure and exposed to the surrounding environment. This portrayal allows understanding the development and efficiency of a city as a series of processes that dictates its potential and highlights its vulnerabilities. As economic activities affect the welfare of the society, whether it is an industrial or service-based economy, the capacity of the community to adapt to the changing environment and demands of the outside world indicates the robustness and vulnerability of the city and subsequently its resilience [15, 16]. However, the existence of resilient infrastructure networks directs the overall effectiveness of any urban society, such as transportation networks facilitating the community's mobility throughout the city, inputs and output streams, and emergency response teams during disasters [6]. Other essential infrastructure networks include ensuring the availability of safe and clean potable water, powering all electrical equipment (for almost all systems in the built environment), and ensuring effective communication within the area and with the outer world through secure connections with minimum delays. However, these networks are not totally independent; in fact, they rely on each other to increase their efficiency and improve the society's life quality as well as the outputs of the economic activities. Furthermore, the integration should increase the city's overall resilience against all possible disturbances and prevent severe degradation caused by failure propagation between various interdependent infrastructures [4, 14].

Urban infrastructures are complex networks (CNs), varying in nature, consisting of many components, and increasingly becoming more interdependent. Each type of urban infrastructure enjoys a set of distinguishing characteristics related to the nature of the flow that uses it as well as the main interdependencies with other networks, as shown in Table 6.1. Transportation networks have a flow of discrete nature, enabling the movement of various units that carry shipments or people, such as vehicles and wagons. Ideally, each moving unit can change direction and is relatively independent of other moving units. Electrical networks carry a continuous flow of electrons, representing the electrical current; this movement powers all the types of connected equipment with the complete connection of the whole circuit essential for it to work. The flow is obstructed by impedance in the carrying lines, causing loss in the form of heat. Water distribution networks also have a continuous flow, with the benefit of holding tanks used to store the potable water, helping

Table 6.1 Urban infrastructure flow types and their interdependencies.

Urban infrastructure	Flow	Type of flow	Main dependencies
Transportation networks	Vehicles, wagons	Discrete, requires physical connections	Electrical networks (for metro)
Water distribution networks	Potable water	Continuous, with some storage capacity	Electrical networks
Electrical networks	Electricity	Continuous	Transportation networks
Communication networks	Data	Discrete, could be wireless	Electrical networks
Sewer systems	Wastewater	Continuous	Electrical networks for pumping and lifting stations

to regulate production and create a better service condition; however, they are exposed to pollution and quality degradation and require maintaining a specific range of pressure to reach customers without damaging the pipes. Communication networks rely on several types of connectivity because some parts are connected by cables and others by wireless connections. These connections generally are not limited to data flow in one direction. Still, it is mandatory to go through a series of hubs or relay points, exposing connections to disturbance if points fail or are damaged, or the probability of intangible attacks through jamming or hacking. The sewer system also has a continuous flow, but it mainly carries foul water and rainwater in case of mixed networks that rely on gravity. While it is essential for sanitation and healthy living of any community, overflow during flooding events and leakage due to degradation during normal service conditions can lead to serious pollution issues affecting the water sources in an area and cause severe contamination and environmental damage. Thus, evaluating the resilience of these infrastructures while considering their unique characteristics is critical to understanding their behavior and performance under all possible disturbances [14].

This chapter starts by presenting the characteristics of several urban infrastructures from a resilience point of view, namely transportation, electricity, and water. Then it continues with one case study focused on a single infrastructure resilience assessment and a second focused on the resilience of interdependent systems in an assessment of interdependent infrastructures. The chapter finishes with concluding remarks.

Individual Network Resilience

The assessment of resilience in each individual system is essential for understanding its performance under possible disturbances and is an important step in the development of holistic resilience evaluation of interdependent systems. In the following subsections, we will present the resilience characteristics of three critical urban infrastructures and conclude with a case study about individual system resilience assessment to facilitate the reader's understanding.

Transportation Network Resilience

The resilience of transportation networks is critical for the efficiency of cities, both during regular daily service and during times of disaster. For example, transportation networks facilitate movement and determine the efficiency of cities by reducing the daily time wasted on congestion. Furthermore, during a disaster, networks provide much-needed mobility for recovery efforts and evacuation plans to minimize losses [50]. The resilience in transportation networks could be defined as preserving commute time due to daily peak hours (considered transient disturbances), limiting service degradation, including extensive delays or inaccessibility in the network during extreme events, and ensuring a fast return to normal, or acceptable, flow and accessibility in various parts of the network. However, the exact definition varies in the literature, as some researchers focus on network vulnerability while others focus on robustness or other characteristics that contribute to network resilience [6, 11]. Moreover, different assessment approaches rely on different performance metrics depending on the stage of interest, whether it is damage reduction or recovery capacity, which is eventually affected by the needed accuracy, resource availability, and nature of the disturbance [6, 11].

Transportation networks are characterized by their discrete flow of vehicles and generally distributed generation of trips with origin–destination pairs, as well as the capacity for rerouting. Almost any point in a city can be a source or goal for trips during different parts of the day, with some sort of storage capacity throughout the network and at parking points. The discrete nature of the flow allows for employing approaches like agent-based modeling (ABM), while the importance of accessibility and connectivity requires the use of CN approaches, using various connectivity metrics. Service metrics such as the number of completed trips and the average trip duration can also be used in performance assessment and thus represent the system's resilience. In the literature, several other approaches are also applied, and the reader can refer to [6, 12] to expand on such approaches. In addition, transportation networks have several modes, such as buses, metro lines, and private cars. These modes have distinctive characteristics, can all contribute to mobility and accessibility throughout the city, and respond differently to various disturbances [17]. Understanding these characteristics and approaches allows for choosing the most suitable assessment approach based on the disturbance scenario, the focus of the assessment, and available resources, thus facilitating an accurate assessment of the resilience of the transportation system.

Electrical Network Resilience

Electrical networks provide the power needed for all modern systems and activities in the city. The availability of a reliable and continuous electricity supply is a must in any modern community, as it powers almost all the equipment, such as HVAC (heating, ventilation, and air conditioning) systems, lighting, appliances, and even other critical infrastructures, such as water treatment and pumping. Furthermore, the demand for electricity supply is expected to increase due to the accelerated adoption of electrical cars caused by increased environmental awareness, which will eventually affect the transportation network and increase its dependency on the electrical network. Electrical networks consist of three components that are generally considered in resilience studies: generation facilities, transmission lines, and distribution networks, with most disruptions happening in the transmission section of the network, thus most studies focusing on it. Due to demand changes and fluctuations throughout the day, a reserve spinning capacity is provided on standby to address any changes in demand.

Furthermore, urban areas are typically connected in a ring system, so if a fault occurs on the main cable, the electricity supply is maintained through another main cable, and the failure is isolated from the network until the fault is repaired. The primary weak points in this system are the substations and transmission towers, which can affect the connected customers or destabilize the network. In recent years, the makeup of the electrical grid has been changing due to the rapid development and implementation of distributed generation through renewable sources like solar panels and wind turbines. Renewable sources can provide electricity supply to the main grid during outages, improving the network's resilience, especially if integrated with storage capacity as in microgrids. Nevertheless, the limitations of storage technology and the impact of atmospheric conditions on renewable energy production limit the efficiency of such approaches.

Electrical networks require a continuous flow of energy while balancing between production and demands. Due to limitations in energy storage technology, a continuous balancing process of production is crucial to satisfy and follow electricity demand, as presently no cost-effective solution is available to store the energy. Additionally, electrical current, unlike in a transportation network, is not easily rerouted. Despite the adaptation of a design mentality of extra safety, such as an $n+1$ rule, several events that have occurred over the past decades have proven that there is a need to take extra measures to ensure network resilience [18–21]. For example, the resilience of the electrical network can be improved using several approaches that focus on reinforcing or relocating the vulnerable assets and taking precautions that can prevent damage propagation during wildfires and flooding events [14, 18, 22]. Additional improvements can be achieved by introducing microgrids with islanding potential, providing input to the grid during normal conditions, and supporting the local demands during main grid failures [23]. Resilience assessment of electrical networks follows several approaches depending on the stage of focus, disturbance, and types of connected components; some focus on the physical laws of electrical currents while others focus on the connectivity metrics. Readers can expand on the assessment approaches in [4, 14, 24, 25].

Water Network Resilience

Securing clean and healthy water is always essential for all human settlements; In modern cities, this process is made more complicated not only because huge quantities need to be treated, transported, and delivered but also because of the need to deal with disposed water in the sewer system. The components of the water distribution network are pipes, to carry the water; treatment plants, responsible for filtering and purifying the water to acceptable quality; pumps, to provide the necessary pressure throughout the network and compensate for the energy lost due to fractions and leaks; and storage tanks, to support the network and control the pressure in surrounding areas. Water networks are exposed to a wide array of possible disruptions or degradations affecting their various elements. In addition to providing the required quantities of water to the customer, water networks need to meet pressure requirements at consumption points, allowing for ease of distribution without exceeding upper thresholds, which may damage the fixtures. Furthermore, water quality is of extreme importance, both at the pumping station and at the customer faucet/tap; thus, water quality is tested at plants and the pressure in the pipes is monitored to prevent any contamination from entering water networks. Additionally, in tanks, settled water is exposed to degradation; thus, it is important to size tanks properly as well as ensure the circulation and consumption of the water. Intuitively, a bigger tank supports consumption during long outages; however, larger tanks are more exposed to contamination and aging [26, 27]. Additionally, advanced monitoring, control, and contamination detection to sense any damage or leak are very important; however, such digitalization of the network exposes it to cyberattacks that must be addressed using deep learning approaches or other cybersecurity measures [26, 28].

Water supply networks focus on delivering sufficient, acceptable-quality and contamination-free water; all these aspects should be reflected in the resilience assessment. On the upside, water has the capacity to be stored; however, on the downside, prolonged storage can degrade the quality of water. In the literature, different assessments and resilience metrics are suggested, focusing on the main objectives of water networks. Some researchers used the fraction of satisfied water demands during a disturbance at different network nodes as a metric for quantifying the resilience of the water network [29]. Other researchers used the safety of water served to the customers during a disruptive event as an indicator for system resilience [30]. Still, other researchers focused on the capacity of the network to maintain the needed hydraulic pressure at various outlets or nodes in the networks during disturbances [31–33]; this assessment method was used to measure the network resilience based on the reserve capacity of the node, with the least difference between the head and demand, and designating this node as the most critical or vulnerable node [34]. Furthermore, connectivity metrics that are based on graph theory are well adopted for resilience assessment [35–38], as well as some applications of graph theory, such as weighting, to reflect the hydraulic properties of the network [39, 40].

Case Study About Individual System Resilience: Transportation Resilience During Mega Sport Events

Mega sport events (MSEs) present a critical case under which the failure of a transportation network can have extensive consequences in terms of the economy and reputation of the host country. During MSEs, all infrastructures are expected to satisfy the increase in demand due to the influx of fans while continuing to present an acceptable level of performance during any potential disturbance. Furthermore, the transportation network plays a critical role in the fans' experience by providing mobility to the venues and different parts of the host city. The tight schedule of events shifts attention to the transportation network's capacity to withstand a disturbance rather than the effectiveness of recovery efforts. Serdar and Al-Ghamdi (2021) developed a resilience assessment framework to assess road network resilience during MSEs and used the FIFA World Cup 2022 in Qatar as a case study [41]. The framework focused on assessing the performance of the network on two levels using different metrics:

1) Critical link level: using average trip duration (T) between the central fan zone and various stadia as a metric.
2) Network level: using total betweenness C_B to represent network connectivity and mobility throughout the city.

Furthermore, several disturbance scenarios that represent extreme situations were considered during the assessment:

1) *Natural hazard (N)*: represented by 100 year ARI flood hazards.
2) *Intentional attack (I)*: targeting the central point and creating an inaccessible 3 km radius buffer zone.
3) *Widespread accidents (A)*: represented by 10 randomly scattered failure points with a 500 m radius buffer zone.

Additionally, the framework included several weighting factors to reflect decision-makers' preferences, including level weighting (α) and disturbance weighting (β). Finally, the framework calculated the resilience index of the road networks during MSEs (MSERRI), as presented in Figure 6.1.

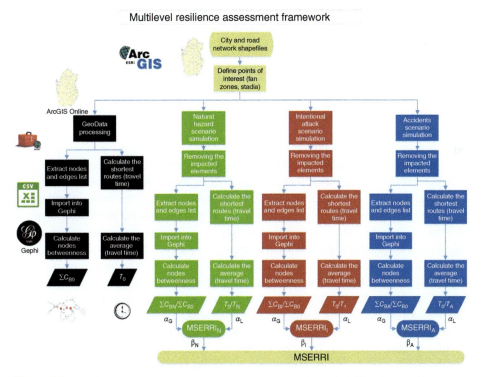

Figure 6.1 Multilevel resilience assessment framework developed for MSEs. *Source:* Adapted from Serdar and Al-Ghamdi [41].

Table 6.2 Doha road network assessment results and resilience index.

	Betweenness		Travel time (min.)		
Case	Max (% change)	Sum (% change)	Max (% change)	Average (% change)	MSERRI
Baseline	1085	37,194	43.5	16	–
Scenario 1	103 (−90%)	914 (−97.5%)	–	–	0.00
Scenario 2	1242 (+14.5%)	34,360 (−7.6%)	43.5	18 (+12.5%)	0.91
Scenario 3	1228 (+13.2%)	31,523 (−15.2%)	46.1 (+5.8%)	17 (+6.25%)	0.9

Source: Adapted from Serdar and Al-Ghamdi [41].

The application of this framework required using multiple software, including ArcGIS, Gephi [42], and an extension to export the features to the edge list [43]. As presented in Table 6.2 and Figure 6.2, the application of the framework highlighted the severe vulnerability of road networks to flood hazards, while showing that minor degradation was suffered during the other scenarios: intentional attacks and accidents. The final MSERRI value of almost 0.59 was due to the low performance toward flood hazards.

The assessment results highlight the road network's low resilience toward flooding hazards. Previous development disregarded such disturbances due to the local arid climate. However, in recent years, Qatar has witnessed several unusual precipitation events, which led to costly flooding in the underprepared urban areas of Qatar; these events are attributed to the impacts of climate change [9, 44].

Infrastructures Interdependencies and Resilience

Leveraging technological developments and the drive to achieve higher performance and service quality levels, urban infrastructures are becoming more interdependent. Although each urban infrastructure serves a specific goal, such as providing mobility through a transportation system or delivering clean potable water through the water distribution network, they collectively contribute to one common goal: facilitating the development of a thriving community with a continuous supply of electricity, clean water, and unrestricted access to communication. Furthermore, some of these infrastructures, like electrical networks, are essential for the functionality of other networks and thus create a sort of interrelation between them, or what is typically referred to in the literature as interdependency. Understanding the interdependencies between various infrastructures is critical for forecasting performance under various disturbances, as failure could propagate from one network to another based on the level of dependency and the resilience of each infrastructure [14]. This interdependency has increased in recent years, especially with the drive toward smart infrastructures, advanced monitoring and management of infrastructures enabled through advanced developments in sensor and communication technology, as well as the development of electrical and autonomous vehicles, which can impact transportation and electrical network resilience [6, 45, 46].

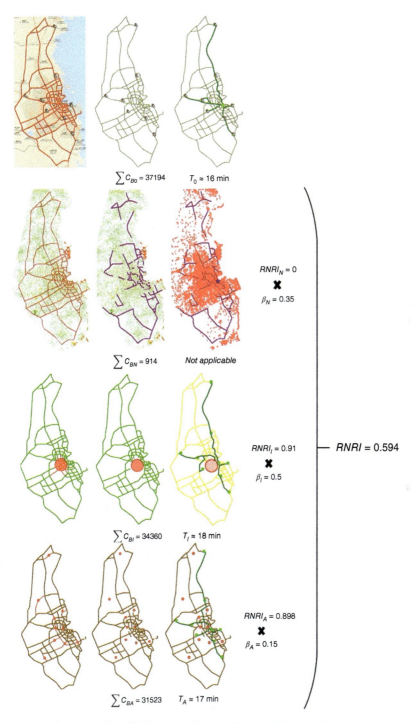

Figure 6.2 Results of Doha road resilience assessment. *Source:* Serdar et al. [41]/from MDPI/ CC BY 4.0.

Several sources constitute the relationship between various critical infrastructures. The relationship between different pairs of infrastructures results in several benefits and challenges, especially failure propagation [47]. Failure propagation and the relationship between critical infrastructures can be attributed to several sources [48]:

1) Physical flow, where the flow from one network feeds into the other, as in the case of water to wastewater network flow or the flow of electricity to power pumps in water networks.
2) Cyber domain, where information about one infrastructure helps calibrate the performance of the other, as in the case of smart metering to calibrate energy and water production and detect losses.
3) Proximity, where the geographical existence of some infrastructures is in close proximity to other infrastructures, as in the case of an explosion in one of the water pipes under a road section that could impact the road and neighboring electrical and gas pipes in the same section of the road.
4) Logical relation, where all dependencies that do not fit in previous categories fall, such as the mobility of recovery teams using road networks to repair damage in the electrical network after a disaster.

These aspects constitute the interdependencies between critical infrastructures and help identify the severity of failure in one network for other dependent ones. The quantification of resilience, considering the complexity of interdependency due to variation in the nature of infrastructures, remains an under-investigated topic [6, 14]. We will now present a case study focusing on the assessment of the resilience of a system of interdependent infrastructures.

Case Study About Interdependent Systems Resilience

Evaluating the system resilience of several interconnected infrastructures is a complex challenge due to the variation of flow and nature of different lifelines. This study proposed a framework to assess the resilience of interdependent infrastructures and demonstrated its application to three interconnected networks: electricity, water, and transport networks [49]. The framework used CN theory to model the networks with nodes assigned to specific data such as the number of users, and later employed to reflect system functionality at each timestep to calculate the resilience of the system.

The developed abstract consisted of 21 nodes and 25 links, including 20 internal links and five interdependent links. The nodes and links refer to a different meaning for each infrastructure, as highlighted in Table 6.3.

Also, each node was assigned several attributes, such as location, importance, recovery initiation time, served users, and cost, where these attributes were based on a mix of information and expert opinion. The developed abstract is demonstrated in Figure 6.3.

An example of failure can be seen in node 3E, from the electrical network, which demonstrated the resulting degradation in performance of independent nodes (Figure 6.4).

A similar simulation was conducted on each node in the network that considered a delay time for failure propagation from one node to the subsequent node. Additionally, the

Table 6.3 Examples of network component.

Infrastructure	Node	Edge
Transportation networks	Junctions	Roads, rail lines, and bridges
Water networks	Pumping stations and reservoirs	Water mains
Electrical networks	Generators, switches, and breakers	Transmission lines

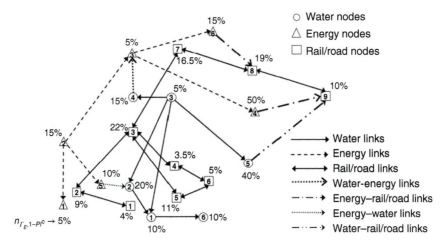

Figure 6.3 Case study abstract. The percentage represents the performance indicator at the base case. Source: Imani and Hajializadeh [49]/IWA Publishing.

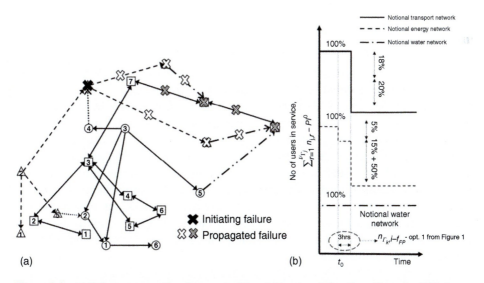

Figure 6.4 (a) Failure propagation from node 3E and (b) network functionality under 3E failure. Source: Imani and Hajializadeh [49]/IWA Publishing.

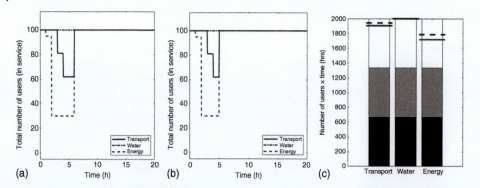

Figure 6.5 (a) Interdependent network performance during (2E) failure scenario with three hours recovery duration; (b) two hours recovery duration; and (c) bar chart of resilience levels. *Source:* Imani and Hajializadeh [49]/IWA Publishing.

recovery time of each node was calculated based on assumptions of recovery resources, such as three hours of recovery time and one hour needed to initiate the recovery effort. For example, node (2E) was calculated under two recovery strategies to demonstrate the impact of the allocation of different recovery resources. This is reflected in the functionality over time, as shown in Figure 6.5, which facilitates the calculation of resilience through mathematical integration.

The application was expanded to consider the impact of the failure of each node on the overall system, which supported the fact that electricity is at the core of the functionality of all other infrastructures. Furthermore, various network metrics were calculated to compare with the direct system resilience assessment result, and it was determined that betweenness presented the closest metric to the assessment results [49].

Conclusion

Modern societies rely on complex networks of interconnected and interdependent infrastructures to provide essential services and facilitate productivity. However, most infrastructure design approaches deal with each infrastructure separately, ignoring the interrelationship between critical infrastructures and the threat of failure propagation. This chapter demonstrated the difference between three critical infrastructures, transportation, water, and electricity, as well as the main considerations during resilience assessment. We provided a case study to demonstrate the resilience assessment of the individual networks. Furthermore, we explained the importance of considering the impact of interdependencies between infrastructures and the sources of such interrelations. Finally, we provided a case study on the resilience assessment of multilayer systems, consisting of the same three critical infrastructures, demonstrating the assessment method's application to facilitate the reader's understanding.

References

1 Takano, T. (2005). Developing urban infrastructure supportive to health: the healthy Cities approach. *Jpn. Med. Assoc. J.* 48: 458–461.
2 United Nations (2019). *Department of Economic and Social Affairs PD. World Urbanization Prospects: The 2018 Revision*. New York. https://doi.org/10.18356/b9e995fe-en.
3 Serdar, M.Z., Koc, M., and Al-Ghamdi, S.G. (2021). Urban infrastructure resilience assessment during mega sport events using a multi-criteria approach. *Front. Sustainability* 2. https://doi.org/10.3389/frsus.2021.673797.
4 Liu, W. and Song, Z. (2020). Review of studies on the resilience of urban critical infrastructure networks. *Reliab. Eng. Syst. Saf.* 193: 106617. https://doi.org/10.1016/j.ress.2019.106617.
5 Butry, D., Davis, C.A., Malushte, S.R. et al. (2021). *Hazard-Resilient Infrastructure*. Reston, VA: American Society of Civil Engineers. https://doi.org/10.1061/9780784415757.
6 Serdar, M.Z., Koç, M., and Al-Ghamdi, S.G. (2022). Urban transportation networks resilience: indicators, disturbances, and assessment methods. *Sustain. Cities Soc.* 76: 103452. https://doi.org/10.1016/j.scs.2021.103452.
7 Serdar, M.Z., Çağlayan, B.Ö., and Al-Ghamdi, S.G. (2022). A special characteristic of an earthquake response spectrum detected in Turkey. *Mater. Today Proc.* https://doi.org/10.1016/j.matpr.2022.04.407.
8 Rus, K., Kilar, V., and Koren, D. (2018). Resilience assessment of complex urban systems to natural disasters: a new literature review. *Int. J. Disaster Risk Reduct.* 31: 311–330. https://doi.org/10.1016/j.ijdrr.2018.05.015.
9 Tahir, F., Ajjur, S.B., Serdar, M.Z. et al. (2021). *Qatar Climate Change Conference 2021*. Hamad bin Khalifa University Press (HBKU Press). https://doi.org/10.5339/conf_proceed_qccc2021.
10 Holling, C.S. (1973). Resilience and stability of ecological systems. *Annu. Rev. Ecol. Syst.* 4: 1–23. https://doi.org/10.1146/annurev.es.04.110173.000245.
11 Hosseini, S., Barker, K., and Ramirez-Marquez, J.E. (2016). A review of definitions and measures of system resilience. *Reliab. Eng. Syst. Saf.* 145: 47–61. https://doi.org/10.1016/J.RESS.2015.08.006.
12 Sun, W., Bocchini, P., and Davison, B.D. (2020). Resilience metrics and measurement methods for transportation infrastructure: the state of the art. *Sustain. Resilient Infrastruct.* 5: 168–199. https://doi.org/10.1080/23789689.2018.1448663 .
13 Serdar, M.Z., Macauley, N., and Al-Ghamdi, S.G. (2022). Building thermal resilience framework (BTRF): a novel framework to address the challenge of extreme thermal events, arising from climate change. *Front. Built Environ.* 8. https://doi.org/10.3389/fbuil.2022.1029992.
14 Serdar, M.Z. and Al-Ghamdi, S.G. (2021). Preparing for the unpredicted: a resiliency approach in energy system assessment. In: *Green Energy and Technology* (ed. J. Ren), 183–201. Cham: Springer International Publishing https://doi.org/10.1007/978-3-030-67529-5_9.
15 Magoni, M. (2017). Resilience thinking and urban metabolism in spatial planning: which possible integrations. *City, Territ Archit* 4: 19. https://doi.org/10.1186/s40410-017-0074-0.

16 Kuznecova, T., Romagnoli, F., and Rochas, C. (2014). Energy metabolism for resilient urban environment: a methodological approach. *Procedia Econ. Financ.* 18: 780–788. https://doi.org/10.1016/S2212-5671(14)01002-8.

17 Ahmed, S. and Dey, K. (2020). Resilience modeling concepts in transportation systems: a comprehensive review based on mode, and modeling techniques. *J. Infrastruct. Preserv. Resil.* 1: 8. https://doi.org/10.1186/s43065-020-00008-9.

18 PG&E. Expanding our efforts to prevent wildfires 2020. https://www.pge.com/en_US/safety/emergency-preparedness/natural-disaster/wildfires/community-wildfire-safety.page.

19 Helman, C. (2019). As $30B In Wildfire Claims Bankrupt PG&E, California Wonders Who Will Pay After The Next Conflagration. Forbes. https://www.forbes.com/sites/christopherhelman/2019/01/21/as-30b-in-wildfire-claims-bankrupt-pge-california-wonders-who-will-pay-after-the-next-conflagration/#7d36267a2699 (accessed 20 October 2020).

20 Kamali, S. and Amraee, T. (2017). Blackout prediction in interconnected electric energy systems considering generation re-dispatch and energy curtailment. *Appl. Energy* 187: 50–61. https://doi.org/10.1016/j.apenergy.2016.11.040 .

21 Zio, E. and Aven, T. (2011). Uncertainties in smart grids behavior and modeling: What are the risks and vulnerabilities? How to analyze them? *Energy Policy* 39: 6308–6320. https://doi.org/10.1016/j.enpol.2011.07.030.

22 USA NG (2016). National Grid 2016 Climate change Resilience planning. Climate Change Adaptation Report by National Grid Electricity Transmission plc September 2010, pp. 1–72.

23 Rosales-Asensio, E., de Simón-Martín, M., Borge-Diez, D. et al. (2019). Microgrids with energy storage systems as a means to increase power resilience: an application to office buildings. *Energy* 172: 1005–1015. https://doi.org/10.1016/j.energy.2019.02.043.

24 Abedi, A., Gaudard, L., and Romerio, F. (2019). Review of major approaches to analyze vulnerability in power system. *Reliab. Eng. Syst. Saf.* 183: 153–172. https://doi.org/10.1016/j.ress.2018.11.019.

25 Cuadra, L., Salcedo-Sanz, S., Del Ser, J. et al. (2015). A critical review of robustness in power grids using complex networks concepts. *Energies* 8: 9211–9265. https://doi.org/10.3390/en8099211.

26 Shin, S., Lee, S., Judi, D. et al. (2018). A systematic review of quantitative resilience measures for water infrastructure systems. *Water* 10: 164. https://doi.org/10.3390/w10020164.

27 Diao, K., Sweetapple, C., Farmani, R. et al. (2016). Global resilience analysis of water distribution systems. *Water Res.* 106: 383–393. https://doi.org/10.1016/j.watres.2016.10.011.

28 Taormina, R. and Galelli, S. (2017). Real-time detection of cyber-physical attacks on water distribution systems using deep learning. In: *World Environmental and Water Resources Congress 2017* (ed. C.N. Dunn and B. Van Weele), 469–479. Reston, VA: American Society of Civil Engineers. https://doi.org/10.1061/9780784480625.043.

29 Zhuang, B., Lansey, K., and Kang, D. (2013). Resilience/availability analysis of municipal water distribution system incorporating adaptive pump operation. *J. Hydraul. Eng.* 139: 527–537. https://doi.org/10.1061/(ASCE)HY.1943-7900.0000676.

30 Klise, K., Murray, R., and Walker, L.T. (2015). *Systems Measures of Water Distribution System Resilience*. Albuquerque, NM, and Livermore, CA (United States): https://doi.org/10.2172/1177373.

31 Todini, E. (2000). Looped water distribution networks design using a resilience index based heuristic approach. *Urban Water* 2: 115–122. https://doi.org/10.1016/S1462-0758(00)00049-2.

32 Greco, R., Di Nardo, A., and Santonastaso, G. (2012). Resilience and entropy as indices of robustness of water distribution networks. *J. Hydroinform.* 14: 761–771. https://doi.org/10.2166/hydro.2012.037.

33 Creaco, E., Franchini, M., and Todini, E. (2016). Generalized resilience and failure indices for use with pressure-driven modeling and leakage. *J. Water Resour. Plan Manag.* 142: 04016019. https://doi.org/10.1061/(ASCE)WR.1943-5452.0000656.

34 Wright, R., Herrera, M., Parpas, P., and Stoianov, I. (2015). Hydraulic resilience index for the critical link analysis of multi-feed water distribution networks. *Procedia Eng.* 119: 1249–1258. https://doi.org/10.1016/j.proeng.2015.08.987.

35 Soldi, D., Candelieri, A., and Archetti, F. (2015). Resilience and vulnerability in urban water distribution networks through network theory and hydraulic simulation. *Procedia Eng.* 119: 1259–1268. https://doi.org/10.1016/j.proeng.2015.08.990.

36 Yazdani, A., Otoo, R.A., and Jeffrey, P. (2011). Resilience enhancing expansion strategies for water distribution systems: a network theory approach. *Environ. Model Softw.* 26: 1574–1582. https://doi.org/10.1016/j.envsoft.2011.07.016.

37 Pandit, A. and Crittenden, J.C. (2016). Index of network resilience for urban water distribution systems. *Int. J. Crit. Infrastruct.* 12: 120. https://doi.org/10.1504/IJCIS.2016.075865.

38 Farahmandfar, Z., Piratla, K.R., and Andrus, R.D. (2017). Resilience evaluation of water supply networks against seismic hazards. *J. Pipeline Syst. Eng. Pract.* 8: 04016014. https://doi.org/10.1061/(ASCE)PS.1949-1204.0000251.

39 Herrera, M., Abraham, E., and Stoianov, I. (2015). Graph-theoretic surrogate measures for analysing the resilience of water distribution networks. *Procedia Eng.* 119: 1241–1248. https://doi.org/10.1016/j.proeng.2015.08.985.

40 Herrera, M., Abraham, E., and Stoianov, I. (2016). A graph-theoretic framework for assessing the resilience of sectorised water distribution networks. *Water Resour. Manag.* 30: 1685–1699. https://doi.org/10.1007/s11269-016-1245-6.

41 Serdar, M.Z. and Al-Ghamdi, S.G. (2021). Resiliency assessment of road networks during mega sport events: the case of FIFA world cup Qatar 2022. *Sustainability* 13: 12367. https://doi.org/10.3390/su132212367.

42 Bastian, M., Heymann, S., and Jacomy, M. (2009). Gephi: an open source software for exploring and manipulating networks. BT - International AAAI Conference on Weblogs and Social. *Int. AAAI Conf. Weblogs Soc. Media* 3 (1): 361–362.

43 Karduni, A., Kermanshah, A., and Derrible, S. (2016). A protocol to convert spatial polyline data to network formats and applications to world urban road networks. *Sci. Data* 3: 160046. https://doi.org/10.1038/sdata.2016.46.

44 Serdar, M.Z., Ajjur, S.B., and Al-Ghamdi, S.G. (2022). Flood susceptibility assessment in arid areas: a case study of Qatar. *Sustainability* 14: 9792. https://doi.org/10.3390/su14159792.

45 Ahmed, S., Dey, K., and Fries, R. (2019). Evaluation of transportation system resilience in the presence of connected and automated vehicles. *Transp. Res. Rec. J. Transp. Res. Board* 2673: 562–574. https://doi.org/10.1177/0361198119848702.

46 Brown, M.A. and Soni, A. (2019). Expert perceptions of enhancing grid resilience with electric vehicles in the United States. *Energy Res. Soc. Sci.* 57: 101241. https://doi.org/10.1016/j.erss.2019.101241.

47 Grafius, D.R., Varga, L., and Jude, S. (2020). Infrastructure interdependencies: opportunities from complexity. *J. Infrastruct. Syst.* 26: 04020036. https://doi.org/10.1061/(asce)is.1943-555x.0000575.

48 Rinaldi, S.M., Peerenboom, J.P., and Kelly, T.K. (2001). Identifying, understanding, and analyzing critical infrastructure interdependencies. *IEEE Control Syst. Mag.* 21: 11–25. https://doi.org/10.1109/37.969131.

49 Imani, M. and Hajializadeh, D. (2020). A resilience assessment framework for critical infrastructure networks' interdependencies. *Water Sci. Technol.* 81: 1420–1431. https://doi.org/10.2166/wst.2019.367.

50 Serdar, M.Z. and Al-Ghamdi, S.G. (2023). Resilience-oriented recovery of flooded road networks during mega-sport events: a novel framework. *Frontiers in Built Environment*, 9. https://doi.org/10.3389/fbuil.2023.1216919

7

Assessing Infrastructure Resilience: Approaches and Considerations

Mohammad Zaher Serdar[1] and Sami G. Al-Ghamdi[1,2,3]

[1] *Division of Sustainable Development, College of Science and Engineering, Hamad Bin Khalifa University, Qatar Foundation, Doha, Qatar*
[2] *Environmental Science and Engineering Program, Biological and Environmental Science and Engineering Division, King Abdullah University of Science and Technology (KAUST), Thuwal, Saudi Arabia*
[3] *KAUST Climate and Livability Initiative, King Abdullah University of Science and Technology (KAUST), Thuwal, Saudi Arabia*

Introduction

Today we face unseen challenges while living in a world characterized by a high standard of living, supported by technological advancement, and a high level of optimization. Technological advancements and the drive to optimize all processes have allowed better resource management and delivery, providing a wide array of services and infrastructures for the world's population, especially in urban areas. However, these services and infrastructures are increasingly suffering from disruptions caused by unforeseen disasters such as climate-change-induced disasters, the COVID-19 pandemic, cybersecurity attacks. As of 2020, more than 50% of the world's population is concentrated in urban areas; thus, these challenges are becoming costlier as the reliance on integrated and interdependent infrastructures increases [1].

In recent years, the world has witnessed an unparalleled increase in natural hazards, in both intensity and ratio, mainly attributed to climate change effects. Throughout past years, humans across the world have experienced unprecedented local events, such as deadly heatwaves in Canada [2], low temperature in Texas [3, 4], and intense rainfalls in the Arabian Peninsula [5–7]. These events are evidence of the alarming increase in the impacts of climate change. Furthermore, these events expose the lack of preparedness to such a rapid succession of natural hazards and the resulting cascading failures of interdependent infrastructures [3].

Modern infrastructures are complex, interconnected, integrate many smart technologies, and face unprecedented challenges that require an unconventional design approach to limit disruptions and provide rapid recovery, in other words, resilience-based design [8].

Sustainable Cities in a Changing Climate: Enhancing Urban Resilience, First Edition.
Edited by Sami G. Al-Ghamdi.
© 2024 John Wiley & Sons Ltd. Published 2024 by John Wiley & Sons Ltd.

The complexity and interdependency of modern infrastructures mean that failure in one system can propagate to others, increasing and multiplying the cost of disasters, as experienced in recent events [3, 4, 9, 10]. Subsequently, we can see that modern infrastructures are interconnected and require different design approaches to address the emerging challenges and ensure robustness and rapid recovery, mainly ensuring their resilience [11].

A resilient system typically enjoys multiple characteristics that contribute to its resilience and can be used as a proxy approach to resilience assessment. Some researchers see resilience as a combination of multiple characteristics like vulnerability, robustness, and reliability; this allows well-established assessment approaches of these characteristics to be used to assess resilience. These approaches are quite common, especially when focusing on a specific part of system performance, such as damage absorption or recovery phases [12, 13].

While presenting a holistic design process is not part of the scope of this chapter, we are providing an introduction for the reader to gain an understanding of the main approaches in assessing the resilience of different infrastructures and systems, an essential step toward resilience-based design. In the following sections, we present complex networks, simulation, and statistical approaches. The choice of suitable networks depends on the scale of the assessed system, the type of disturbance, and the inputs available.

Complex Networks

Complex networks theory originated in social science studies but later gained momentum in computer sciences and resilience engineering. The complex networks theory was introduced in the social sciences to describe the connection and centrality of individuals to their colleagues or social network [14]. However, complex networks found further application in computer networks with the rise of social network sites, such as Twitter and Facebook, as they can provide important characteristics of the behavior patterns of people as well as marketing directions. The capacity for complex networks to reflect the relationship between different elements in a system allows for its implementation in resilience studies.

Generally, complex networks consist of many nodes connected through links, creating a complex network of relationships that can be hard to interpret through simple observation. Complex networks consist of nodes that can resemble individuals, airports, generation facilities, etc., and these nodes are connected by links such as social relationships, roads, and transmission lines. To focus on network properties, networks also refer to graphs that can provide an abstract of the relationships and connections between the nodes, regardless of their physical distribution. Table 7.1 shows examples of graph theory applications in several research fields.

Types of Graphs

The complex network method, using connectivity metrics, is effective in presenting and reflecting on the performance of a system. The use of an abstract simplifies assessing a

Table 7.1 Examples of graph theory and resilience in several research fields.

Field of study	Graph theory examples
Social networks	[15–17]
Transportation networks	[12, 18–23]
Electrical networks	[11, 24–28]
Water networks	[29–34]
Computer science	[35–39]

network or infrastructure by focusing on the structural properties and the availability of alternatives. However, several considerations can fundamentally affect the assessment results as well as the representativeness of the graphs and should be clearly identified at the beginning of the abstracting process. These considerations, mainly direction and weighting, are discussed in the following subsections.

Directed and Undirected Graphs
The main advantage of graph theory is that it focuses on the relationships between system nodes, such as people in a community or generation facilities and substations in an electrical grid, with these relations represented as edges between associated nodes. However, these relations are different in their meaning and nature, and subsequently, they can be divided into directed and undirected graphs.

Undirected graphs refer to networks where the relationship between the nodes has no flow, is interchangeable, or the flow between nodes can go in both directions. For example, in a graph representing a community, the friendship between its members, the nodes, is represented in the form of edges; these edges (relationships) are interchangeable as both the start and end nodes are friends. Another example is a graph describing a highway network, where nodes are intersections and roads between them are edges. We can consider the roads in both directions as a single edge, with the flow going in both directions. In resilience literature, undirected graphs are widely applied as the size of an urban highway system is massive, and thus, it is very important to simplify the assessment process [18, 40, 41].

On the other hand, directed graphs refer to networks where the direction of flow between nodes is physically meaningful, such as in the case of infrastructures, or represents a hierarchy in a workplace or society. Electricity and water distribution networks are a clear example of these types of graphs, where the direction of the flow is critical to the nature of the network. Subsequently, edges have directions indicating the flow of resources from a source to a destination. Figure 7.1 shows these two types of graphs.

Weighted and Unweighted Graphs
Standard or unweighted graphs focus on the structural properties of the network and how its nodes are connected, which can yield different metrics, regardless of the physical arrangement or the nature of the network. This approach allows the abstracting process

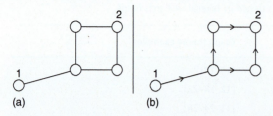

Figure 7.1 Directed and undirected graphs. (a) An undirected graph, how travelers can go from point 1 to point 2 and return in the opposite direction. (b) A directed graph representing how the flow can go from point 1 to point 2, but how it is impossible to go in the opposite direction.

and the performance assessment to be simplified. However, ignoring the physical properties could yield a confusing or inaccurate picture of the network.

To improve the accuracy of graphs, we can introduce some of their physical properties in the form of weighting. For example, weighting can account for road length in a transportation network, pipe diameter in a water distribution network, or impedance in an electrical network [11, 42]. Introducing weighting increases accuracy, as metric calculations such as betweenness measure the actual shortest paths rather than the least number of steps between two pairs of nodes.

Main Applications in Resilience Assessment

Betweenness Centrality

Betweenness centrality expresses the role of a certain element, node or edge, in connecting any pair of nodes in the graph. It calculates the number of shortest paths between all pairs of nodes in a network that pass through an element. Betweenness can identify the most critical nodes in a network, which can be considered bottle necks. Changes in betweenness can also detect disturbance and the loss of connectivity, which can be used to assess a network's resilience [12, 40]. However, in some situations where a network experiences limited damage, the maximum betweenness can rise rather than decrease due to the redirection of shortest paths to other critical nodes; this situation can be addressed by considering the summation of the betweenness values of all the nodes, providing a more consistent and robust metric [23].

Typically, betweenness is calculated using the following formula suggested by [43]:

$$C_B(i) = \sum_{a \neq b \neq i \in N} \frac{\sigma_{ab}(i)}{\sigma_{ab}}$$

where

$C_B(i)$ is betweenness centrality of element (i),
σ_{ab} is the count of shortest routes between nodes (a) and (b), and
$\sigma_{ab}(i)$ is the portion of these routes that include element (i).

Generally, betweenness is used to assess infrastructures under different hazards, including natural, human-induced, and accidents [40].

Graph Percolation
Percolation is used for assessing the infrastructure's robustness and resilience to random or targeted attacks. The percolation concept relies on the graph's capacity to sustain the removal of a portion of the nodes or links before suffering from fragmentation. Here, the term "fragmentation" refers to the drop in the size of the largest connected component below a certain threshold, known as the "percolation threshold." This threshold marks the phase change of a well-connected graph due to rapid decay in the size of its largest connected component. However, the interpretation of this threshold differs, depending on the context of the investigation, for example, between community virus protection to transportation networks. Take the case of a virus spread within a community. If a small portion of the connections is removed, as in the case of a low threshold, limiting or preventing the spread of the virus means a high resilience level; in other words, we can protect the community by vaccinating a very small percentage of the population. On the other hand, in the case of a transportation network, if the network accommodates the loss of a large portion of its components before suffering fragmentation, it would be resilient.

The percolation process can follow different approaches during the application based on the simulated scenario. The process of removal can take different paths; it can be based on random removal, targeted removal, or targeted randomized removal [22]. Furthermore, the selection of the targeted nodes or edges can be based on different centrality measures or metrics, such as node degree or betweenness centrality. Generally, the application of random removal of network elements is less impactful compared to the targeted removal, especially those based on betweenness centrality [22, 44].

Strengths and Limitations of Complex Networks
The main advantages of applying complex network approaches include:

1) Simplifying the network by focusing on topological properties and the relations between the elements.
2) Introduction of weighting to reflect some important characteristics of the network.
3) Ease of application and scaling up compared to other approaches, as they require limited resources.
4) Solutions and approaches in one research field or infrastructure could be used to improve other networks.

The main drawbacks of complex network approaches include:

1) The actual flow properties and dynamics could be lost due to simplification.
2) This approach ignores disturbance as a progressive event and focuses on the total impact on the network.
3) Generally, reflects network vulnerability and robustness, but ignores recoverability.

Simulation Approaches

Simulation approaches enjoy the capacity to predict the system performance during future scenarios and the efficiency of possible development alternatives. In simulation approaches, we focus on developing a numerical simulation of the system components and flow, thus allowing for testing of different hypothetical disturbance scenarios or evaluation of development plans.

System Simulation

Simulating system performance and measuring its performance, and subsequently, its resilience, under different scenarios relies on the nature of that system and the accuracy of the developed models. Different infrastructures have different specifications, nature, and performance metrics expressed by different equations. Subsequently, the resilience assessment for each of the critical infrastructures relies on a different set of indicators and tools.

Transportation networks have a discrete nature of the flow, allowing redirection and queuing of vehicles. Additionally, the networks have definitive origin–destination locations, as trips generated during a certain time of the day are reversed during another time of the day, for example, between residential complexes and several workplaces. This nature requires the use of trip generation approaches, based on land use, and allocating such trips, based on different models, before calibrating and simulating the network performance under any specific scenario. These steps are typically used during the design phase but can also be used for resilience assessments, using different computer software such as PTV VISSUM, which can perform multimodal traffic simulations. Nevertheless, performing large-scale simulation and assessment is resource-intensive and requires carefulness to calibrate the model accurately.

Electrical and water distribution infrastructures are similar, in the sense of having a specific origin–destination scheme, yet they have very distinctive natures. Electrical networks have the nature of continuous flow. However, considering the immaturity of battery and energy storage technologies, generation facilities need to continuously meet and balance generation with demand while considering the physical laws of electrical current. However, technologies, such as solar panels and wind turbines, enable on-site energy production and conversion to apply islanding modes, which can improve the resilience of specific areas or buildings. Modeling of an electrical system can follow a deterministic approach or a probabilistic one, and the selection of the suitable one depends on the scale and the accuracy needed for the assessment. To further expand on the resilience assessment of electrical systems through modeling and simulation, readers can refer to [45–48].

Water distribution networks also have a semicontinuous flow, but, unlike electrical networks, water can be stored and has different types of disturbances. There are several aspects of concern regarding water network resilience assessment, including quality, contamination, aging, and pressure. Water can be stored at elevated tanks to preserve the pressure and regulate the distribution, but the age of stored water and the potential for contamination can affect the water's suitability for consumption. Water pressure deviation is among the main metrics used for resilience assessment as an indicator of network performance. However, the reliance on pumps to provide the needed pressure can expose the network to disturbances caused by failures in the electrical network, creating a sort of dependency. Generally, several software or open-source tools can be used to simulate the network performance, including contamination, aging, and pressure, such as WaterGEMS and EPANET; readers wishing to expand on the details can refer to [49].

Agent-Based Modeling

Agent-based modeling (ABM) facilitates measuring system performance in response to relational or behavioral changes in its components. ABM focuses on programming a particular type of behavior or set of rules for the agents, such as cars, and assessing the resulting impact on the system due to their interactions. In this sense, ABM allows for the investigation of complex systems by programming a set of rules and agent behavior for the surrounding environment. This allows one to decompose the complex systems, of the real world, and simulate the impacts of certain changes or events, thus assessing the resilience of the networks. ABM is not very common in the field of infrastructure resilience but has the potential to improve and assess the impact of emerging technologies and their integration, especially by reflecting the interdependence of different infrastructures. A good example of such an assessment is investigating the feasibility of using electric cars for energy storage and distribution during electrical network disruption, as reported in [50]. Among the most suitable software for ABM are NetLogo and AnyLogic.

GIS-Based Approaches

The development of remote sensing and the capacity of new geoprocessing tools allow different disaster scenarios to be projected on already existing networks, identifying exposed parts and the potential damage level. Geographic Information Systems (GIS) enable processing of geographic information such as land-use maps, infrastructure, and elevations so that valuable information can be extracted such as flood potential, heat island effect, and the impact of any other disturbance scenarios. By overlaying the disturbance impacts or hazards over the infrastructure layers, impacted areas can be identified and eliminated from the performance assessment. In some cases, the performance assessment can be conducted within a GIS environment, such as measuring trip duration between two points, while in other cases, it can be inferred during the development of the simulation model of the damaged network [23]. Thus, GIS can be used as a supporting technology that enables one to conduct approaches on a realistic and city-scale level. The most popular software used for GIS applications are ArcGIS (developed by ESRI) and QGIS (a free and open-source tool).

Strengths and Limitations of Simulation Approaches

The main advantages of applying simulation approaches include:

1) The ability to simulate the impacts of unprecedented events on the infrastructure.
2) The ability to assess the effectiveness of development options, including smart infrastructure, on network resilience.
3) The capability to assess the impact of the behavior and interactions of individual components on the resilience of complex systems, as in ABM.
4) The capacity to provide the basis for applying other assessment approaches, through GIS applications, which consider realistic situations and disturbance scenarios.

The main drawbacks of applying simulation approaches include:

1) The need for accurate modeling and calibration; otherwise, the overall results can be misleading.
2) Scalability, as extensive processing capacity is required, especially for large-scale assessments.
3) Data availability, as a complete understanding of the system, its functionality, and the relationships between its components is required.

Other Approaches

Some approaches are less common and are primarily used to complement the aforementioned assessment methods. These include statistical approaches to reflect on previous events and optimization approaches to better manage available resources in both pre- and post-disaster operations.

Statistical Approaches

The main applications of statistics in resilience assessment are to reflect on the records of previous events and real-time monitoring of a system's performance using big data streams. Reflectiveness is applied to analyze infrastructure performance and resilience under previous events, as it can provide the actual measure of a system's resilience as well as the recovery time for a certain event [51]. The capacity to reflect on previous events requires detailed records of the event's progress and infrastructure response. The data can be harvested through the increasingly integrated sensors found in all modern products and infrastructure components, like Global Positioning System (GPS) devices in cars or smart meters and appliances in homes [52]. Through the application of big data concepts, used to filter and extract valuable performance metrics, we can identify trends and detect a system's performance deviations in previous and real-time events using available streams of data. Based on the trend projections, integrated early warning and emergency response efforts can be initiated automatically [4, 52, 53].

The statistical approaches, combined with the big data applications, can be considered a promising field with huge potential to grow as digitalization of infrastructures, such as smart infrastructures, is accelerated and smart devices and sensors become widespread. Considering the accelerated environmental changes and emerging threats, using records to reflect on previous events can facilitate learning from such experiences, even if they happened in other countries, which is crucial. However, these approaches still cannot reflect the interaction between different infrastructures, require huge investment to make smart infrastructure development viable, and are highly exposed to cybersecurity challenges.

Optimization Approaches

The core concept of engineering is developing an efficient design using minimum resources, in other words, optimizing resource consumption. Optimization approaches aim to strike a balance between resource usage and infrastructure performance to maximize resilience while meeting budget and performance constraints. The main application of such approaches focuses on recovery efforts and post-disaster response, aiming to boost recovery using a limited budget [59]. For example, investigating the optimum repair sequence of damaged bridges in order to maximize a transportation network's performance and resilience [54], or determining the benefits of introducing new health centers to post-disaster recovery [55].

Another field for optimization approaches is presented by game theory application. The capacity for game theory to model the interests of different parties allows for its application in different contexts, especially in an attacker–defender scenario. The concept in this competitive game scenario is to identify the best resource allocation, from a defender perspective, that minimizes resulting damage in the network and thus, maximize resilience by reinforcing vulnerabilities [56, 57]. On the other hand, in cooperative game scenarios, different parties or infrastructures allocate their resources to achieve maximum shared performance and thus maximize network resilience. Examples of such games in the electrical network are provided in [58].

Optimization approaches are important for recovery planning and modeling the interests and interactions of different parties, facilitating the choices that maximize infrastructure resilience. However, the need to accurately model the system mathematically while also identifying the objectives and constraints makes it susceptible to errors and hard to scale up.

Conclusion

Most people today live in urban areas, and this trend is expected to continue increasing. These urban areas are supported by complex and interconnected infrastructure networks. Additionally, in recent years, the world has been witnessing an increase in the impact and rate of disasters, attributed both to unprecedented population concentration and climate change. Subsequently, urban infrastructures are required to support the continuous productivity of urban areas while protecting the inhabitants under these challenging situations. To achieve this, resilience should be incorporated into infrastructure design, and the resilience of current infrastructure should be assessed. This chapter briefed the reader on the main assessment approaches of infrastructure resilience as well as their feasibility and limitations. Furthermore, it provided readers with some sources, tools, and examples to expand their knowledge regarding the application of such approaches. Finally, we hope this chapter has facilitated the readers understanding and stimulated their interest in promoting resilience as a core concept in the design of future infrastructures, a necessary and integral part of sustainable development.

References

1 United Nations, Department of Economic and Social Affairs PD (2019). *World Urbanization Prospects: The 2018 Revision*. New York: UN-iLibrary, United Nations Department of Economic and Social Affairs, US. https://doi.org/10.18356/b9e995fe-en.

2 Vaughan, A. (2021). The heat is on out west. *New Sci.* 250: 10–11. https://doi.org/10.1016/S0262-4079(21)01169-6.

3 Doss-Gollin, J., Farnham, D.J., Lall, U., and Modi, V. (2021). How unprecedented was the February 2021 Texas cold snap? *Environ. Res. Lett.* 16: 64056. https://doi.org/10.1088/1748-9326/ac0278.

4 Serdar, M.Z., Macauley, N., and Al-Ghamdi, S.G. (2022). Building thermal resilience framework (BTRF): a novel framework to address the challenge of extreme thermal events, arising from climate change. *Front. Built Environ.* 8. https://doi.org/10.3389/fbuil.2022.1029992.

5 Salimi, M. and Al-Ghamdi, S.G. (2020). Climate change impacts on critical urban infrastructure and urban resiliency strategies for the Middle East. *Sustain Cities Soc* 54: 101948. https://doi.org/10.1016/j.scs.2019.101948.

6 Tahir, F., Ajjur, S.B., Serdar, M.Z. et al. (2021). *Qatar Climate Change Conference 2021*. Hamad bin Khalifa University Press (HBKU Press) https://doi.org/10.5339/conf_proceed_qccc2021.

7 Serdar, M.Z., Ajjur, S.B., and Al-Ghamdi, S.G. (2022). Flood susceptibility assessment in arid areas: a case study of Qatar. *Sustainability* 14: 9792. https://doi.org/10.3390/su14159792.

8 Serdar, M.Z., Koc, M., and Al-Ghamdi, S.G. (2021). Urban infrastructure resilience assessment during mega sport events using a multi-criteria approach. *Front. Sustain.* 2. https://doi.org/10.3389/frsus.2021.673797.

9 Rus, K., Kilar, V., and Koren, D. (2018). Resilience assessment of complex urban systems to natural disasters: a new literature review. *Int. J. Disaster Risk Reduct.* 31: 311–330. https://doi.org/10.1016/j.ijdrr.2018.05.015.

10 Serdar, M.Z., Çağlayan, B.Ö., and Al-Ghamdi, S.G. (2022). A special characteristic of an earthquake response spectrum detected in Turkey. *Mater. Today Proc.* https://doi.org/10.1016/j.matpr.2022.04.407.

11 Serdar, M.Z. and Al-Ghamdi, S.G. (2021). Preparing for the unpredicted: a resiliency approach in energy system assessment. In: *Green Energy Technol* (ed. J. Ren), 183–201. Cham: Springer International Publishing https://doi.org/10.1007/978-3-030-67529-5_9.

12 Serdar, M.Z., Koç, M., and Al-Ghamdi, S.G. (2022). Urban transportation networks resilience: indicators, disturbances, and assessment methods. *Sustainable Cities Soc.* 76: 103452. https://doi.org/10.1016/j.scs.2021.103452.

13 Gu, Y., Fu, X., Liu, Z. et al. (2020). Performance of transportation network under perturbations: reliability, vulnerability, and resilience. *Transp. Res. Part E Logist. Transp. Rev.* 133: 101809. https://doi.org/10.1016/j.tre.2019.11.003.

14 Ghoshal, G., Mangioni, G., Menezes, R., and Poncela-Casanovas, J. (2014). Social system as complex networks. *Soc Network Anal. Min.* 4: 238. https://doi.org/10.1007/s13278-014-0238-9.

15 De Meo, P., Ferrara, E., Fiumara, G., and Ricciardello, A. (2012). A novel measure of edge centrality in social networks. *Knowl. Based Syst.* 30: 136–150. https://doi.org/10.1016/j.knosys.2012.01.007.

16 Das, K., Samanta, S., and Pal, M. (2018). Study on centrality measures in social networks: a survey. *Soc Network Anal. Min.* 8: 13. https://doi.org/10.1007/s13278-018-0493-2.

17 Grando, F., Noble, D., and Lamb, L.C. (2016). An analysis of centrality measures for complex and social networks. In: *2016 IEEE Global Communications Conference (GLOBECOM)*, 1–6. IEEE https://doi.org/10.1109/GLOCOM.2016.7841580.

18 Kermanshah, A. and Derrible, S. (2017). Robustness of road systems to extreme flooding: using elements of GIS, travel demand, and network science. *Nat. Hazards* 86: 151–164. https://doi.org/10.1007/s11069-016-2678-1.

19 Furno, A., El Faouzi, N.-E., Sharma, R., and Zimeo, E. (2018). Fast computation of betweenness centrality to locate vulnerabilities in very large road networks. 97th Annual Meeting of the Transportation Research Board, Washington DC, United States (7–11 January 2018), pp. 1–17.

20 Mohmand, Y.T. and Wang, A. (2013). Weighted complex network analysis of Pakistan highways. *Discrete Dyn. Nat. Soc.* 2013: 1–5. https://doi.org/10.1155/2013/862612.

21 Derrible, S. (2012). Network centrality of metro systems. *PLoS One* 7: e40575. https://doi.org/10.1371/journal.pone.0040575.

22 Casali, Y. and Heinimann, H.R. (2020). Robustness response of the Zurich road network under different disruption processes. *Comput. Environ. Urban Syst.* 81: 101460. https://doi.org/10.1016/j.compenvurbsys.2020.101460.

23 Serdar, M.Z. and Al-Ghamdi, S.G. (2021). Resiliency assessment of road networks during mega sport events: the case of FIFA World Cup Qatar 2022. *Sustainability* 13: 12367. https://doi.org/10.3390/su132212367.

24 Boussahoua, B. and Elmaouhab, A. (2019). Reliability analysis of electrical power system using graph theory and reliability block diagram. In: *2019 Algerian Large Electrical Network Conference (CAGRE)*, 1–6. IEEE. https://doi.org/10.1109/CAGRE.2019.8713175.

25 Zahedi, Z., Mawengkang, H., Masri, M. et al. (2019). Mathematical fallacies and applications of graph theory in electrical engineering. *J. Phys. Conf. Ser.* 1255: 12085. https://doi.org/10.1088/1742-6596/1255/1/012085.

26 Wang, Z., Scaglione, A., and Thomas, R.J. (2010). Electrical centrality measures for electric power grid vulnerability analysis. In: *49th IEEE Conf. Decis. Control, IEEE*, 5792–5797. https://doi.org/10.1109/CDC.2010.5717964.

27 Werho, T., Vittal, V., Kolluri, S., and Wong, S.M. (2016). Power system connectivity monitoring using a graph theory network flow algorithm. *IEEE Trans. Power Syst.* 31: 4945–4952. https://doi.org/10.1109/TPWRS.2016.2515368.

28 Lulli, A., Ricci, L., Carlini, E., and Dazzi, P. (2015). Distributed current flow betweenness centrality. *2015 IEEE 9th Int. Conf. Self-Adaptive Self-Organizing Syst., IEEE* 71–80. https://doi.org/10.1109/SASO.2015.15.

29 Zarghami, S.A. and Gunawan, I. (2019). A domain-specific measure of centrality for water distribution networks. *Eng. Constr. Archit. Manag.* 27: 341–355. https://doi.org/10.1108/ECAM-03-2019-0176.

30 Giustolisi, O., Ridolfi, L., and Simone, A. (2019). Tailoring centrality metrics for water distribution networks. *Water Resour. Res.* 55: 2348–2369. https://doi.org/10.1029/2018WR023966.

31 Agathokleous, A., Christodoulou, C., and Christodoulou, S.E. (2017). Topological robustness and vulnerability assessment of water distribution networks. *Water Resour. Manage.* 31: 4007–4021. https://doi.org/10.1007/s11269-017-1721-7.

32 Simone, A., Ciliberti, F.G., Laucelli, D.B. et al. (2020). Edge betweenness for water distribution networks domain analysis. *J. Hydroinf.* 22: 121–131. https://doi.org/10.2166/hydro.2019.030.

33 Mortula, M.M., Ahmed, M.A., Sadri, A.M. et al. (2020). Improving resiliency of water supply system in arid regions: integrating centrality and hydraulic vulnerability. *J. Manage. Eng.* 36: 05020011. https://doi.org/10.1061/(ASCE)ME.1943-5479.0000817.

34 Soldi, D., Candelieri, A., and Archetti, F. (2015). Resilience and vulnerability in urban water distribution networks through network theory and hydraulic simulation. *Procedia Eng.* 119: 1259–1268. https://doi.org/10.1016/j.proeng.2015.08.990.

35 Green, O. and Bader, D.A. (2013). Faster betweenness centrality based on data structure experimentation. *Procedia Comput. Sci.* 18: 399–408. https://doi.org/10.1016/j.procs.2013.05.203.

36 Sariyuce, A.E., Kaya, K., Saule, E., and Catalyurek, U.V. (2013). Incremental algorithms for closeness centrality. *2013 IEEE Int. Conf. Big Data, IEEE* 487–492. https://doi.org/10.1109/BigData.2013.6691611.

37 Elser, B. and Montresor, A. (2013). An evaluation study of BigData frameworks for graph processing. *2013 IEEE Int. Conf. Big Data, IEEE* 60–67. https://doi.org/10.1109/BigData.2013.6691555.

38 Amato, F., Moscato, V., Picariello, A. et al. (2018). Centrality in heterogeneous social networks for lurkers detection: an approach based on hypergraphs. *Concurrency Comput. Pract. Exper.* 30: e4188. https://doi.org/10.1002/cpe.4188.

39 Kaur, R., Kaur, M., and Singh, S. (2016). A novel graph centrality based approach to analyze anomalous nodes with negative behavior. *Procedia Comput. Sci.* 78: 556–562. https://doi.org/10.1016/j.procs.2016.02.102.

40 Kermanshah, A., Karduni, A., Peiravian, F., and Derrible, S. (2014). Impact analysis of extreme events on flows in spatial networks. *2014 IEEE Int. Conf. Big Data (Big Data), IEEE* 29–34. https://doi.org/10.1109/BigData.2014.7004428.

41 Serdar, M.Z., Koc, M., and Al-Ghamdi, S.G. (2021). Public transportation resilience towards climate change impacts: the case of Doha metro network. Proceeding B.

42 Liu, W. and Song, Z. (2020). Review of studies on the resilience of urban critical infrastructure networks. *Reliab. Eng. Syst. Saf.* 193: 106617. https://doi.org/10.1016/j.ress.2019.106617.

43 Freeman, L.C. (1978). Centrality in social networks conceptual clarification. *Soc Networks* 1: 215–239.

44 Duan, Y. and Lu, F. (2013). Structural robustness of city road networks based on community. *Comput. Environ. Urban Syst.* 41: 75–87. https://doi.org/10.1016/j.compenvurbsys.2013.03.002.

45 Saadat, H. (2011). *Power System Analysis Third Edition*, 3e. Boston: WCB/McGraw-Hill: McGraw-Hill.

46 Ibrahim, M. and Alkhraibat, A. (2020). Resiliency assessment of microgrid systems. *Appl. Sci.* 10: 1824. https://doi.org/10.3390/app10051824.

47 Maria Luisa, P., Michele, D., and All, E. (2018). Resilience of Distribution Grids. CIRED's point of view, International Conference on Electricity Distribution, pp. 1–107.

48 Abedi, A., Gaudard, L., and Romerio, F. (2019). Review of major approaches to analyze vulnerability in power system. *Reliab. Eng. Syst. Saf.* 183: 153–172. https://doi.org/10.1016/j.ress.2018.11.019.

49 Shin, S., Lee, S., Judi, D. et al. (2018). A systematic review of quantitative resilience measures for water infrastructure systems. *Water* 10: 164. https://doi.org/10.3390/w10020164.

50 Esteban, M. and Portugal-Pereira, J. (2014). Post-disaster resilience of a 100% renewable energy system in Japan. *Energy* 68: 756–764. https://doi.org/10.1016/j.energy.2014.02.045.

51 Diab, E. and Shalaby, A. (2019). Metro transit system resilience: Understanding the impacts of outdoor tracks and weather conditions on metro system interruptions. *Int. J. Sustain. Transp.* 1–14. https://doi.org/10.1080/15568318.2019.1600174.

52 Donovan, B. and Work, D.B. (2017). Empirically quantifying city-scale transportation system resilience to extreme events. *Transp Res Part C Emerg Technol* 79: 333–346. https://doi.org/10.1016/j.trc.2017.03.002.

53 Ilbeigi, M. (2019). Statistical process control for analyzing resilience of transportation networks. *Int. J. Disaster Risk Reduct.* 33: 155–161. https://doi.org/10.1016/j.ijdrr.2018.10.002.

54 Ye, Q. and Ukkusuri, S.V. (2015). Resilience as an objective in the optimal reconstruction sequence for transportation networks. *J. Transp. Saf. Secur.* 7: 91–105. https://doi.org/10.1080/19439962.2014.907384.

55 Wu, Y. and Chen, S. (2019). Resilience modeling of traffic network in post-earthquake emergency medical response considering interactions between infrastructures, people, and hazard. *Sustain Resilient Infrastruct* 4: 82–97. https://doi.org/10.1080/23789689.2018.1518026.

56 Perea, F. and Puerto, J. (2013). Revisiting a game theoretic framework for the robust railway network design against intentional attacks. *Eur. J. Oper. Res.* 226: 286–292. https://doi.org/10.1016/j.ejor.2012.11.015.

57 Zhang, C. and Ramirez-Marquez, J.E. (2013). Protecting critical infrastructures against intentional attacks: a two-stage game with incomplete information. *IIE Trans.* 45: 244–258. https://doi.org/10.1080/0740817X.2012.676749.

58 He, L. and Zhang, J. (2019). Distributed solar energy sharing within connected communities: a coalition game approach. *2019 IEEE Power Energy Soc. Gen. Meet., IEEE* 1–5. https://doi.org/10.1109/PESGM40551.2019.8973867.

59 Serdar, M.Z. and Al-Ghamdi, S.G. (2023). Resilience-oriented recovery of flooded road networks during mega-sport events: a novel framework. *Frontiers in Built Environment*, 9. https://doi.org/10.3389/fbuil.2023.1216919

8

Enhancing Buildings Resilience: A Comprehensive Perspective on Earthquake Resilient Design

Mohammad Zaher Serdar[1] and Sami G. Al-Ghamdi[1,2,3]

[1] *Division of Sustainable Development, College of Science and Engineering, Hamad Bin Khalifa University, Qatar Foundation, Doha, Qatar*
[2] *Environmental Science and Engineering Program, Biological and Environmental Science and Engineering Division, King Abdullah University of Science and Technology (KAUST), Thuwal, Saudi Arabia*
[3] *KAUST Climate and Livability Initiative, King Abdullah University of Science and Technology (KAUST), Thuwal, Saudi Arabia*

Introduction

People spend much of their time within buildings, and thus the safety and resilience of these buildings are essential for people's comfort and productivity. People spend about 90% of their time in buildings, whether in homes, offices, workshops, restaurants, stadiums, or hotels; thus, the safety of the buildings they occupy affects their productivity and quality of life [1, 2]. However, the safety of these buildings is not limited to the indoor environment but also extends to the structural safety and the functionality of supporting systems during and after disasters [3].

Extreme events and disasters have increased in pace and severity in recent years. These extreme events and their adverse effects have increased annually during the past two decades, fueled by the impacts of climate change [4–6]. These impacts are becoming more alarming, as countries across the globe are exposed to unprecedented events such as extreme precipitation and successive forest wildfires; these extreme events are projected to increase in the coming years [4, 7–9]. The impacts of extreme natural events are not limited to the emerging ones caused by climate change but also to reoccurring natural disasters such as earthquakes [10]. Earthquakes affect millions of people throughout the world and cause wide destruction to the urban centers and their inhabitants; they are projected to increase from more than 50% presently to two-thirds of the world population by 2050 [11–13].

Different disasters have different impacts on the built environment in urban areas, but earthquakes are among the most threatening. Extreme events affect different components of the built environment in urban areas, including buildings and other critical infrastructures, with the degree of each disaster varying on different parts. For example, a heatwave does not

---

Sustainable Cities in a Changing Climate: Enhancing Urban Resilience, First Edition.
Edited by Sami G. Al-Ghamdi.
© 2024 John Wiley & Sons Ltd. Published 2024 by John Wiley & Sons Ltd.

impact a building structurally, but it affects its thermal load and consequently, its electricity consumption and the electrical network throughout the city [3, 10, 14]. Some disasters may limit a building's functionality by impacting the supporting infrastructures directly, whether it is water, electricity, or transportation networks; the latter is significant during mega sport events, or indirectly due to failure propagation between the interdependent infrastructures [15–17]. However, earthquakes are different from other disasters in scale and destruction, impacting all infrastructures and buildings directly and indirectly through related disasters, such as tsunamis and landslides, or due to weak recovery [18–21].

To reduce the damage and cost of these severe events, the focus has turned to a resilience-based design in recent years. The increase in population density and value of urban areas has increased the cost of disasters [10, 16, 22]. The cost of a disaster can be divided into the cost of direct damage to the assets and population and the indirect cost due to downtime. Resilience in systems means reducing the direct and indirect damage to the system caused by disturbances and speeding up the recovery process [23]. Thus, integrating resilience-oriented thinking means designing structures and systems in a way that reduces the impacts caused by disturbances and limits the downtime to an acceptable level [24, 25]. Adopting resilience-based design thinking into structural engineering practices means expanding the structural design boundaries to the repair and recovery process, including the supporting systems, especially in extreme cases like earthquakes [25].

Earthquake resilience in structures should focus on both structural systems and functionality supporting systems. As the resilience concept focuses on preserving the overall performance during and after the earthquake, the design should accurately anticipate the structural performance of the building elements and the resulting drifts and deformations during earthquakes. Furthermore, supporting systems that are essential for the functionality and performance of occupants, such as heating, ventilation, and air conditioning (HVAC), electrical, and IT systems, should be accounted for to anticipate the possible damage and downtime. Due to the importance of reducing downtime and regaining functionality, which translates into cost, a resilience design should be based on a holistic understanding of the building's functionality and needs, with all pieces of equipment designed and fixed in a way that contributes to the desired resilience level. Such interdisciplinary understanding requires following an integrative design approach from the early stages of the project, with the participation of all stakeholders, to achieve resilience in a building.

To simplify the discussion regarding building resilience, we first discuss structural representation for assessing building resilience (from a structural perspective). Next, we discuss performance-based design (PBD) as the cornerstone of structural resilience assessment. Then, we focus on the role of supporting systems in building resilience. Finally, we end with concluding remarks, highlighting the importance of green building rating systems for improving the resilience of buildings.

Structural Resilience Representation

The concept of resilience has many representations in different fields aiming to facilitate measuring the resilience of the system of concern. Many representations are suggested in the literature, plotting the performance against time, most prominently in the work of

Bruneau et al. [26], who introduced the concept of the "resilience triangle" [10, 26, 27]. This concept allows one to quantify the resilience of the system by calculating the area under the performance curve, which is done through mathematical integration over time, from the moment of disaster to the end of recovery [34]. To further expand on the concept, readers can refer to [26].

The resilience of a building is not only related to the structural elements but also extends to the supporting systems; thus, there is a need to expand the resilience representation. Despite the critical importance of the safety and integrity of the structural elements, putting a building back in use also requires all the necessary supporting systems to function at an acceptable level [25]. Thus, we are suggesting a modified representation that reflects the actual requirements for building resilience, as presented in Figure 8.1.

The suggested representation for building resilience toward earthquakes, as shown in Figure 8.1, connects all the contributing systems to the building performance. These systems are divided into two main groups: structural systems and supporting systems. Furthermore, the supporting systems are divided into two categories: the first includes the systems necessary for the building's functionality that exist within the building perimeter, such as mechanical, electrical and plumbing (MEP), and the second consists of the supporting systems that exist outside the building and are critical for its functionality, such as electrical, transportation network, and all other critical infrastructures. The assessment of all these systems varies depending on their nature and scale, but they should be normalized to align with a structural system's main categorization levels, according to PBD: immediate occupancy (IO), life safety (LS), and collapse prevention (CP) performance. Additionally, the repair and recovery should follow a damage assessment process, and to ensure building resilience, the planned recovery efforts should be less than a certain time limit set during the integrative design step. The building performance is related to the lowest-performing system, reflects the building functionality loss, and can be used to quantify the resilience using mathematical integration, similar to the method mentioned in the previous paragraph.

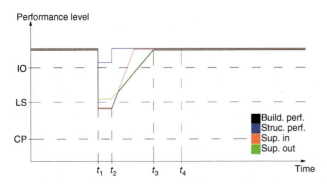

Figure 8.1 Building seismic/earthquake resilience representation. This representation connects various elements that contribute to the building's performance (Build. Perf.) and resilience toward earthquakes, namely structural system performance (Struc. Perf.), supporting systems within the building, such as MEP (Sup. In), and supporting systems outside the building, representing various infrastructures (Sup. Out). The used performance level notations are immediate occupancy (IO), life safety (LS), and collapse prevention (CP). The time-related ones are t_1: earthquake happens; t_2: the end of damage evaluation; t_3: the planned end of repairs; and t_4: repair time limit.

The discussion regarding the performance and assessments of the system is presented in the following section; however, the primary focus is on structural system performance assessment using PBD.

Performance-Based Design (PBD)

Structural resilience is related to structural design and construction, in other words, related to core and shell developments. This aspect should be addressed from a structural engineering perspective and focus on specific threats, primarily earthquakes. Structural resilience is not concerned with the structural safety under regular (bearing) loads, such as dead and live loads, because the capacity of bearing loads is considered ensured under current design codes. The main concern is structural behavior under exceptional events such as earthquakes/seismic loads, where design codes fail to describe the exact expected behavior of buildings. To address this limitation in design codes, the resilience-based approach relies on the application of PBD under several earthquake levels and probabilities to assess buildings' resilience and performance levels [28].

The response spectrum approach adopted for earthquake design fails to represent the actual damage to a building. The response spectrum relies on several properties related to project location and type of soil to generate a diagram of maximum responses that will affect a building based on the building's period [29]. The application of such a method mainly yields a shear force applied to a building that will generate internal forces in its elements and cause drifts. The aim of this design approach is to make sure all structural elements can sustain the resultant forces and keep drifts within allowable service levels. The main drawback of this method is to view a building as a simple oscillator and convert an earthquake, which consists of a series of unpredictable vibrations with numerous frequencies, into a single force that affects a building linearly. This simplification ignores the actual impact of an earthquake and the development of plastic hinges in a building, which can change a building's response and periods under a possible series of vibrations contained in an earthquake.

PBD provides better insights into a building's behavior under earthquake conditions than conventional response spectrum analysis. By leveraging the increased computational power in recent years, structural engineers are capable of simulating building behavior under specific ground shaking records, representing earthquake scenarios, in what is known as time-history analysis. Unlike the response spectrum analysis, which is limited to the application of static forces at story levels, time-history analysis reflects the actual building behavior and allows for monitoring of energy dissipation solutions and the development of plastic hinges throughout an earthquake scenario. Energy dissipation and development of plastic hinges are then used to describe a building's performance level.

PBD can account for the randomness of earthquakes by using several earthquake records. To address the random nature of earthquakes and the vast range of possible frequencies, PBD requires running a series of specifically selected ground shaking records based on the geological properties of the building area. These properties are related to site classification, such as fault type and distance, and project soil, which affects wave velocity.

Synthetic accelerograms can be generated and used when suitable records are lacking. However, a large database of ground motion records from around the world can be accessed through [30]. Although, the records accessed through this database are unscaled, a tool is provided on the website to generate scaling factors that can be applied in the simulation software [30].

Generally, ground motion records require scaling to match earthquake spectrums at different levels and probabilities of reoccurrence. The scaling process is typically done in the time domain and follows two main approaches: scaling over a single period of interest and scaling over a range of interests. The first approach focuses on making the response spectrum of the record at the natural period of the structure (T_n) equal to the code-provided response spectrum at the same point. The second approach focuses on using a scaling factor that considers the average variations between periods over a range from $0.2T_n$ to $1.5T_n$. Alternatively, engineers can apply spectral matching, which manipulates the frequencies, to smooth the differences between the records and the target in the time domain. However, spectral matching is not recommended as it changes the frequency content of the ground motion records.

Several performance objectives, or levels, are assigned to different earthquake probability levels. PBD aims to assess the structural behavior under scaled ground motion records and ensure it does not exceed the performance limit defined by the formation of hinges and development associated with each earthquake level. The main performance objectives are IO, LS, and CP, referring to the development of several levels of plastic hinges in the building; readers can expand on the topic by referring to [25, 28, 31]. These performance criteria are then assigned to specific levels of earthquake probability of exceedance such as 2% in 50 years equal to an exceedance level of once every 2500 years event and 10% in 50 years equal to once every 475 years [25, 28].

Planning for the post-earthquake assessment and repairs is essential to ensure the resilience of the building and its timely recovery process. From the early design process, scenarios of a post-earthquake event should be considered, such as a lack of resources for damage assessment and retrofitting. To accommodate such a situation and to avoid unnecessary downtime or risk the lives of occupants, building owners should have binding contracts with professional experts to assess the damage suffered to the structural elements and with contractors to apply any necessary retrofitting before reoccupying a building. The time needed for such an assessment as well as the mobilization of the contractor should be determined through the contracts and accounted for as part of a building's recovery time. Additionally, the details of these contracts, both in time and cost, should be based on the results of a peer review process of the PBD results.

Supporting Systems

Several aspects can affect building resilience, especially regarding the supporting systems, which are necessary for the building's functionality and achieving performance goals. These systems can be divided into two main categories: within the building, such as MEP, and offsite/external supporting systems, such as critical infrastructures.

Supporting Systems Within the Building

Resilience-based thinking should extend beyond the structural element's performance during an earthquake, as the survivability of nonstructural components is also important in labeling a building as resilient. In a resilient building, it is mandatory to ensure that all nonstructural components suffer acceptable degradation levels that do not prevent the system from meeting its intended performance goal. For example, if the damage due to shaking affects a building's envelope, causing severe cracks and threatening the safety or comfort of occupants, the building may not be suitable for immediate occupancy, even if the structural parts meet the required performance levels; the same concept can be applied to the MEP.

The assessment of these nonstructural systems should include facades, electrical, Internet, HVAC, and plumbing systems. To prevent any impact or loss beyond the scope allowed by the intended performance, the assessment of such systems should include checks on the impacts of oscillation, drifts, and ductile deformation, as well as confirming proper installation and repair of related equipment. Such assessments are mandatory to evaluate possible damage and the recovery of resources and time, thus creating a holistic understanding of building earthquake resilience. Such consideration can allow the application of the suggested representation of Figure 8.1, which facilitates the capture of the recovery plans and the assessment of different system alternatives based on the expected recovery rates, costs, and supply complexities. Furthermore, it is crucial to ensure the availability of maintenance and repair teams and equipment, through pre-disaster contracts, to avoid prolonged downtime due to either a shortage of equipment or resources during the post-disaster period. These considerations are essential when designing a resilient building because a damaged HVAC system, lightning, or any other system may limit or eradicate a building's functionality.

Beyond the Building Limits

Urban infrastructures are essential to a building's functionality and, thus, to its resilience. Critical infrastructures such as transportation, electricity, water supply, and sewer networks are essential to the livelihood of any city. In the aftermath of an earthquake, these infrastructures are crucial for the relief effort. For example, transportation networks facilitate the mobility of repair teams and equipment and allow the return of a building's occupants.

The performance assessment of infrastructures under conditions of earthquakes is different, in scale and nature, from the one done for the structural system; thus, it should be done separately and provided by the governing authorities, such as municipalities. A governing authority should assess and map the performance and damage expected to be suffered by critical infrastructures due to earthquakes. Such an assessment should consider the possible damage caused to each infrastructure at each magnitude of earthquake reoccurrence probability, similar to those applied in PBD, as part of the regulating codes and building standards. Additionally, it is necessary to reflect the degradation and delays caused by the interdependency between different infrastructures, such as inaccessibility to a substation due to loss of a bridge or loss of metro services due to the damage of the electrical network. The results of these assessments should be mapped to reflect the expected damage, serviceability level, expected downtime, and recovery duration; such a process should be done in a Geographic Information System (GIS) environment and provided either publicly or on-demand to the developers.

The use of microgrid applications and on-site energy production can help improve resilience. The application of microgrids and energy production using renewable energy could provide some level of functionality for the related system, especially if the design that is considered works on an islanded mode with enough energy storage capacity [3]. However, due to the debris of an earthquake, these systems are also susceptible to direct and indirect impacts, and their efficiency is related to climate factors and storage capacity. On-site energy production using a backup generator could provide a temporary solution, but its safety and the risk of a lack of fuel supply in the post-earthquake period could limit its reliability.

Conclusion

With the increased concentration of people living in urban areas, it is becoming more critical to reduce the impact caused by any disturbance, especially large-scale ones such as earthquakes. With the large number of cities located in active seismic areas and the increase in population concentrated in them, the impact of any future earthquake is expected to be more disastrous. The impact of an earthquake can span over large areas and affect buildings, infrastructures, and even the sociology of humans. It is crucial to base preparation and risk management planning on two main principles: impact reduction and swift recovery, i.e., resilience-based planning. Resilience-based planning should expand to the design standards and practices of buildings and infrastructures and consider their interactions, especially with high-value facilities. Furthermore, designers should consider all the supporting systems within the building and the earthquake's impact on these systems. Considering all the systems within and outside the building, in addition to its structural system, can provide a holistic understanding of the impact of such an event and facilitate efficient recovery planning, thus protecting human and economic capitals and improving building and city resilience.

Aiming for efficient design and performance of a building by following green building regulations can also improve its resilience to any disturbance, including earthquakes. By following green building rating system recommendations and requirements, such as LEED or BREEAM, engineers can develop an efficient design, in terms of ventilation, natural lighting, and energy and water consumption [32]. Additionally, these rating systems promote on-site energy production and water reuse, extending the survivability of the building and its occupants [32, 33]. Furthermore, following resilience-oriented rating systems, even if not fully adopted, can substantially improve the resilience of a structure. Most notable among these rating systems are REDI, developed by ARUP [25], which focuses on earthquake resilience, and RELI, developed by USGBC [33], which focuses on the resilience of a building to different disturbances.

References

1 Mannan, M. and Al-Ghamdi, S.G. (2021). Indoor air quality in buildings: a comprehensive review on the factors influencing air pollution in residential and commercial structure. *Int. J. Environ. Res. Public Health* 18: 3276. https://doi.org/10.3390/ijerph18063276.

2 Klepeis, N.E., Nelson, W.C., Ott, W.R. et al. (2001). The National Human Activity Pattern Survey (NHAPS): a resource for assessing exposure to environmental pollutants. *J. Exposure Sci. Environ. Epidemiol.* 11: 231–252. https://doi.org/10.1038/sj.jea.7500165.

3 Serdar, M.Z. and Al-Ghamdi, S.G. (2021). Preparing for the unpredicted: a resiliency approach in energy system assessment. In: *Green Energy Technology* (ed. J. Ren), 183–201. Cham: Springer International Publishing https://doi.org/10.1007/978-3-030-67529-5_9.

4 Salimi, M. and Al-Ghamdi, S.G. (2020). Climate change impacts on critical urban infrastructure and urban resiliency strategies for the Middle East. *Sustainable Cities Soc.* 54: 101948. https://doi.org/10.1016/j.scs.2019.101948.

5 Serdar, M.Z. and Al-Ghamdi, S. (2022). Resilience assessment of transportation networks to climate change induced flooding: the case of doha highways network. *2022 7th Asia Conference on Environment and Sustainable Development (ACESD 2022)* (4–6 November 2022), Kyoto, Japan, pp. 9–17.

6 Tahir, F., Ajjur, S.B., Serdar, M.Z. et al. (2021). *Qatar Climate Change Conference 2021*. Hamad bin Khalifa University Press (HBKU Press) https://doi.org/10.5339/conf_proceed_qccc2021.

7 Dupuy, J., Fargeon, H., Martin-StPaul, N. et al. (2020). Climate change impact on future wildfire danger and activity in southern Europe: a review. *Ann. For. Sci.* 77: 35. https://doi.org/10.1007/s13595-020-00933-5.

8 Natole, M., Ying, Y., Buyantuev, A. et al. (2021). Patterns of mega-forest fires in east Siberia will become less predictable with climate warming. *Environ. Adv.* 4: 100041. https://doi.org/10.1016/j.envadv.2021.100041.

9 Serdar, M.Z., Ajjur, S.B., and Al-Ghamdi, S.G. (2022). Flood susceptibility assessment in arid areas: a case study of Qatar. *Sustainability* 14: 9792. https://doi.org/10.3390/su14159792.

10 Serdar, M.Z., Koç, M., and Al-Ghamdi, S.G. (2022). Urban transportation networks resilience: indicators, disturbances, and assessment methods. *Sustainable Cities Soc.* 76: 103452. https://doi.org/10.1016/j.scs.2021.103452.

11 United Nations, Department of Economic and Social Affairs PD. World Urbanization Prospects: The 2018 Revision. New York: 2019. https://doi.org/10.18356/b9e995fe-en.

12 Olson, D.L. and Wu, D.D. (2010). Earthquakes and risk management in China. *Hum. Ecol. Risk Assess.: Int. J.* 16: 478–493. https://doi.org/10.1080/10807031003779898.

13 Food and Agriculture Organization of the United Nations (FAO) (n.d.). Earthquakes. https://www.fao.org/emergencies/emergency-types/earthquakes/en/ (accessed 16 November 2021).

14 Serdar, M.Z., Macauley, N., and Al-Ghamdi, S.G. (2022). Building thermal resilience framework (BTRF): a novel framework to address the challenge of extreme thermal events, arising from climate change. *Front. Built Environ.* 8: https://doi.org/10.3389/fbuil.2022.1029992.

15 Serdar, M.Z. and Al-Ghamdi, S.G. (2021). Resiliency assessment of road networks during mega sport events: the case of FIFA World Cup Qatar 2022. *Sustainability* 13: 12367. https://doi.org/10.3390/su132212367.

16 Serdar, M.Z., Koc, M., and Al-Ghamdi, S.G. (2021). Urban infrastructure resilience assessment during mega sport events using a multi-criteria approach. *Front. Sustainability* 2. https://doi.org/10.3389/frsus.2021.673797.

17 Serdar, M.Z. and Al-Ghamdi, S. (2022). Public transportation resilience towards climate change impacts: the case of doha metro network. *2022 7th Asia Conference on Environment and Sustainable Development (ACESD 2022)* (4–6 November 2022), Kyoto, Japan, pp. 1–8.

18 Ramalanjaona, G. (2011). Impact of 2004 tsunami in the Islands of Indian Ocean: lessons learned. *Emerg. Med. Int.* 2011: 1–3. https://doi.org/10.1155/2011/920813.

19 Panwar, V. and Sen, S. (2019). Economic impact of natural disasters: an empirical re-examination. *Margin J. Appl. Econ. Res.* 13: 109–139. https://doi.org/10.1177/0973801018800087.

20 Hombrados, J.G. (2020). The lasting effects of natural disasters on property crime: evidence from the 2010 Chilean earthquake. *J. Econ. Behav. Organ.* 175: 114–154. https://doi.org/10.1016/j.jebo.2020.04.008.

21 Sheller, M. (2013). The islanding effect: post-disaster mobility systems and humanitarian logistics in Haiti. *Cult. Geogr.* 20: 185–204. https://doi.org/10.1177/1474474012438828.

22 Liu, W. and Song, Z. (2020). Review of studies on the resilience of urban critical infrastructure networks. *Reliab. Eng. Syst. Saf.* 193: 106617. https://doi.org/10.1016/j.ress.2019.106617.

23 Hosseini, S., Barker, K., and Ramirez-Marquez, J.E. (2016). A review of definitions and measures of system resilience. *Reliab. Eng. Syst. Saf.* 145: 47–61. https://doi.org/10.1016/J.RESS.2015.08.006.

24 Butry, D., Davis, C.A., Malushte, S.R. et al. (2021). *Hazard-Resilient Infrastructure*. Reston, VA: American Society of Civil Engineers https://doi.org/10.1061/9780784415757.

25 Almufti, I. and Willford, M. (2013). REDi™ rating system, resilience-based earthquake design initiative for the next generation of buildings, Version 1, pp. 1–133.

26 Bruneau, M., Chang, S.E., Eguchi, R.T. et al. (2003). A framework to quantitatively assess and enhance the seismic resilience of communities. *Earthquake Spectra* 19: 733–752. https://doi.org/10.1193/1.1623497.

27 Ayyub, B.M. (2014). Systems resilience for multihazard environments: definition, metrics, and valuation for decision making. *Risk Anal.* 34: 340–355. https://doi.org/10.1111/risa.12093.

28 Manohar, S. and Madhekar, S. (2015). *Performance-Based Seismic Design*, 417–437. https://doi.org/10.1007/978-81-322-2319-1_11.

29 Serdar, M.Z., Çağlayan, B.Ö., and Al-Ghamdi, S.G. (2022). A special characteristic of an earthquake response spectrum detected in Turkey. *Mater. Today Proc.* https://doi.org/10.1016/j.matpr.2022.04.407.

30 The University of California Pacific Earthquake Engineering Research Center (n.d.). PEER Ground Motion Database. https://ngawest2.berkeley.edu/.

31 (2017). *Seismic Evaluation and Retrofit of Existing Buildings*. Reston, VA: American Society of Civil Engineers https://doi.org/10.1061/9780784414859.

32 Champagne, C.L. and Aktas, C.B. (2016). Assessing the resilience of LEED certified green buildings. *Procedia Eng.* 145: 380–387. https://doi.org/10.1016/j.proeng.2016.04.095.

33 USGBC (2018). RELi 2.0 rating guidelines for resilient design and construction. U.S. Green Building Council. https://www.usgbc.org/sites/default/files/RELi_Dec_2018-FINAL.pdf (accessed 1 August 2020).

34 Serdar, M.Z. and Al-Ghamdi, S.G. (2023). Resilience-oriented recovery of flooded road networks during mega-sport events: a novel framework. *Frontiers in Built Environment*, 9. https://doi.org/10.3389/fbuil.2023.1216919

9

Enhancing Built Environment Resilience: Exploring Themes and Dimensions

Mohammed M. Al-Humaiqani[1] and Sami G. Al-Ghamdi[1,2,3]

[1] Division of Sustainable Development, College of Science and Engineering, Hamad Bin Khalifa University, Qatar Foundation, Doha, Qatar
[2] Environmental Science and Engineering Program, Biological and Environmental Science and Engineering Division, King Abdullah University of Science and Technology (KAUST), Thuwal, Saudi Arabia
[3] KAUST Climate and Livability Initiative, King Abdullah University of Science and Technology (KAUST), Thuwal, Saudi Arabia

Introduction

Historically, the concept of resilience can be traced back to the mid-twentieth century (the 1950s) when a group of scholars studied environmental stability with an emphasis on biodiversity. Crawford Stanley Holling [1] made the first systematic discussion on the resilience concept in 1973 and pointed out the need for persistence and stability of ecological systems, which was considered an extension of the stability view [2]. Later, Holling expanded the concept of resilience to cover management and engineering perspectives [1]. Resilience has become popular due to increasing concerns about natural, environmental, and climate threats and impacts. The concept of resilience has progressively complemented sustainability [3]. Different multidisciplinary resilience studies have also been conducted by many scholars over the last several years. The studies have been conducted to respond to the increasing number of climate and natural challenges, including the environmental instability faced by the world [4]. The global challenges and critical issues highlighted by the World Economic Forum in the Global Risks Report of 2019 include environmental fragilities, economic vulnerabilities, societal and political strains, geopolitical tensions, and technological instabilities [5]. The concept of resilience has evolved over the last two decades and is increasingly used by various practitioners, academics, scholars, and policy-makers across many disciplines [6–8]; however, the term "resilience" can have different meanings depending on the boundaries of the discipline it is used for [8].

Climate resilience is the ability of a system to reduce vulnerability and sensitivity to climatic changes, withstand extreme climate events, rapidly recover after undergoing stress and adverse impacts, and learn from climatic issues and events to reflect on future decisions [9]. The ability includes risk reduction, response, absorptive (coping) and

Sustainable Cities in a Changing Climate: Enhancing Urban Resilience. First Edition.
Edited by Sami G. Al-Ghamdi.
© 2024 John Wiley & Sons Ltd. Published 2024 by John Wiley & Sons Ltd.

Table 9.1 Resilience concepts.

Resilience perspective	Area of focus	Attributes	Resilience status
Engineering (material) resilience	Stability and recovery	Efficiency and recovery time	The status is about the vicinity of a stable equilibrium
Management (organizational) resilience	Anticipation, preparedness, responses, and adaptation	Prioritize reliability, recognize complexity, strong leaders, acknowledging risk, and decentralized decision-making	To survive and prosper
Ecological (ecosystem) resilience	Stability and robustness	Redundant capacity, maintained function, and ability to withstand shock	Multiple equilibria, stable green lands, and landscapes
Social-ecological resilience	Adaptive and transformative capacities, knowledge, and creativity	Reorganization, sustaining, developing, and thriving	Dynamic interactions and feedback from the integrated system

Source: Adapted from [13].

adaptive capacity, and improving the environmental, physical, social, institutional, and governance structures. The definition can also include the degree of disturbance a system can withstand before its primary function is operational. The system can be a single system or a group of systems forming an urban development. Urban climate resilience includes several components and measures, starting with assessing the level of vulnerability at the urban level by measuring the systems' sensitivity, exposure, and response capacity. They also include learning from past and former ineffective responses and actions, developing a responsive action plan, and building risk reduction capacity [10]. Overall, urban resilience is the capacity of cities to function regardless of encountered climatic stresses or shocks, enabling people to survive and thrive safely [9, 11, 12]. Table 9.1 outlines the resilience concepts, including the area of focus, attributes, and resilience perspective and status.

Uncertainty

Uncertainty is a general term used differently depending on the context of the debate. Generally, uncertainty is applied to anticipate future events and unknown measurements [14]. According to Grebici et al. , uncertainty has no consistent definition in the literature [15]. In engineering, two types of uncertainties are known and unknown, as distinguished by Clarkson and Eckert [16]. The known uncertainties can be depicted and adequately handled based on cases experienced in the past. In contrast, the unknown uncertainties are related to unforeseen events and circumstances, e.g., wildfires in different countries during 2021 due to severe heat waves and their impact on communities and the built environment [17, 18]. Apart from these types of uncertainty, engineering analysis classifies uncertainty into two categories: reducible (epistemic) and irreducible (stochastic) [19].

The difference between reducible and irreducible uncertainty is that reducible uncertainty can be decreased through measurements, further studies, and expert consultation, while irreducible uncertainty cannot. The reason is that irreducible uncertainty is inherent to a system and cannot be easily modified [15]. In practice, the separation of both uncertainty types is not always possible. Overall, the uncertainty can be caused by several reasons, including conflicting evidence, abundance, ignorance, lack of information, belief, ambiguity in the description of collected measurements, and improper measurements [20]. The lack of knowledge is interpreted as either unknown or vaguely known, and the lack of definition means not yet decided or specified. There is another classification of uncertainty called content and context classification. Content-related uncertainty is related to uncertainty in the information, such as errors, unspecificity, incorrectness, and fuzziness. Context-related uncertainty is related to context complications, such as the instability of a system [20]. Overall, the uncertainty level is determined by assessing the readiness of information, represented by several metrics, including unreliability, invalidity, and inconsistency.

Risk Identification and Assessment

Risk is described as the likely impact or impacts due to the vulnerability of a system and the consequence of uncertainty. Consequently, risk management in engineering is required to convert uncertainty into useful sources of information. Risk management is critical for the planning, engineering design, operation, and maintenance of different built environment systems [21]. In the built environment, understanding the risks is significant for making investment decisions. The frequency and magnitude of climate events and their changes over time in the present and future are necessary to inform urban infrastructure. Understanding potential future changes through realizing current weather patterns, disasters, and climate risks can help predict disasters and reduce climate impacts or reoccurring cyclic events [22].

Resilience Capacities

The three themes of resilience in a community (reducing disaster, increasing adaptability, and rapid recovery) can be named and described in four capacities, as follows [23, 24]:

a) Absorptive (resistive or coping) capacity is defined as a system's ability to mitigate the impacts of disruption caused by an adverse event and recover from the problem with predetermined responses that help preserve and restore essential basic functions and structures. Examples of adaptive capacities are savings, trained disaster risk personnel, installed early warning systems, climate insurance systems, and dike systems to reduce susceptibility to floods [25–27].

b) Adaptive capacity is defined as a system's ability to adapt, alter, or change its characteristics to respond better to both present and anticipated future climatic shocks and environmental stresses, as well as take advantage of given opportunities. Adaptive capacities include improved ecological resource management, awareness events regarding climate change, modified planting performance, and diversification of early warning systems [25, 28, 29].

Table 9.2 Urban resilience key capacities to achieve resilience.

No.	Urban resilience characteristics	The role played and relationship with the resilience phases
1	Preparedness	Preparing the system to withstand an event before it occurs
2	Absorption or coping (persistence) → stability	Make the system able to absorb the occurring event
3	Recovery	The system recovers after having undergone an event
4	Adaptability → flexibility	The ability to adjust to a change and cope with consequences during and after an event
5	Transformability (transformational responses) → change	Transformation in policies/regulations, governance mechanisms, community networks, infrastructure, etc., after an event

c) Restorative (recovery) capacity is the extent to which a system can be restored from a disturbing effect over a specific period of time. Restorative capacity can enhance the recoverability of a system undergoing a disruptive event or stress [30].

d) Transformative capacity [26] is the fundamental change a system can make to its characteristics when the current situation turns out to be unsound in the face of climatic shocks and environmental stresses. Examples of transformative capacities are migrating people from rural areas to cities, business transformation (e.g., a transformation from corn farmer to fish farmer), and using renewable energy systems instead of systems based on fossil fuel energy [25, 26, 31].

Table 9.2 presents the five characteristics of urban resilience discussed by Tong. The characteristics reveal the level of resilience of any urban system and may inform future actions to improve its resilience; hence, monitoring their values is very important.

Resilience Components

Resilience can be described in several thematic areas, where every area contains components that differ from one theme to another. Table 9.3 describes the resilience components for each thematic area [32].

Types of Resilience

The capacity of a system to react to unanticipated risks, changes, and events by mitigating or preventing their effects is known as resilience. This concept is more precisely defined from one field to another [33]. As defined by the United Nations, "Resilience is the ability of a system, community or society exposed to hazards to resist, absorb, accommodate and recover from the effects of a hazard in a timely and efficient manner, including through the preservation and restoration of its essential basic structures and functions [34–36]."

Table 9.3 Resilience components and theme areas.

	Resilience theme areas				
	Knowledge (education)	Assessment (risk)	Management (risk and vulnerability)	Preparedness (disaster)	Governance (policies and regulations)
Resilience components	Awareness dissemination, information management, attitudes, learning, and research	Data collection, risk assessment, vulnerability assessment, and capacity assessment	Resources management, sustainable businesses, financial tools, and technical measures	Warning systems, infrastructure and emergency resources, response and recovery, and organizational capacities	Policies integration, planning, regulations, emergency response, responsibilities, partnerships, and community participation

Another definition as stated by the National Academies Press is, "the ability to prepare and plan for, absorb, recover from and more successfully adapt to adverse events [37, 38]." The types of resilience covered in this section include ecological and engineering resilience, community and social resilience, specified and general resilience, critical infrastructure resilience, and technical systems, products, and production resilience.

Ecological and Engineering Resilience

In ecology, resilience has been defined differently, with contrasting meanings, each referring to a distinct connotation. The first connotation is the rapidity of returning a system, after being disrupted, to its initial status (equilibrium), which is usually measured in time [39]. Ecological resilience, also known as ecosystem resilience, is based on multiple alternate states. According to Gunderson, ecological resilience is defined as "the amount of disturbance that an ecosystem could withstand without changing self-organized processes and structures [28]." It is also defined as the magnitude of disturbance that a system can withstand and absorb before it causes a change to the system's stable states, and it refers to the stability domain limit [29].

Resilience was initially defined by Holling in 1973 as the ability to absorb disturbance while maintaining the primary function and structure and not changing to a qualitatively different state. This type of resilience emphasizes change in a system, persistence, and unpredictability [40]. Although several scholars and researchers have considered the time required by a system to return to its steady-state or equilibrium status as a measure of stability [41], Holling [40] considered the return time as "engineering resilience."

The existence of multiple stable states is the critical assumption behind ecological resilience. It is also known as regimes, multiple equilibria, or basins of attraction [39]. Figure 9.1 illustrates some fundamental concepts of ecological resilience that occur due to disturbances: (1 and 2) several stable states and (3 and 4) changes in resilience and regime shifts.

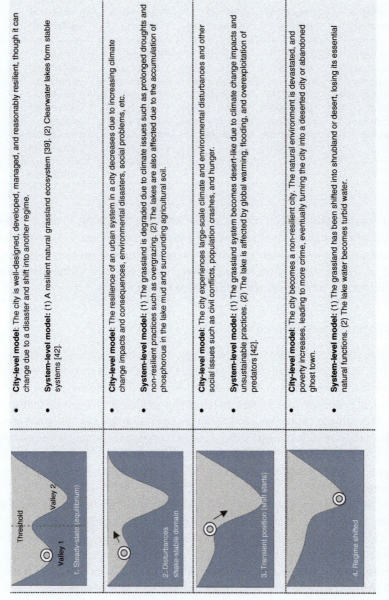

Figure 9.1 Key concepts of ecological resilience. Valleys, balls, and arrows represent the stability domains, system, and disturbances, respectively. The slope in the stability domain (valley) edges can determine the engineering resilience (also called shape). Similarly, the width of the domain can determine the system's ecological resilience. Consequently, the system's adaptive capacity is determined by its ability to persist in the stability domain (valley). *Source:* The ideas for plotting this figure were adapted from different sources, including (2013) Wu and Wu [39] and Folke et al. 2004 [42].

- **City-level model**: The city is well-designed, developed, managed, and reasonably resilient, though it can change due to a disaster and shift into another regime.
- **System-level model**: (1) A resilient natural grassland ecosystem [39]. (2) Clearwater lakes form stable systems [42].

- **City-level model**: The resilience of an urban system in a city decreases due to increasing climate change impacts and consequences, environmental disasters, social problems, etc.
- **System-level model**: (1) The grassland is degraded due to climate issues such as prolonged droughts and non-resilient practices such as overgrazing. (2) The lakes are also affected due to the accumulation of phosphorous in the lake mud and surrounding agricultural soil.

- **City-level model**: The city experiences large-scale climate and environmental disturbances and other social issues such as civil conflicts, population crashes, and hunger.
- **System-level model**: (1) The grassland system becomes desert-like due to climate change impacts and unsustainable practices. (2) The lake is affected by global warming, flooding, and overexploitation of predators [42].

- **City-level model**: The city becomes a non-resilient city. The natural environment is devastated, and poverty increases, leading to more crime, eventually turning the city into a deserted city or abandoned ghost town.
- **System-level model**: (1) The grassland has been shifted into shrubland or desert, losing its essential natural functions. (2) The lake water becomes turbid water.

Adaptive capacity is introduced as a new term describing the process of ecological resilience modification [41].

Engineering resilience is defined as the return time that measures the stability of a system. In other words, engineering resilience is the time that a system takes to return to its stable state [40].

Community and Social Resilience

Community resilience is defined as the ability of the community to utilize resources and strategies to absorb, respond, withstand, and recover from adverse situations [43]. Community resilience expresses the interconnected systems that directly impact human society on ecological, socioeconomic, and built environment levels [44]. According to Arbon [45], a community is resilient if its members are connected and work together, are able to sustain critical systems under stress, and can adapt to social, physical, environmental, and economic changes. Also, it is self-sustaining if limited resources can be managed and if it has the ability to keep improving itself by learning from the past to nurture future decision-making. Communities are prominent actors that can play a significant role in creating resiliency through preparedness (learning capacity), the ability to persist (robustness), and the ability to adapt and innovate changes (transformation capacity) [46].

Social resilience is another topic of resilience and is defined as the human ability to withstand and recover from external shocks, such as socioeconomic and environmental stresses and political perturbations, within a reasonable time [39, 47]. Social resilience includes how people come together after a disaster to help and support each other as needed. It also involves learning to adapt to a new team's work environment and building and establishing a positive workplace culture. Social resilience faces several challenges and problems on different levels (major, situational, and daily).

Worldwide, many cities and communities experience significant yearly damage due to climate and natural hazards, such as heavy precipitation, floods, tornados, earthquakes, hurricanes, and temperature rises [9]. The community functions might be suspected of disturbance or disruption due to the hazardous event, resulting in permanent changes [15]. Many examples have been witnessed worldwide, with extensive damage across several communities. However, many communities are now focusing on mitigating vulnerabilities and threats and reducing the potential exposure to hazards. The activities that make a community resilient may include additional actions and out-of-the-box strategies depending on the location and type of the community, as well as the type and magnitude of potential threats and vulnerabilities. Community resilience requires plans to maintain acceptable system functionality levels during and after the event. The plans must include strategies for rapid system recovery within the shortest possible time.

Apart from adopting resilience codes and standards and enforcing relevant regulations, additional activities are still needed to develop resilient communities. A community resilience plan, including performance goals, mitigation strategies, response actions, and recovery activities, needs to be set to address anticipated climate and natural hazards to the different systems. Several aspects would inform the resilience plan's performance. The aspects include emergency response, community and welfare protection, human health, security and safety, business continuity, temporary and permanent measures, land use policy, and other resilience strategies.

Community resilience consists of four dimensions: technical, organizational, social, and economic. These dimensions are sometimes referred to by the acronym TOSE. The dynamic properties of the four dimensions are measured by applying quantitative indicators. According to Stevenson et al. [13], the properties include robustness, redundancy, resourcefulness, and rapidity. However, eight functions within a city are critical to resilience, including health, well-being, economy, society, urban systems, services, leadership, and strategy. These functions and the 12 key themes contribute to city resilience. The eight qualities of resilient cities were systematically studied by Al-Humaiqani and Al-Ghamdi [9, 12] and differently represented in different resilient city frameworks, such as the one developed by Rockefeller and Arup in 2014 [48].

Specified and General Resilience

Cities can face known threats and vulnerabilities by designing and incorporating specified resilience systems. The specified resilience is designed and built depending on the target city's location and the potential impacts expected to attack it. For example, cities located in a hot region, where maximum temperatures are reached, causing huge impacts on utilities and other infrastructure, or cities located on coastal lines, where storm surges, tsunamis, and hurricanes occur, resulting in infrastructure damage and collapse. As a response, the infrastructure systems should be subjected to further improvements to mitigate the impact and withstand such occurrences [39]. Similarly, security protocols should be developed to hedge terrorist dangers to political and financial centers.

Critical Infrastructure Resilience

Critical infrastructure is highly vulnerable to natural disasters and climate change impacts. The exposure of critical infrastructure assets to such disasters can result in major consequences and damages after facing a disaster. Consequently, repairing or replacing infrastructure elements or assets can sometimes become costlier and difficult, eventually increasing a community's suffering. For example, Australia has recently faced several natural disasters, such as floods, cyclones, storms, and bushfires resulting in destructive and damaging consequences [49].

Adaptation and implementation of suitable standards, codes, and regulations and other activities, such as enforcing emergency management and encouraging standards-based design, are also being implemented. Regardless of the implementation and enforcement of best practices and standards, and the advancement of regulations and codes of practice, significant damages and losses continue to harm communities. Independent design and development of different systems within a community, such as buildings, utilities, and other infrastructure systems, are not enough to achieve the performance goal of community resilience. The best way to achieve higher resiliency within a community is through interdependent design and development, which address the interrelationships between systems. The achievement of the integrated performance of urban built environment systems, including buildings and infrastructure, cannot be limited to adopting codes and regulations, especially where existing buildings and infrastructure systems are built following different standards. This is because the hazard level at the time of development could have been different compared to the most recent developments, in which climate and natural hazards are substantially different.

Although the requirements of infrastructure planning refer to resilience in some cases, there is a lack of guidelines to achieve the target level of resilience. Building codes and standards, as well as land use planning systems, can provide suitable requirements for developing resilient infrastructure; however, some infrastructure systems or assets still require a higher level of resilience [49]. Determining resilience measures is challenging, especially before a disaster event or critical infrastructure is built. It requires a detailed assessment of the probability of a potential hazard that may affect a planned infrastructure asset or system. It also requires an analysis of potential resilience options to protect systems and mitigate disaster impacts.

Economically, evaluating resilience before any approvals can change decisions on infrastructure investment. Services such as electricity and telecommunications in risk areas can directly benefit from any resilience measures [49]. For example, electricity service was lost in Australia in 2007 due to the Victoria bushfire; however, they could have avoided this by investing $11M/Km in building underground transmission lines. Also, in 2011, the Brisbane floods cost users $1M/day and amounted to an overall $1M for the Optus telecommunications company. Under the same circumstances, future costs could amount to $9M; however, since 2011, Optus has invested about $4M to improve telecommunications infrastructure resilience, resulting in benefits that exceed the cost of the implemented measures.

Similarly, traffic disruptions caused by flooding highway bridges in New South Wales (NSW) resulted in a considerable cost. The estimated future cost of the same event is about $75M, while the replacement cost would only be $7.4M. The cost of future events is estimated at $75M, totaling about $92M (in present value terms) over the projected life of the asset.

Technical Systems, Products, and Production Resilience

The resilience of technical systems, products, and production is essential. It is achieved through anticipating, preventing, and mitigating risks and threats. It should be considered part of the industry and economy's decision-making processes [50]. The production system is to be designed in accordance with the specific task while increasing its productivity. Some frameworks propose the proper methods for considering resilience in the supply chain, production, execution, control, and automation systems [50]. However, various changes should be taken into consideration to integrate resilience into the design of production systems effectively.

Resilience Dimensions and Capitals

A given community consists of several capitals that describe its assets and resources. As discussed in the literature, capitals are classified into seven types [51]: natural (ecological), physical (built), economic (financial), human, social, cultural, and political. The seven capitals are interrelated and interact with each other. These capitals can be described as follows and are graphically represented in Figure 9.2 [51, 52]:

1) Natural (ecological) capital includes natural resources (mainly water, land, soil, air, oil, minerals, and biological resources such as biodiversity and trees), weather, and amenities. Human management significantly impacts the productivity of such resources, which can improve or degrade them.
2) Physical (built) capital includes all the infrastructure systems and buildings located within a community. The infrastructure systems under this capital support the other capital's activities.
3) Financial capital includes income, financial savings, and investments. The financial resources are utilized to build the capacities of the communities and develop businesses that eventually accumulate wealth for the community and result in better social and civic entrepreneurship. It also consists of stocks of money, income levels, financial savings, disposable assets such as livestock (sometimes considered natural capital), savings in liquid form, and societal distribution.
4) Human capital includes skills, knowledge, health, and physical ability of the community members. It also includes the ability of people to access and develop resources needed to increase their level of understanding and knowledge, as well as encouraging procedures to be identified and community capacities to be built. Human capital addresses the ability of the community to focus on assets and resources, act proactively, and lead effectively toward shaping a promising future for the community.
5) Social capital includes associations, social networks, and connections among people. It also includes assets such as rights or claims, the ability to ask for the help of others when needed, seek support from professional associations and businesses, and the ability to approach politicians or chiefs to assist.
6) Cultural (institutional) capital includes mannerisms, attitudes, language, symbols, competencies, and traditions. It also includes governance and mitigation policies.
7) Political capital includes accessing resources, influencing the distribution of resources, and engaging external entities to achieve goals. It also reflects the community's connection to power and resource agents and access to organizations. Furthermore, it refers to the ability of community members to engage in actions that contribute to the community's well-being.

The four capacities (absorption, adaptation, restoration, and transformation) can be combined with the resilience dimensions (capitals) in a single urban climate change resilience matrix (Table 9.4). The matrix can be used to represent how the system can deal with climate change stresses and shocks. It also helps identify the important factors contributing to the system's resilience against present and future potential impacts and risks.

Resilience Measuring

Why do we measure resilience? According to Martin-Breen and Andries [54], resilience must be measured to gain control over it, and resilience must be related to other properties that could be ascertained through observation [13]. Resilience is not directly observable.

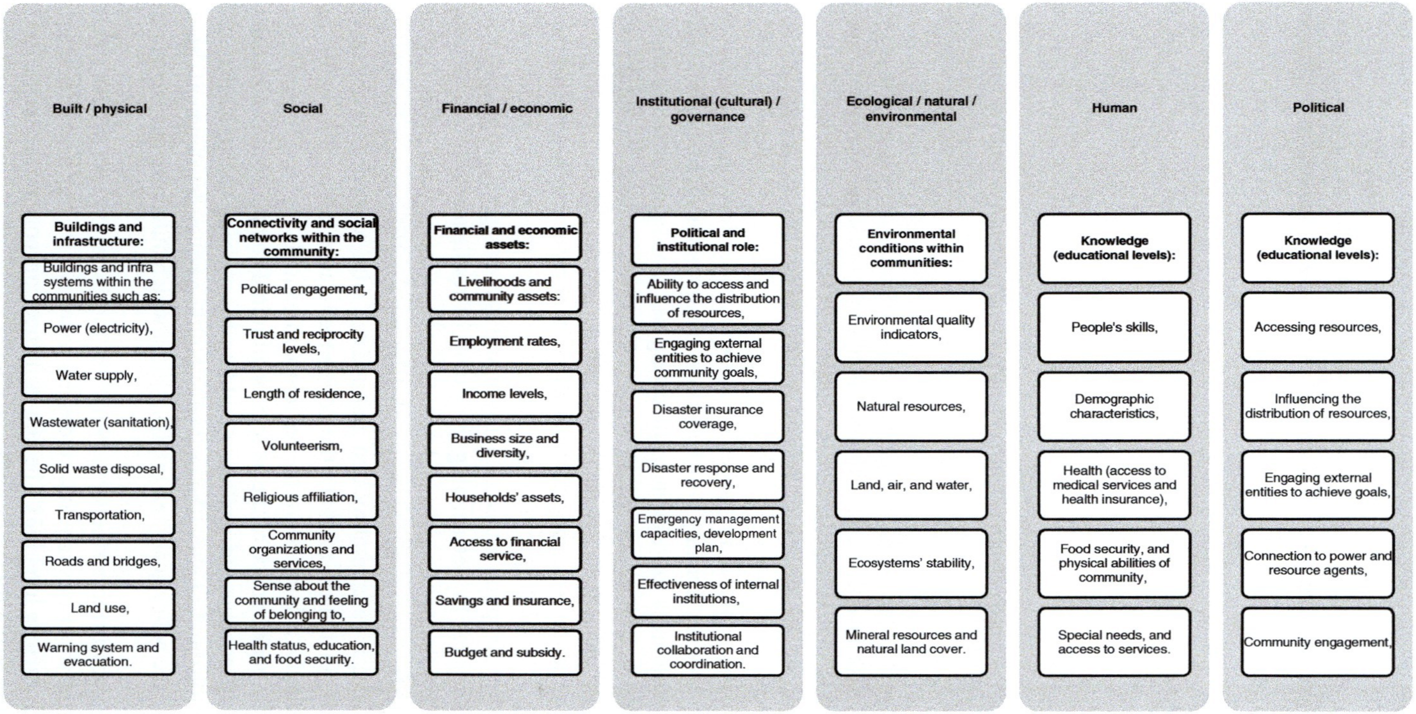

Built / physical	Social	Financial / economic	Institutional (cultural) / governance	Ecological / natural / environmental	Human	Political
Buildings and infrastructure:	**Connectivity and social networks within the community:**	**Financial and economic assets:**	**Political and institutional role:**	**Environmental conditions within communities:**	**Knowledge (educational levels):**	**Knowledge (educational levels):**
Buildings and infra systems within the communities such as:	Political engagement,	Livelihoods and community assets:	Ability to access and influence the distribution of resources,	Environmental quality indicators,	People's skills,	Accessing resources,
Power (electricity),	Trust and reciprocity levels,	Employment rates,	Engaging external entities to achieve community goals,	Natural resources,	Demographic characteristics,	Influencing the distribution of resources,
Water supply,	Length of residence,	Income levels,	Disaster insurance coverage,	Land, air, and water,	Health (access to medical services and health insurance),	Engaging external entities to achieve goals,
Wastewater (sanitation),	Volunteerism,	Business size and diversity,	Disaster response and recovery,	Ecosystems' stability,	Food security, and physical abilities of community,	Connection to power and resource agents,
Solid waste disposal,	Religious affiliation,	Households' assets,	Emergency management capacities, development plan,	Mineral resources and natural land cover.	Special needs, and access to services.	Community engagement.
Transportation,	Community organizations and services,	Access to financial service,	Effectiveness of internal institutions,			
Roads and bridges,	Sense about the community and feeling of belonging to,	Savings and insurance,	Institutional collaboration and coordination.			
Land use,	Health status, education, and food security.	Budget and subsidy.				
Warning system and evacuation.						

Figure 9.2 Resilience capitals (dimensions) and their variables (indicators).

Table 9.4 Community resilience against climate change.

Community dimensions (capitals)	Capacities			
	Absorptive	Adaptive	Restorative	Transformative
Natural	X	X	X	X
Physical (built)	X	X	X	
Financial (economic)	X	X	X	X
Human	X	X	X	X
Social	X	X	X	X
Cultural	X	X		X
Political	X	X	X	X

Source: Adapted from [26].

There is a potential need for assigning quantitative values to complicated social issues [13]. According to Risk Frontiers 2015, systematic and quantitative assessment processes are crucial for measuring resilience. Prior and Hagmann [55] explain that measuring resilience has several reasons and purposes. These purposes can be summarized as follows:

1) Characterizing resilience in a specific context includes developing a measure for resilience in multidimensionality, which requires an expression of the elements of resilience.
2) Allocating resources for resilience: allocating risk management resources by allowing for quantitative comparison measurements. This step includes the fund method using the resilience index to ensure high transparency.
3) Building resilience: managing disruption and its consequences in low-resilient entities is significant. Measuring resilience is crucial to assess the implementation of best practices and the disruption impacts on a system. Risk management agencies can consider the best measures by determining the extent of resilience. Also, the development of resilience thresholds can assist in determining the relevant good policy decisions.
4) Raising public awareness: the scope of disseminating awareness includes the at-risk entities that can be communicated and assisted with resilience measures. Assistance requires the managers to share the resilience-related information with the entities whose resilience is lying below the resilience threshold.
5) Monitor the performance of the policy: longitudinal comparisons can be used to determine the effectiveness of resilience building by applying the resilience index to the targeted entity by the policy. The resilience-building policy development should include integrating the policy goals and targets against which outcomes need to be assessed.

In general, measuring resilience depends on the infrastructure system's performance after an external shock or stress. It includes the recovery time that a system takes to resume its initial performance level [53]. The key resilience concepts are explained in Figure 9.3; accordingly, the resilience index, robustness, and rapidity are defined in Equation 9.1–9.3, respectively.

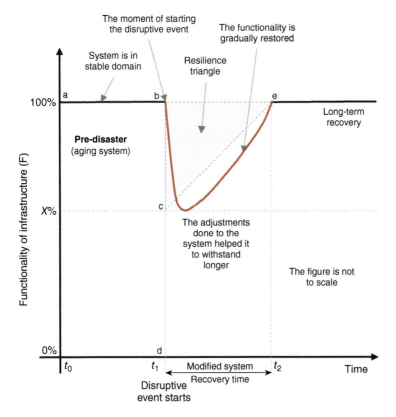

Figure 9.3 Key concepts of the resilience triangle. *Source:* Adapted from Ayyub [53].

$$\text{Resilience}(R) = \frac{\int_{t_1}^{t_2} F(t) \, dt}{100(t_1 - t_2)} \tag{9.1}$$

$$\text{Robustness}(\text{in percentage}) = c - d \tag{9.2}$$

$$\text{Rapidity}(\text{in percentage per time}) = \frac{b - c}{t_1 - t_2} \tag{9.3}$$

Equation 9.1 measures the performance per time, where F represents the performance or quality of the infrastructure system, t_1 is the time of disturbance incidence, and t_2 represents the time when the system is fully recovered. The equations are adapted from [53].

Conclusion

Climatic shocks and stresses have been experienced on various levels in recent years. Heat waves, more frequent and unusual heavy rains, floods, sea-level rise, storm surges, cyclone activities, land droughts and windstorms, and other impacts are all becoming more common. The stress they put on urban systems can eventually put communities at potential risk. These risks need to be reduced to protect the communities. The built environment systems should also be able to quickly and effectively resist, absorb, adapt, and recover from the effects of hazards. The requirements of resistance, absorption, adaptation, and recovery of community systems can be expressed in a single term called "resilience," which has become popular due to the increasing concern over natural, environmental, and climate threats and impacts. The concept of resilience has progressively complemented sustainability. Over the last two decades, it has evolved and is widely used by practitioners, academics, scholars, and policymakers across many disciplines. It can also have different meanings, depending on the boundaries of the discipline it is used for. Climate resilience represents the ability of the system to reduce vulnerability and sensitivity to climatic change, withstand extreme climate events, rapidly recover after undergoing stress and adverse impacts, and learn from the climatic issues and events to reflect on future decisions. It also includes risk reduction, response, absorptive, and adaptive capacities. Furthermore, it covers improving the environmental, social, physical, institutional, and governance structures.

References

1 Holling, C.S. (1973). Resilience and stability of ecological systems. *Annu. Rev. Ecol. Syst.* 4 (1): 1–23.
2 He, X., Lin, M., Chen, T.L. et al. (2020). Implementation plan for low-carbon resilient city towards sustainable development goals: challenges and perspectives. *Aerosol Air Qual. Res.* 20 (3): 444–464.
3 Verlinghieri, E. (2020). Learning from the grassroots: a resourcefulness-based worldview for transport planning. *Transp. Res. Part A Policy Pract.* 133: 364–377.
4 Tahir, F., Ajjur, S.B., Serdar, M.Z. et al. (2021). Proceedings of the Qatar Climate Change Conference 2021. In: *Qatar Climate Change Conference 2021*, 44.
5 World Economic Forum (2019). Global Risks Report 2019. Geneva Switzerland. Geneva. https://www.weforum.org/publications/the-global-risks-report-2019.
6 Fitzpatrick, T. (2016). Community disaster resilience. In: *Disasters and Public Health*, 2e, 57–85. Butterworth-Heinemann.
7 McAslan, A. (2010). The concept of resilience. Understanding its origins, meaning and utility. *Inf. Secur. J.* 50.
8 Cork, S. (2010). *Resilience and Transformation: Preparing Australia for Uncertain Futures*, 205. CSIRO Publishing.
9 Al-Humaiqani, M.M. and Al-Ghamdi, S.G. (2022). The built environment resilience qualities to climate change impact: concepts, frameworks, and directions for future research. *Sustainable Cities Soc.* 80: 103797.
10 U.S. EPA (2017). *Evaluating Urban Resilience to Climate Change: A Multi-Sector Approach (Final Report)*, vol. 10. Washington, DC: Environmental Protection Agency.
11 Asian Development Bank (2014). Urban climate change resilience: a synopsis. https://www.adb.org/publications/urban-climate-change-resilience-synopsis.

12 Al-Humaiqani, M.M. and Al-Ghamdi, S.G. (2023). Assessing the built environment's reflectivity, flexibility, resourcefulness, and rapidity resilience qualities against climate change impacts from the perspective of different stakeholders. *Sustainability* 15 (6): 5055.

13 Stevenson, J.R., Vargo, J., Ivory, V. et al. (2015). Resilience benchmarking & monitoring review. Resilience to Nature's Challenges.

14 Ayyub, B.M. (2004). From dissecting ignorance to solving algebraic problems. *Reliab. Eng. Syst. Saf.* 85 (1–3): 223–238.

15 Grebici, K., Goh, Y.M., and McMahon, C. (2008 (Section 2)). Uncertainty and risk reduction in engineering design embodiment processes. In: *Proceedings DESIGN 2008, the 10th International Design Conference*, 143–156.

16 Clarkson, J. and Eckert, C. (2005). Design process improvement. In: *Design Process Improvement: A Review of Current Practice* (ed. J. Clarkson and C. Eckert), 1–560. London: Springer London.

17 Center for Disaster Philanthropy. (2021) International wildfires – Center for Disaster Philanthropy. https://disasterphilanthropy.org/disaster/2021-international-wildfires/?gclid =Cj0KCQjwnJaKBhDgARIsAHmvz6dtYFW7nSOR3DcWLxTLnbQ9BeVkp4A2NaUG oH9tusvm0Up_-iuyyAsaAsEEEALw_wcB (cited 18 September 2021).

18 Serdar, M.Z., Macauley, N., and Al-Ghamdi, S.G. (2022). Building thermal resilience framework (BTRF): a novel framework to address the challenge of extreme thermal events, arising from climate change. *Front. Built Environ.* 8: 1029992.

19 Oberkampf, W.L., DeLand, S.M., Rutherford, B.M. et al. (2002). Error and uncertainty in modeling and simulation. *Reliab. Eng. Syst. Saf.* 75 (3): 333–357.

20 Zimmermann, H.J. (2000). An application-oriented view of modeling uncertainty. *Eur. J. Oper. Res.* 122 (2): 190–198.

21 Behera, P., Dwre, P.E., Hill, R., Haghani, S., and Design, M. (2019). Board 16 : Work in Progress : Design of "Risk and Resilience" Focused Courses for Undergraduate Engineering Education Towards a Hazard-Resilient Built Environment.

22 Gallego-Lopez C, Essex J. Understanding risk and resilient infrastructure investment. *EOD Resil. Resour.* 2016;1–29. https://doi.org/10.12774/eod_tg.july2016.gallegolopezessex3.

23 Norris, F.H., Stevens, S.P., Pfefferbaum, B. et al. (2008). Community resilience as a metaphor, theory, set of capacities, and strategy for disaster readiness. *Am. J. Community Psychol.* 41 (1–2): 127–150.

24 Weilant, S., Strong, A., and Miller, B. (2019). *Incorporating Resilience into Transportation Planning and Assessment. Incorporating Resilience into Transportation Planning and Assessment*. RAND Corporation.

25 Béné, C., Wood, R.G., Newsham, A., and Davies, M. (2012). Resilience: new utopia or new tyranny? reflection about the potentials and limits of the concept of resilience in relation to vulnerability reduction programmes. *IDS Work Pap.* 2012 (405): 1–61.

26 GIZ (2014). Assessing and monitoring climate resilience: from theoretical considerations to practically applicable tools – a discussion paper. From Theory Considerations to Practical Application Tools – A Discuss Paper.

27 Cutter, S.L., Barnes, L., Berry, M. et al. (2008). A place-based model for understanding community resilience to natural disasters. *Global Environ. Change* 18 (4): 598–606.

28 Brooks, N. (2003). *Vulnerability, Risk and Adaptation: A Conceptual Framework*. Tyndall Centre for Climate Change Research.
29 IPCC (2012). Managing the risks of extreme events and disasters to advance climate change adaptation. https://www.ipcc.ch/report/managing-the-risks-of-extreme-events-and-disasters-to-advance-climate-change-adaptation/. (cited 30 Jun. 2020).
30 Morshedlou, N., Barker, K., and Sansavini, G. (2019). Restorative capacity optimization for complex infrastructure networks. *IEEE Syst. J.* 13 (3): 2559–2569.
31 Walker, B., Holling, C.S., Carpenter, S.R., and Kinzig, A. (2004). Resilience, adaptability and transformability in social-ecological systems. *Ecol. Soc.* 9 (2): 5.
32 Twigg, J. (2009). Characteristics of a disaster-resilient community: a guidance note.
33 Büyüközkan, G., Ilıcak, Ö., and Feyzioğlu, O. (2022). A review of urban resilience literature. *Sustainable Cities Soc.* 77: 103579.
34 Harrison, C.G. and Williams, P.R. (2016). A systems approach to natural disaster resilience. *Simul. Model Pract. Theory* 65: 11–31.
35 Kusumastuti, R.D., Viverita, Husodo, Z.A. et al. (2014). Developing a resilience index towards natural disasters in Indonesia. *Int. J. Disaster Risk Reduct.* 10: 327–340.
36 De Groeve, T., Poljansek, K., and Vernaccini, L. (2014). Index for Risk Management - InfoRM: Concepts and Methodology. Luxembourg.
37 Sharifi, A. (2016). A critical review of selected tools for assessing community resilience. *Ecol. Indic.* 69: 629–647.
38 Sharifi, A. and Yamagata, Y. (2016). On the suitability of assessment tools for guiding communities towards disaster resilience. *Int. J. Disaster Risk Reduct.* 18: 115–124.
39 Wu, J. and Wu, T. (2014). Ecological resilience as a foundation for urban design and sustainability. In: *The Ecological Design and Planning Reader*, 541–556. Washington, DC: Island Press/Center for Resource Economics.
40 Holling, C.S. (1996). Engineering resilience versus ecological resilience. In: *Engineering Within Ecological Constraints*, 214.
41 Gunderson, L.H. (2000). Ecological resilience—in theory and application. *Annu. Rev.* 31: 425–439.
42 Folke, C., Carpenter, S., Walker, B. et al. (2004). Regime shifts, resilience, and biodiversity in ecosystem management. *Annu. Rev. Ecol. Evol. Syst.* 35 (1): 557–581.
43 CQ Net (2021). Organizational resilience: what is it and why does it matter? p. 1. https://www.ckju.net/en/dossier/organizational-resilience-what-it-and-why-does-it-matter-during-a-crisis (cited 16 Sep. 2021).
44 Fitzpatrick, T. (2016). Community disaster resilience. In: *Disasters and Public Health*, 2e, 57–85. Elsevier.
45 Arbon, P. (2014). Developing a model and tool to measure community disaster resilience. *Aust. J. Emerg. Manag.* 29 (4): 12–16.
46 Daniere, A.G. and Garschagen, M. (2017). *Urban resilience Urban Climate Resilience in Southeast Asia*. Springer.
47 Adger, W.N. (2000). Social and ecological resilience: are they related? *Prog. Hum. Geogr.* 24 (3): 347–364.
48 Rockefeller and Arup (2014). City resilience framework: city resilience index.
49 Deloitte Access Economics (2016). Building resilient infrastructure.

50 Ihlenfeldt, S., Wunderlich, T., Süße, M. et al. (2021). Increasing resilience of production systems by integrated design. *Appl. Sci.* 11 (18): 8457.
51 Flora, C.B., Emery, M., Fey, S., and Bregendahl, C. (2004). *Community Capitals: A Tool for Evaluating Strategic Interventions and Projects.* North Central Regional Center for Rural Development (NCRCRD).
52 Elasha, B.O., Elhassan, N.G., Ahmed, H., and Zakieldin, S. (2005). Sustainable livelihood approach for assessing community resilience to climate change: case studies from Sudan. Second International Conference on Climate Impacts Assessment (SICCIA). Higher Council for Environment and Natural Resources, Sudan: AIACC Working Paper.
53 Ayyub, B.M. (2018). *Climate-Resilient Infrastructure: Adaptive Design and Risk Management*, 311. American Society of Civil Engineers (ASCE).
54 Martin-Breen, P. and Anderies, J.M. (2011). Resilience: a literature review' Bellagio initiative, Brighton:IDS. https://opendocs.ids.ac.uk/opendocs/handle/20.500.12413/3692.
55 Timothy Prior & Jonas Hagmann (2014) Measuring resilience: methodological and political challenges of a trend security concept, *J. Risk Res.*, 17:3, 281-298, https://doi.org/10.1080/13669877.2013.808686.

10

Unveiling Urban Resilience: Exploring the Qualities and Interconnections of Urban Systems

Mohammed M. Al-Humaiqani[1] and Sami G. Al-Ghamdi[1,2,3]

[1] Division of Sustainable Development, College of Science and Engineering, Hamad Bin Khalifa University, Qatar Foundation, Doha, Qatar
[2] Environmental Science and Engineering Program, Biological and Environmental Science and Engineering Division, King Abdullah University of Science and Technology (KAUST), Thuwal, Saudi Arabia
[3] KAUST Climate and Livability Initiative, King Abdullah University of Science and Technology (KAUST), Thuwal, Saudi Arabia

Introduction

Climate extremes are significantly changing over time, as informed by the scientific community. The changes are considerably driven by the emissions of greenhouse gases (GHGs) generated by human activity; however, predicting future climate models has become difficult due to the future uncertainties of GHGs needed for extreme climate and weather statistics and calculations [1]. Human influence is dominant in the different ways the climate is changing, as stated by the Intergovernmental Panel on Climate Change (IPCC) [2]. The changes include an increase in temperature, extreme precipitation, and sea level rise (SLR). These changes are projected to continue, placing more stress on the environment.

The resilience of the built environment against climate change impacts has become a hot topic that is receiving increasing attention. It describes the capacity of the community, system, or structure to rapidly and efficiently withstand, absorb, adapt, transform, and recover from a climate change shock or stress [3]. The importance of resilience is reflected as a significant feature in a range of the United Nations (UN) Sustainable Development Goals (SDGs) [4]. Hence, resilient building design and construction must be considered to predict extreme climate shocks and disruptions that impact the built environment [3, 5].

Resilience is required in different community sectors to mitigate the long-term economic impacts caused by disruptions and threats. For example, there is an urgent need for resilient solutions in terms of buildings to maintain acceptable levels of comfort against extreme weather events [3]. Also, it is needed to protect the infrastructure from SLR, storm surges, and other events [6, 7], manage the flooding of communities, and protect businesses and individuals [8] from other interacting stresses. Emerging studies on resilience qualities raise many questions regarding the politics of resilience expertise and knowledge, the

---

Sustainable Cities in a Changing Climate: Enhancing Urban Resilience. First Edition.
Edited by Sami G. Al-Ghamdi.
© 2024 John Wiley & Sons Ltd. Published 2024 by John Wiley & Sons Ltd.

contribution of communities to built environment resilience, and how resilience qualities circulate to integrate systems and engage all relevant sectors. This chapter defines the different resilience qualities and highlights their boundaries and scope of consideration. It also helps readers understand why resilience qualities need to be considered in the planning and design stages of city systems. Furthermore, this chapter facilitates identifying the interrelations between the resilience qualities as well as encourages adaptive, absorptive, and transformative resilient urban environments.

Urban Resilience to Climate Change

The concept of resilience was formally introduced in ecology by Holling [9]. It ranges from ecology to anthropology and social research to infrastructure systems; it is measured by the ability of a system to absorb shocks and stresses and continue functioning. More than 120 definitions for resilience have been accumulated by researchers from both peer-reviewed and gray literature [10]. According to Alexander [11], the term resilience was passed on from mechanics to ecology and psychology and then adopted by sustainability science and social research. The evolution of resilience is schematically represented in a diagram, as shown in Figure 10.1. In terms of operationalizing resilience, the relationship between disaster risk reduction concepts and resilience must be appropriately interpreted because the interpretation is challenging [12]. An appropriate interpretation would help make resilience usable and useful for management and policy purposes. In addition, applying some steps to move from theory to practical action is significant. These steps include the following [10]:

1) Defining the resilience of the system/element under study.
2) Creating a suitable vision that includes an operational definition and criteria.
3) Defining system drivers and their contribution to resilience. The drivers include resources, processes, and people.
4) Benchmarking resilience and examining the efficiency of interventions by identifying the relevant components and behaviors.
5) Defining the context of the operation boundary of resilience-specific features. This step would include interpreting features that function differently and comparing them with other systems, as applicable.
6) Considering the most appropriate assessment approaches in the context of the resilience operation. The evaluation method could be selected based on engagement and consultation in this step.

Table 10.1 outlines the resilience-specific definitions in different contexts of systems.

Table 10.2 summarizes the general definitions of resilience according to different disciplinary boundaries.

Climate Change Impacts on Built Environment Systems

According to Rapoport [31], the concept of a built environment was introduced by social scientists. The relatively recent concept recognizes the interests, modeling, and management

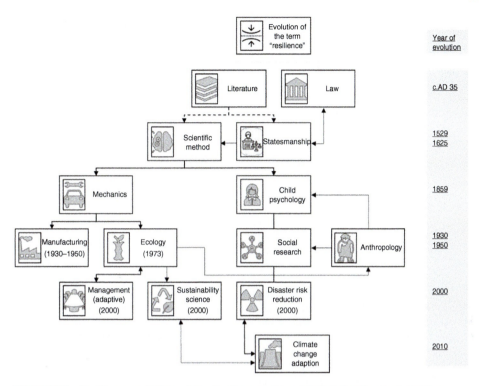

Figure 10.1 Resilience evolution schematic diagram. *Source:* Adapted and reproduced from Alexander [11] with written permission from both the author and the publisher via email on August 28 and September 06, 2021.

Table 10.1 Resilience-specific definitions in different contexts.

Context	Definition	Reference
Ecological systems	Ecological resilience measures the systems' ability to withstand and absorb a disturbance and maintain the relationships between the variables	[9]
	It is also defined as the system's capacity to absorb disruption and rearrange during a crisis and retain its structure, function, and identity	[13]
Materials/physical systems	Material can absorb and release energy within the elastic range	[14]
Socio-ecological systems	It is explained as the amount of disturbance a system can stand before changing to a different state	[15]
Resilience engineering	Resilience engineering is expressed as the ability of a system to recognize, adapt and absorb disruptions, disturbances, changes, and shocks	[16]
Organizational	It depicts the organization's capacity to survive, thrive, and reconstruct	[17]

Source: Adapted from [18].

Table 10.2 General definitions of resilience.

Disciplinary boundary	Definition	Author(s) name	Reference
Ecosystem resilience	Ecosystem resilience is the ability to absorb changes and continue functioning after being attacked	Holling (1973)	[9]
Resilience	Resilience should be considered with adaptability and transformability	Cork (2010)	[19]
	Resilience is the system's ability to withstand or recover from adverse circumstances	USEPA (2017)	[20]
	Resilience reduces the probability of a disaster in a community, increases its ability to absorb and resist a potential shock, and increases the systems' adaptability by maintaining their functions during the shock and enabling rapid recovery	Sarah Weilant et. al. (2019)	[21]
	"Resilience means the ability to prepare for and adapt to changing conditions and withstand and recover rapidly from disruptions"	The White House	[22]
Urban resilience	Urban resilience is the capacity of businesses, communities, organizations, individuals, and urban systems to maintain their function during a shock or stress regardless of its magnitude, rapidly recover, and continue thriving. Urban resilience includes urban ecological infrastructure, seismic resilience, flood resilience, energy resilience, and urban food systems resilience	Frantzeskaki (2016)	[23]
Community resilience	Community resilience is a term that describes the interconnected systems' networks that have a direct impact on human society, including the ecological, socio-economic, and built environments. The community is resilient if people are connected and can manage and sustain critical systems under stress, adapt to changes that may occur in social, economic, and physical environments, learn from the past, and be self-reliant when external resources are cut-off.	Arbon et al. (2012)	[24, 25]
Engineering resilience	Engineering resilience is a concept that incorporates resiliency into engineering practices. It is principally application-oriented, with diverse applications in engineering sectors.	Yodo and Wang (2016)	[26]
Socio-ecological resilience	The social-ecological system is able to absorb disturbance and change without being changed to a different regime or transferring to a new system of different structures and processes.	Garmestani et al. (2019), Walter and Salt (2006)	[27, 28]

Table 10.2 (Continued)

Disciplinary boundary	Definition	Author(s) name	Reference
Social resilience	The capacity of communities and people to deal with external shocks and stresses. It is related to the contribution to the preparedness of the communities, response to the disaster, and recovery after being undergone a disaster. Social resilience comprises three dimensions, including coping capacities, adaptive capacities, and transformative capacities [29]. Coping capacities include the capacity of the social actors to cope with different stresses, while adaptive capacities include the ability of the communities to learn from past experiences and adapt to projected stresses and challenges. Finally, transformative capacities are related to fostering the robustness of individuals and a sustainable society against future crises	Keck and Sakdapolrak (2013), Kwok et al. (2016)	[29, 30]

of urban systems and structures. According to Steiniger et al. [32], urban structures are classified into five categories based on the characterization of buildings:

1) Inner-city areas,
2) Commercial and industrial areas,
3) Urban areas (which contain dense buildings),
4) Suburban areas (dispersed buildings), and
5) Rural areas (single buildings).

The continuous modernization of urban areas with new developments and remodeling of existing ones to meet today's and future community needs has resulted in the introduction of new systems, technologies, and methods. The diversity and the wide range of development of multifunctional systems and structures in developed countries lead to strong interactions between the different systems that may not fall directly under a specific category. Thus, the authors have considered the latest ideas and ways of classifying the systems of the built environment and classified them under four categories as follows:

- Shelter systems mainly include buildings such as residential and commercial buildings and centers.
- Life support systems include utilities such as energy and water supply, telecommunications, human health, and welfare.
- Movement systems related to transportation infrastructure include roads, bridges, highways, tunnels, and airports.
- Open space systems include the utility for park recreation, such as public and semipublic parks and natural systems.

Resilience in the built environment is not only considered for new infrastructure but can also be managed for existing ones. Infrastructure resilience includes planning, designing, and interaction within the infrastructure system, as well as its contribution to the overall impacts on the community, society, and economy resilience [33]. A natural event is considered a disaster event if it disrupts the typical functioning of the community, causing losses and damages exceeding the capacity of the disturbed community to cope. Worldwide, several natural and climate hazards have attacked different cities and countries, resulting in the loss of lives, as well as damage to the environment and infrastructure systems. For example, some of the major climate impacts in the Middle East (ME) are increases in temperature, changes in precipitation patterns, SLR, floods, and droughts [34–36]. At the same time, potential natural disasters range from tsunamis to storm surges and earthquakes [37]. A "disaster" is a sudden event that disrupts the functioning of a system, or group of systems, in a community or society. The combination of natural and anthropogenic causes can form disasters that result in mass displacement and have an enormous impact.

Worldwide, the frequency of natural disasters in many countries has been increasing since 2004. The number of people exposed to risk due to these disasters frequently increases, especially in developing countries. The Asian and ME countries have become more exposed to natural disasters with a high damage rate. In 2014, about 50% of disasters were recorded in Asia and the Pacific, affecting nearly eighty million people and resulting in about USD 60 billion in economic losses [38]. In the Middle East and North Africa (MENA), the natural disasters in 2014 almost tripled from those witnessed 35 years ago. The average annual loss in the MENA region due to natural disasters reaches USD $1 billion [39]. In summary, resilience focuses on the ability of a system to withstand and adapt to such effects, stresses, and disasters and cope with unanticipated threats [37]. Figure 10.2 shows the characteristics of a climate change-resilient system and desired outcomes adapted from GIZ [40].

Temperature Rise

The temperature rise is a severe issue for human health, causing damage to environmental (ecological) systems, and can result in the failure of other systems, such as machinery in operation. Worldwide, a significant number of regions have experienced an increase in temperature above 1.5 °C [41]. Heat stress intensity increases due to the conversion of lands into developments, infrastructure, and other uses. In turn, this causes a lack of water and green lands, impacting the quality and quantity of habitats [36]. Similarly, traffic jams occur due to road and bridge system failures caused by high temperatures [23].

Sea Level Rise (SLR)

Over the last century, people in different parts of the world have been observing a warming climate and rising sea levels. Intense precipitation extremes result from increases in the atmosphere's water-holding capacity [1]. Several reports indicate that extreme precipitation events have been witnessed in several parts of the world, including Australia, Central Africa, Central America, parts of Eastern Europe, and Southwest Asia [1, 38, 39, 42–44].

Interacting Stresses

Stresses due to the overuse of energy and water, those caused by longer seasons, effects on terrestrial ecosystems, and stresses due to population shifts are examples of interacting

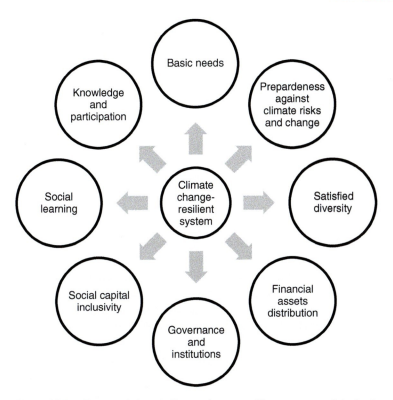

Figure 10.2 Characteristics of climate change-resilient system and desired outcomes.

stresses. Such stresses result in many direct and indirect effects on the environment. The effects may include deteriorating building materials, plants' susceptibility, disasters, crop and livestock productivity reductions, and water stress for agriculture and cooling. Also, stresses derived from human behavior, such as overconsumption of natural species on land or at sea, are another source of trouble. Another example of the interacting stresses is the stress placed by the Covid-19 pandemic. Similarly, eight resilience strategies are also reported that could be considered to enhance the response to the health workforce's needs [45]. Hence, understanding the interacting stresses on communities and built environment systems is crucial, given the growth of city expansions and the dependence of some economies on real estate developments [46].

The interacting stresses must be managed to protect communities, businesses, and individuals [8]. Heat stress should not be allowed to exacerbate, allowing for less influence on people's health and comfort. The environment and social dimensions can be combined to enhance heat stress resilience [47]. For example, the conceptual typo-morphological framework, characteristics, and measures of public spaces, semiprivate/semipublic spaces, and accessibility to social infrastructure are examples of the multidimensional approach to studying heat-stress resilience. A quantitative Geographic Information System (GIS)- and Machine Learning (ML)-based method is appropriate for classifying the social dimension of heat-stress resilience [47]. Other climate stresses like anthropogenic stressors affect fungal and bacterial communities [48], food production [49], livestock supply chain systems and transformations [50, 51], fish health [52], multiple stressors on mountain communities [53], job stressors and chronotype [54], and many more.

Major Uncertainties and Interrelations

Stress could be caused by uncertainty or unpredictability, such as hazard strength, timing, and financial insecurity [55]. Climate injustice can happen due to the vulnerability of the natural and human systems to the adverse effects of uncertainty, specifically when communities have insufficient coping capacity [56]. Although science provides an extent of predictability for climate change scenarios and projections, substantial uncertainty might still be considered higher in the projections of major social dimensions (changes in demography, economics, and the political incentives of adopting new technology) [57, 58].

Since uncertainty prevails regarding the impacts' spatial distribution, variability in time, and change in climate conditions, impacts of climate change are more difficult to identify and quantify [59]. The lack of knowledge about the climate impacts and hazards on different levels (local and regional) can increase the uncertainty of socioeconomic scenarios impacting the spatial distribution and frequency magnitude of future climate hazards [59, 60].

In summary, the uncertainty level in future conditions and climate models can be considered a motivator toward prioritizing infrastructure disruption scenarios by applying the existing modeling tools of infrastructure systems and vulnerability analysis approaches. Planning for risk, crisis, and uncertainty in urbanization is needed to reduce cities' vulnerability to shocks and stresses. In the opinion of Glaeser, uncertainty is among the major challenges humans have to respond to when considering environmental changes, which all together form the notion of risk assessment [61].

Resilience Qualities

Built environment resilience consists of several qualities that have already been demonstrated by different resilience frameworks and assessment methods. The resilience qualities include reflectivity (learning from the past), robustness, redundancy, flexibility, resourcefulness, inclusivity, integration, and rapidity. This section discusses the essential qualities of a resilient built environment, focusing on the key definitions and themes.

Reflectivity

Learning from the past is a philosophy that humans have practiced since life started on Earth. The things humans experience daily are accumulated, contribute to building knowledge, and improve societies and skills of individuals. As such, accumulative learning should help improve performance over time and enhance lifestyle. It is an important measure that promotes good future practice, helps progression on a community level, and measures the future [62].

In terms of the built environment, the necessity to withstand today's climate change impact emphasizes deploying the best methods, and innovative measures to help urban systems survive under such conditions. So, knowing how societies have faced similar disasters in the past is crucial [63]. It helps to realize the ways in which past societies dealt with the environment and enables an understanding of the changing relationship between them. It also allows for identifying the risks associated with these changes to inform the present and future decision-makers on better protecting the built environment [63]. The reflective environment consists of systems that accept the change and uncertainty inherited from the past or generated over time. The systems can potentially evolve continuously and have mechanisms that allow them to deal with such uncertainties. This would require modifying standards, norms, or regulations by considering collected information and data from the past rather than relying permanently on the current status [64].

Robustness

Robustness is defined as the ability of a system to withstand external stresses and shocks while maintaining its primary performance without significant losses [1]. It is also known as the amount of environmental uncertainty against which a system can maintain specified functionality. Robustness is also known as the persistence of the system [65]. It could also be defined as the ability of a system to withstand external shocks and be stable despite uncertainty [66]. The factors that weaken a system's robustness eventually lower an urban system's resilience on both environmental and social scales [67].

Redundancy

The redundancy of a system or a structure has become crucial in numerous design philosophies as designers and engineers have recognized its importance. The term "redundancy" means that the system can undergo minor deterioration in performance when a system component fails. The term redundancy refers to the extent to which systems and elements can sustain and satisfy functionality requirements during the disturbance event [1]. At the same time, the redundancy of a structure refers to the degradation that a structure can sustain before losing its functionality or specified elements. Overall, the degradation might not be predictable during the design stage; hence, structural redundancy can be related to robustness against uncertainty [68].

Flexibility

Due to climate change, accepting changes, uncertainty, and risks faced by urban systems has become a challenge. Nevertheless, it acknowledges the need for flexibility, allowing the system to adopt alternative strategies to absorb, respond to, and recover from a sudden shock [40]. Flexibility means that systems are designed and prepared to allow them to change, adapt, and evolve while undergoing climate change or naturally changing circumstances. According to the Rockefeller and Arup resilience framework, flexibility means converting traditional knowledge and practices to new techniques as needed. This assumption emphasizes that new knowledge, technologies, and obligations must be introduced [64].

Resourcefulness

Resourcefulness is defined as the ability of a system to find alternative ways, in a time of crisis, to achieve goals or meet needs [69]. It is also known as the capacity of a system to manage the needed services and resources during emergencies promptly and correctly [70]. In other words, resourcefulness means prioritizing problems and introducing solutions by identifying and monitoring all resources that make a system flexible, including human, technological, economic, technical, infrastructure, and informational [1]. The term "resourcefulness" has also been used as an alternative approach to foster resilience by community environmental groups [71]. The resourcefulness idea was introduced by MacKinnon and Derickson as a practical critique of the concept [71]. It is considered a complementary concept to sustainability, so building resourcefulness means distributing and accessing all resources. Resourcefulness is a concept that indicates where to tackle and deal with crises or shocks and can be used to link academia, policymaking, and the public sector [72]. As considered by Verlinghieri [72], the core resources that constitute resourcefulness can be classified into the following three categories:

1) Material resources include environmental conditions, health, food, housing, tools, infrastructures, and financial resources.
2) Intellectual resources include social capital, networking, ecological knowledge, culture, time, education, and science.
3) Civic resources focus on citizenship as an idea that enables people to participate in the public domain.

Rapidity of Recovery

Rapidity is a measure considered, besides resourcefulness, in determining recovery speed. Rapidity is defined as the ability of the system to recover from the crisis encountered, even with some losses, and avoid future disruptions [1].

Inclusivity

Broad consultation and community engagement are needed to create a sense of mutual ownership. Community engagement includes all groups, including vulnerable groups, collectively addressing the stresses and shocks faced by the different sectors and attempts to eliminate community-to-community isolation. The inclusion approach ensures a joint vision and shared responsibility to build city resilience [64].

Integration

Alignment and integration between systems and across their operational scales support the outcome and maintain consistency in decision-making. It involves information exchange among systems to collectively function and rapidly respond to a threat or disaster within a city [64].

Table 10.3 outlines the built environment system and services and the applicable indicators against each resilience quality sourced from the resilience framework [64, 73].

Table 10.3 Categorization of built environment systems and services showing the applicability of the relevant indicators in terms of resilience qualities.

	Category: Built environment systems and services		
	Indicators		
Resilience quality	Reliable mobility and communications	Continuity of critical services	Reduced physical exposure
Reflectivity	•	•	
Robustness		•	•
Redundancy	•	•	•
Flexibility	•	•	•
Resourcefulness	•	•	
Rapidity			
Inclusivity	•	•	
Integration	•	•	

The table covers the most well-known eight resilience qualities: reflectivity, robustness, redundancy, flexibility, resourcefulness, rapidity, inclusivity, and integration [73, 74].

Interrelation of Resilience Qualities

Built environment systems are interrelated and influence each other, principally through physical proximity and operational interaction. Damage to one infrastructural system or component can rapidly cascade into damage to adjacent components or systems [75]. The contiguous infrastructural systems or components are subject to increased risk. The proximity of old and exhausted systems to other vital systems or facilities, such as oil mains and hydroelectric power stations, increases the possibility of unexpected catastrophes for which no provisions have been established.

In resilient built environments, the characteristics such as planning, efficiency, redundancy, adaptability, interrelation, diversity, strength, and flexibility need to be combined [73, 76]. Accordingly, resilience requirements need to be integrated into built environment systems, considering the strategies focusing on the interrelations of resilience qualities [73, 74].

Conclusion

The term resilience was passed from mechanics to ecology and psychology and then adopted by sustainability science and social research. Resilience is required to mitigate the long-term climate impacts caused by disruptions and threats. Resilient solutions in different sectors vary from one region to another depending on the urgency of climate impacts. An acceptable comfort level against extreme weather events is to be maintained in buildings. Likewise, infrastructure systems must be protected from SLR, storm surges, and other events affecting the coastal environment. Also, flood management within the communities and businesses and individual protection are very important.

Urban climate resilience consists of several components and measures. The components include the assessment of the level of vulnerability of urban systems by measuring their three major capacities: sensitivity, exposure, and response. In addition, the climate resilience of urban systems considers learning from past and ineffective responses and actions, building risk reduction capacity, and developing a responsive action plan. The term "urban climate resilience" always means that people can survive and thrive safely, and the built environment systems can function irrespective of the encountered climatic stresses or shocks.

Resilience is not only considered for the new infrastructure but can also be managed for the existing ones. Infrastructure resilience includes planning, designing, and interacting within the infrastructure system and contributing to the overall impacts on the community, society, and economic resilience. Risk, crisis, and uncertainty are significant challenges that require communities to respond. Stress could be caused by uncertainty or unpredictability, such as hazard strength and timing and financial insecurity. Uncertainty could also cause climate injustice, specifically when communities have insufficient coping capacity. The lack of knowledge about the climate impacts and hazards on local and regional levels can increase the uncertainty of socioeconomic scenarios impacting the spatial distribution and frequency magnitude of future climate hazards.

There are many questions about the politics of resilience expertise and knowledge, community involvement in building resilience, and how resilience qualities are integrated into systems and sectors. Resilience qualities are needed to be considered in the planning and design stages of city systems. The relationship between disaster risk reduction concepts and resilience is challenging and needs to be appropriately interpreted to make it usable for policy and management purposes. Also, moving from theory to practical action is very important. This may include the definition of the resilience of the system under study, creating a suitable vision, defining drivers and contributions, benchmarking resilience, examining the efficiency of interventions, defining the context of the operation boundaries of specific features, and considering the appropriate assessment approaches.

References

1 Ayyub, B.M. (2018). *Climate-Resilient Infrastructure: Adaptive Design and Risk Management*. American Society of Civil Engineers (ASCE). 311 p.
2 IPCC (2012). Managing the risks of extreme events and disasters to advance climate change adaptation. https://www.ipcc.ch/report/managing-the-risks-of-extreme-events-and-disasters-to-advance-climate-change-adaptation/ (cited 30 Jun. 2020).
3 Attia, S., Levinson, R., Ndongo, E. et al. (2021). Resilient cooling of buildings to protect against heat waves and power outages: key concepts and definition. *Energy Build* 239: 110869.

4 Nations U (2015). THE 17 GOALS – sustainable development. https://sdgs.un.org/goals (cited 17 Sep. 2021).

5 Attia, S. (2018). *Net Zero Energy Buildings (NZEB): Concepts, Frameworks and Roadmap for Project Analysis and Implementation*, 382. Elsevier.

6 Abou-Mahmoud, M.M.E. (2021). Assessing coastal susceptibility to sea-level rise in Alexandria, Egypt. *Egypt. J. Aquat. Res.* 47 (2): 133–141.

7 Al-Mutairi, N., Alsahli, M., El-Gammal, M. et al. (2021). Environmental and economic impacts of rising sea levels: a case study in Kuwait's coastal zone. *Ocean Coastal Manag.* 205 (May 2020): 105572.

8 Zevenbergen, C., Gersonius, B., and Radhakrishan, M. (2020). Flood resilience. *Philos. Trans. R. Soc. London, Ser. A* 378 (2168): 20190212.

9 Holling, C.S. (1973). Resilience and stability of ecological systems. *Annu. Rev. Ecol. Syst.* 4 (1): 1–23.

10 Stevenson, J.R., Vargo, J., Ivory, V. et al. (2015). Resilience benchmarking & monitoring review. Resilience to Nature's Challenges.

11 Alexander, D.E. (2013). Resilience and disaster risk reduction: an etymological journey. *Nat. Hazards Earth Syst. Sci.* 13 (11): 2707–2716.

12 Serdar, M.Z. and Al-Ghamdi, S.G. (2021). Preparing for the unpredicted: a resiliency approach in energy system. In: *Energy Systems Evaluation*, vol. 1, 1–19.

13 Walker, B., Holling, C.S., Carpenter, S.R., and Kinzig, A. (2004). Resilience, adaptability and transformability in social-ecological systems. *Ecol. Soc.* 9 (2): 5.

14 Gere, J.M. and Goodno, B.J. (2013). Mechanics of materials. 8th ed. Stamford, CT : Cengage Learning; 2013. 1130 p.

15 Carpenter, S., Walker, B., Anderies, J.M., and Abel, N. (2014). From metaphor to measurement: resilience of what to what? *Ecosystems* 4 (8): 765–781.

16 Hollnagel, E., Woods, D.D., and Leveson, N. (2006). *Resilience Engineering: Concepts and Precepts*. Ashgate, 397.

17 Hamel, G. and Välikangas, L. (2003). The quest for resilience. *Harvard Bus. Rev.* 81 (9): 52–63, 131.

18 Hassler, U. and Kohler, N. (2014). Resilience in the built environment. *Build Res. Inf.* 42 (2): 119–129.

19 Cork, S. (2010). *Resilience and Transformation: Preparing Australia for Uncertain Futures*, 205. CSIRO Publishing.

20 U.S. EPA (2017). *Evaluating Urban Resilience to Climate Change: A Multi-Sector Approach (Final Report)*, vol. 10. Washington, DC: Environmental Protection Agency.

21 Weilant, S., Strong, A., and Miller, B. (2019). *Incorporating Resilience into Transportation Planning and Assessment. Incorporating Resilience into Transportation Planning and Assessment*. RAND Corporation.

22 The White House (2013). Presidential policy directive – critical infrastructure security and resilience. USA.

23 Frantzeskaki, N. (2016). Urban resilience. A concept for co-creating cities of the future, pp. 51.

24 Paul, A. and Gebbie, K. (2012). Developing a model and tool to measure community disaster resilience: final report. *Aust. J. Emerg. Manage.* 29.

25 Fitzpatrick, T. (2016). Community disaster resilience. In: *Disasters and Public Health*, 57–85. Elsevier.

26 Yodo, N. and Wang, P. (2016). Engineering resilience quantification and system design implications: a literature survey. *J. Mech. Des.* 138 (11): 111408.

27 Garmestani, A., Craig, R.K., Gilissen, H.K. et al. (2019). The role of social-ecological resilience in coastal zone management: a comparative law approach to three coastal nations. *Front. Ecol. Evol.* 7 (October): 1–14.

28 Walter, B. and Salt, D. (2006). *Resilience Thinking: Sustaining People and Ecosystems in a Changing World*. Washington, DC, USA: Island Press.

29 Keck, M. and Sakdapolrak, P. (2013). What is social resilience? Lessons learned and ways forward. *JSTOR* 67 (1): 5–19.

30 Kwok, A.H., Doyle, E.E.H., Becker, J. et al. (2016). What is 'social resilience'? Perspectives of disaster researchers, emergency management practitioners, and policymakers in New Zealand. *Int. J. Disaster Risk Reduct.* 19: 197–211.

31 Rapoport, A. (1976). *The Mutual Interaction of People and Their Built Environment* (ed. A. Rapoport). The Hague: Chicago: De Gruyter Mouton 505 p.

32 Steiniger, S., Lange, T., Burghardt, D., and Weibel, R. (2008). An approach for the classification of urban building structures based on discriminant analysis techniques. *Trans. GIS* 12 (1): 31–59.

33 Foltz-Gray, D. (2016). Designing for dementia. *Contemp. Longterm Care* 28–30.

34 Tahir, F., Ajjur, S.B., Serdar, M.Z. et al. (2021). Proceedings of the Qatar Climate Change Conference 2021. In: *Qatar Climate Change Conference 2021*, 44.

35 Tahir, F. and Al-Ghamdi, S.G. (2023). Climatic change impacts on the energy requirements for the built environment sector. *Energy Rep.* 9: 670–676.

36 Salimi, M. and Al-Ghamdi, S.G. (2020). Climate change impacts on critical urban infrastructure and urban resiliency strategies for the Middle East. *Sustainable Cities Soc.* 54 (October 2019): 101948.

37 Institute ME (2019). Responding to natural disasters in the MENA region and Asia: rising to the challenge?https://www.mei.edu/publications/responding-natural-disasters-mena-region-and-asia-rising-challenge (cited 2021 Sep 3)

38 UNESCAP (2015). Disasters in Asia and the Pasific: 2014 year in review.

39 Banerjee, A., Bhavnani, R., Burtonboy, C.H. et al. (2014). *Natural Disasters in the Middle East and North Africa: A Regional Overview*. Washington, DC: World Bank Group.

40 GIZ (2014). Assessing and monitoring climate resilience: from theoretical considerations to practically applicable tools – a discussion paper.

41 Wigley, T.M.L., Jones, P.D., and Kelly, P.M. (1981). Global warming? *Nature* 291 (5813): 285–285.

42 Zheng, F., Westra, S., and Leonard, M. (2015). Opposing local precipitation extremes. *Nat. Clim. Change* 5 (5): 389–390.

43 Wasko, C., Sharma, A., and Westra, S. (2016). Reduced spatial extent of extreme storms at higher temperatures. *Geophys. Res. Lett.* 43 (8): 4026–4032.

44 (2014). Climate change impacts in the United States: the third national climate assessment. GlobalChange.gov. https://www.globalchange.gov/browse/reports/climate-change-impacts-united-states-third-national-climate-assessment-0 (cited 4 Sep. 2021).

45 Tebes, J.K., Awad, M.N., Connors, E.H. et al. (2022). The stress and resilience town hall: a systems response to support the health workforce during COVID-19 and beyond. *Gen. Hosp. Psychiatry* 77 (February): 80–87.

46 World Finance (2016). Real estate could become the driving force behind Indonesia's economic development. https://www.worldfinance.com/infrastructure-investment/real-estate-could-become-the-driving-force-behind-indonesias-economic-development. (cited 6 Jul. 2022).

47 Eldesoky, A.H., Gil, J., and Pont, M.B. (2022). Combining environmental and social dimensions in the typomorphological study of urban resilience to heat stress. *Sustainable Cities Soc.* 83 (January): 103971.

48 Juvigny-Khenafou, N.P.D., Zhang, Y., Piggott, J.J. et al. (2020). Anthropogenic stressors affect fungal more than bacterial communities in decaying leaf litter: a stream mesocosm experiment. *Sci. Total Environ.* 716: 135053.

49 Bazzana, D., Foltz, J., and Zhang, Y. (2022). Impact of climate smart agriculture on food security: an agent-based analysis. *Food Policy* 111: 102304.

50 Godde, C.M., Mason-D'Croz, D., Mayberry, D.E. et al. (2021). Impacts of climate change on the livestock food supply chain; a review of the evidence. *Global Food Sec.* 28: 100488.

51 Blekking, J., Giroux, S., Waldman, K. et al. (2022). The impacts of climate change and urbanization on food retailers in urban sub-Saharan Africa. *Curr. Opin. Environ. Sustainable* 55: 101169.

52 Murdoch, A., Mantyka-Pringle, C., and Sharma, S. (2020). The interactive effects of climate change and land use on boreal stream fish communities. *Sci. Total Environ.* 700: 134518.

53 Roxburgh, N., Stringer, L.C., Evans, A. et al. (2021). Impacts of multiple stressors on mountain communities: insights from an agent-based model of a Nepalese village. *Global Environ. Chang.* 66 (November 2020): 102203.

54 Togo, F., Yoshizaki, T., and Komatsu, T. (2022). Interactive effects of job stressor and chronotype on depressive symptoms in day shift and rotating shift workers. *J. Affect Disord. Rep.* 9: 100352.

55 Wamsler, C., Reeder, L., and Crosweller, M. (2020). The being of urban resilience. In: *The Routledge Handbook of Urban Resilience*, 1e (ed. M.A. Burayidi, A. Allen, J. Twigg, and C. Wamsler), 47–58. New York, NY: Routledge.

56 Cheng, C. (2020). Climate justicescape and implications for urban resilience in American cities. In: *The Routledge Handbook of Urban Resilience*, 1e (ed. M.A. Burayidi, A. Allen, J. Twigg, and C. Wamsler), 85–96. New York, NY: Routledge.

57 Shen, S. (2020). Critical infrastructure and climate change. In: *The Routledge Handbook of Urban Resilience*, 1e (ed. M.A. Burayidi, A. Allen, J. Twigg, and C. Wamsler), 117–129. New York, NY: Routledge.

58 Adger, W.N., Dessai, S., Goulden, M. et al. (2009). Are there social limits to adaptation to climate change? *Clim. Change* 93 (3–4): 335–354.

59 Durand, F.A. (2020). Building urban resilience to climate change: the case of Mexico City Megalopolis. In: *The Routledge Handbook of Urban Resilience*, 1e (ed. M.A. Burayidi, A. Allen, J. Twigg, and C. Wamsler), 143–157. New York, NY: Routledge.

60 Thomalla, F., Downing, T., Spanger-Siegfried, E. et al. (2006). Reducing hazard vulnerability: towards a common approach between disaster risk reduction and climate adaptation. *Disasters* 30 (1): 39–48.

61 Glaeser, E.L. (2022). *Urban Resilience: Planning for Risk, Crisis, and Uncertainty*. Urban Studies. vol. 59, 3–35.

62 Borie, M., Pelling, M., Ziervogel, G., and Hyams, K. (2019). Mapping narratives of urban resilience in the global south. *Global Environ. Change* 54 (January): 203–213.

63 Garber, D. (1986). Learning from the past: reflections on the role of history in the philosophy of science. *Synthese* 67 (1): 91–114.

64 Rockefeller and Arup (2014). City resilience framework: city resilience index.

65 Daniere, A.G. and Garschagen, M. (2017). *Urban Climate Resilience in Southeast Asia*. Springer.

66 Bankes, S. (2010). Robustness, adaptivity, and resiliency analysis. *Complex Adaptive Systems—Resilience, Robustness, and Evolvability: Papers from the AAAI Fall Symposium* (FS-10-03) pp. 2–7.

67 Wu, J. and Wu, T. (2014). Ecological resilience as a foundation for urban design and sustainability. In: *The Ecological Design and Planning Reader*, 541–556. Washington, DC: Island Press/Center for Resource Economics.

68 Kanno, Y. and Ben-Haim, Y. (2011). Redundancy and robustness, or when is redundancy redundant? *J. Struct. Eng.* 137 (9): 935–945.

69 Smart City Hub (2020). Resilient cities. http://smartcityhub.com/resilient-cities/ (cited 21 Jan. 2021).

70 Diego, S. (2015). Disaster resilience framework 75% draft for San Diego, CA Workshop. CA Workshop.

71 MacKinnon, D. and Derickson, K.D. (2013). From resilience to resourcefulness: a critique of resilience policy and activism. *Prog. Hum. Geogr.* 37 (2): 253–270.

72 Verlinghieri, E. (2020). Learning from the grassroots: a resourcefulness-based worldview for transport planning. *Transp. Res. Part A Policy Pract.* 133 (August 2019): 364–377.

73 Al-Humaiqani, M.M. and Al-Ghamdi, S.G. (2022). The built environment resilience qualities to climate change impact: concepts, frameworks, and directions for future research. *Sustainable Cities Soc.* 80: 103797.

74 Al-Humaiqani, M.M. and Al-Ghamdi, S.G. (2023). Assessing the built environment's reflectivity, flexibility, resourcefulness, and rapidity resilience qualities against climate change impacts from the perspective of different stakeholders. *Sustainability* 15 (6): 5055.

75 Cerns, T.D.O. (2007). Critical infrastructure, interdependencies, and resilience. *Bridg. Link Eng. Soc.* 37 (1): 22–29.

76 Godschalk, D.R. (2003). Urban hazard mitigation: creating resilient cities. *Nat. Hazards Rev.* 4 (3): 136–143.

11

Quantifying Urban Resilience: Methods and Approaches for Comprehensive Assessment

Mohammed M. Al-Humaiqani[1] and Sami G. Al-Ghamdi[1,2,3]

[1] *Division of Sustainable Development, College of Science and Engineering, Hamad Bin Khalifa University, Qatar Foundation, Doha, Qatar*
[2] *Environmental Science and Engineering Program, Biological and Environmental Science and Engineering Division, King Abdullah University of Science and Technology (KAUST), Thuwal, Saudi Arabia*
[3] *KAUST Climate and Livability Initiative, King Abdullah University of Science and Technology (KAUST), Thuwal, Saudi Arabia*

Introduction

Various scientific tools and technical mechanisms are commonly used during the planning and development of sustainability and resilience strategies. Building resilience can be informed by applied practices, such as conducting vulnerability and risk assessments as well as implementing action plans [1]. Although quantitative and qualitative approaches and tools are used to quantify resilience in the built environment, quantitative tools are considered more reliable and generate more authority than other approaches [2]. According to Kovacic [2], qualitative and creative approaches may be used alongside quantitative approaches [3]. Using creative methods and deliberative mapping, in addition to conventional tools, can result in values and diverse perspectives [4, 5]. In practice, this can help develop more approaches toward mapping more policy options for the future [3]. Various resilience assessment tools, frameworks, indices, and programs that quantify and measure community resilience progress have emerged [6]. In general, the emerging community resilience programs are diverse, including international programs (large scale), national programs (medium scale), and local efforts (small scale). These three scales of resilience include many examples of programs and tools. On the international level, the United Nations Office for Disaster Risk Reduction Making Cities Resilient Campaign, 100 Resilient Cities (international), and the Z Zurich Foundation (Germany) are good examples. On the national level, there are examples of good assessment programs such as ICLEI, RISE Resilience Innovations, the National Academies Resilient America Program, and the National Institute of Standards and Technology's Community Resilience Program. Lastly, an example of resilience efforts

---

Sustainable Cities in a Changing Climate: Enhancing Urban Resilience. First Edition.
Edited by Sami G. Al-Ghamdi.
© 2024 John Wiley & Sons Ltd. Published 2024 by John Wiley & Sons Ltd.

at the local/community level is the Charleston Resilience Network and Sustainable Seattle in the United States.

A targeted community must take various vital actions to build and measure its resilience. The actions include (1) developing goals and priorities through community engagement; (2) identifying resilience needs and challenges through accounting of the various dimensions of the community discussed in Chapter 9: (natural (ecological), physical (built), financial, human, social, cultural (institutional); (3) encouraging resilience measurement, through multi-beneficial actions by introducing incentives; and (4) linking the measurement of community resilience to decision-making processes [6].

Urban Resilience

A complex urban system is made up of a series of interconnected networks of various physical and social elements. Individual components have dynamic interactions that allow the system to work efficiently [7–9]. It is impossible to assess a city's resilience without considering the interdependence of several vital systems. Therefore, before conducting an urban resilience assessment, it is vital to understand the concept and attributes of a resilient urban system and how to build and manage resilient cities [10].

A comprehensive evaluation of urban system resilience should answer questions concerning a system's preparedness as well as its ability to absorb disturbances, respond quickly, and adapt to new conditions. In this chapter, different ways to evaluate resilience are discussed. The qualitative resilience assessment approach is summarized in the following sections. Alternative attempts at quantitative resilience evaluation are provided and classified according to the observable system components.

Resilience Strategies

Cities and communities may experience stressors and shocks due to climate change; therefore, measuring resilience is significant to help communities survive and thrive by ensuring the following strategies [6, 11, 12]:

1) Defining resilience means for each community or city.
2) Disseminating awareness among stakeholders about the significance of resilience in the built environment.
3) Establishing resilience baselines for the communities to continue monitoring progress.
4) Prioritize the needs and identify the risks.
5) Identify resources for building resilience.
6) Prioritize investments to enhance resilience and quantify accompanied anticipated returns.

At the same time, different indicators and effective measures could be used to assist a community in improving its response as well as its recovery planning (to shocks and stressors). Overall, measuring resilience must be perceived within the relevant past conditions and historical environmental, social, and economic contexts.

Urban and Community Resilience Assessment

Various qualitative, quantitative, and mixed methods and approaches have been developed to measure urban resilience formally. Qualitative assessment enables ranking, while qualitative methods could be limited to producing an index. Qualitative questions also determine capabilities, capacities, and geospatial representations that illustrate the capacity intersections used for comparing places. Yet, there is no validation of every effort to measure resilience [6]. Some methods include a conceptual framework, while others include numerical methods, index methods or indicators, and surveys [13].

The numerical method is considered a quantitative approach to representing resilience through an analysis of multi-criteria, scenario simulation, and performance curve [9, 13, 14]. The conceptual framework method relies mainly on experts assessing various qualitative aspects of resilience according to best practices [13, 15]. The indicator method is either quantitative or qualitative, representing a goal, achievement, change, and performance after being measured. The mapping method includes selected resilience attributes and data related to the studied area [9]. The interview method entails the collection of views, opinions, and judgments of an individual about specific concerns or subjects that eventually enable weighing indicators or variables to be used to quantify resilience. The Delphi method allows for weighing different resilience indicators (through interactive and systematic forecasting) collected judgments made by a panel of experts. The survey method is designed to collect data by distributing questionnaires among targeted people. Table 11.1 outlines the urban resilience methods and their use, matrices, and limitations.

Table 11.1 Applied methods in the assessment of urban resilience, use, matrices, and limitations (Adapted from Tong [13] and modified by authors).

No.	Method	Use	Metric	Limitation
1	Numerical method	The numerical method is used to solve mathematical problems obtained in urban resilience assessment. Examples of the problems are simulation, fuzz, and optimization [13, 14]	The numerical method computes resilience as the function, performance, and quality of network systems	The framework mainly depends on urban data consistency and integrity
2	Conceptual/ theoretical framework	The framework defines urban resiliency, realizes the threats and hazards, and identifies priorities and needs to build resilience into urban ecosystems. Also, it indicates methods to quantify urban resilience [13, 16, 17]	The theoretical framework approach helps identify and analyze resilience	The framework mainly depends on the individuals' knowledge and experience
3	Indicators or Indices	The indicators or indices are used to assess the performance and achievement of resilience. They also record characteristics, changes, etc. [13]	The dimensions and characteristics of resilience are computed using various metrics	The analysis and interpretation require precise and realistic methods

(Continued)

Table 11.1 (Continued)

No.	Method	Use	Metric	Limitation
4	Mapping method	The mapping method is mainly designed for data related to geographic information. It collects, visualizes, and analyzes relevant data [13]	It focuses on the relationship between the components of a single system in which the quality of resilience is quantified	Experience and skills in mapping software, as well as updated data, are required
5	Interview method	The interview method helps collect extensive and intensive data from individuals. It also helps exchange experience and knowledge and obtains evidence [13]	The interview method is a qualitative approach that gathers, studies, and interprets resilience	Several limitations constrain this method: it consumes more time, accompanying costs are high, and it lacks standardized representativeness
6	Delphi method	The Delphi method aims at collecting particular information from experts	The method is a theoretical approach to defining and selecting resilience-relevant variables and assigning weights	-
7	Survey method	The survey method is an excellent way to gather a wide range of data from individuals or inhabitants, about their opinions, on a particular subject	The method helps quantify the resilience, based on the provided information, by the inhabitants, such as their behavior and life quality level [18]	This method is limited by potential bias

The increase in the impact of climate change on different aspects of life makes the resilience of cities crucial [13, 19]. Due to the daily interactions between humans and nature that eventually impacts cities, scholars have emphasized this type of resilience [20]. Different methods are used to assess urban resilience; more than 85% of studies use indices or indicator methods; however, a combination of data and methods is suggested [13]. Infrequently considered are survey and interview methods; however, since climate change directly impacts communities, urban resilience assessment may need to include the perspectives of humans [13, 21]. Table 11.2 outlines practical examples of programs used to quantify community resilience.

Table 11.2 Community resilience and example programs.

Program level	Example of resilience programs	Key actions to build and measure resilience
International programs	United Nations Office for Disaster Risk Reduction Making Cities Resilient Campaign, 100 Resilient Cities (International), and Z Zurich Foundation are good examples [6]	1. Community participation and engagement (efforts on building and measuring resilience)
National programs	ICLEI, RISE Resilience Innovations, the National Academies Resilient America Program, and the National Institute of Standards and Technology's Community Resilience Program [6]	2. Resilience design and measuring using multiple community dimensions 3. Usable and relatable collected, synthesized, and integrated data used to make decisions
Local programs	Charleston Resilience Network and Sustainable Seattle in the United States [6]	4. Incentivizing the resilience measurement

Resilience Assessment Approaches

Assessing and modeling the resilience of potentially complex and large-scale systems has recently piqued the interest of researchers' and practitioners and has resulted in several definitions of resilience techniques, as well as resilience being measured across various application domains [15]. Resilience quantification approaches are different from one discipline to another. They are categorized into qualitative and quantitative approaches (indicators) with subcategories. The category of qualitative methods is comprised of two subcategories [15]:

i) Conceptual frameworks: This subcategory includes best practices.
ii) Semiquantitative indices: This subcategory includes an expert's assessments of various qualitative aspects of resilience.

The quantitative methods category also contains two subcategories:

i) General resilience approaches: This subcategory includes applying domain-agnostic measures to quantify resilience across applications.
ii) Structural-based modeling approaches: This subcategory models include domain-specific representations of resilience components.

Qualitative assessments are required specifically for those aspects that cannot be quantified or need to be explained without a scale [22]. The quantitative assessment indicators identify characteristics that shape community resilience, and their use may be limited by some non-extent indicators that prevent further comparisons. Good examples of qualitative indicators include learning, risk awareness, reorganization, and readiness. These indicators are used because the resilience of a community may go beyond the views and capacities of the community and could recognize interests [23]. The qualitative indicators are classified based on several factors: knowledge and education, risk assessment and management, vulnerability reduction, preparedness, disaster response, and governance [22]. The factors include coordination and organization, updated data on vulnerability, a budget to invest in risk reduction, and reducing risk on infrastructure. They also include enforcing risk-related regulations, protecting natural ecosystems, using early warning systems, achieving education programs, and ensuring the needs of the affected people.

There is a question that many practitioners ask: "Is it possible to make qualitative indicators quantifiable?" To answer this question, a practitioner may need to refer to good examples of coding schemes, rating scales, and structured subjective methods. Most qualitative approaches define communities by considering societal and social factors, such as community values and interests. Qualitative indicators contain aspects considered essential factors. Aspects include diversity, participation, communication, efficacy, coordination, and equity. These indicators are considered a suitable tool for decision-making because they are values added to a composite indicator.

Quantitative indicators may depend on other indicators covering specific aspects of community resilience. The indicators can be aligned into different resilience domains: economic, social, institutional, community capital, housing, environmental, and infrastructure. The environmental resilience domain covers risk and exposure, sustainability, and protective resources, while the infrastructure resilience domain covers healthcare, housing, transportation networks, communication services, etc. [24]. Another good example of quantitative indicators used to assess the resilience of the built environment is the one used to determine the resilience index for urban water distribution networks, proposed by Cimellaro et al. [25]. The index is based on three indicators describing three different features: water quality, water demand (number of users), and capacity (water tank height) [22]. These indicators are used in assessing the functionality of a water distribution system and leveraging the decision-making process. These indicators feed water distribution system functionality, providing solid metrics that cover different community resilience perspectives (Figure 11.1).

Qualitative Resilience Assessment

Urban designers and spatial planners may prefer using a qualitative approach for assessing resilience [9]. Conceptual frameworks are used to examine the spatial qualities of various qualitative components of resilience and best practices, based on the judgments of experts. Detailed descriptions and discussions of apparent urban systems allow for different (subjective) interpretations and strategies to improve resilience. A qualitative assessment method for planning and maintaining resilience was proposed by Lu and Stead [26]. The proposed framework consists of six attributes of resilience (Table 11.3) in relation to climate disturbance. The framework evaluated the application and importance of the concept of resilience across a wide range of planning documents from the local, regional, and national levels based on policy reviews and interviews with planning professionals.

The qualitative framework to assess a community's resilience may consider numerous characteristics, including four main categories: qualities, capacities, assets, and external resources. A set of indicators can be evaluated for each category, as similarly performed by Woolf et al. [30]. The indicators that describe each category are given in Table 11.4. The framework assessment relies on user judgment and perception; however, the objectivity

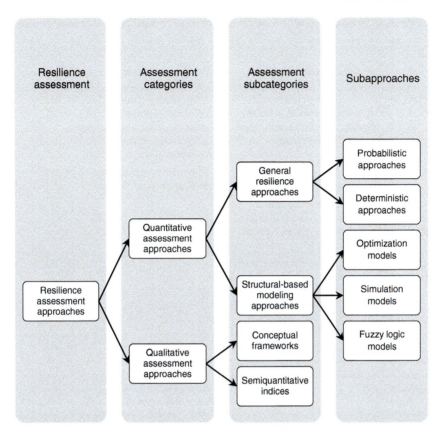

Figure 11.1 Classification of resilience assessment approaches. *Source:* Adapted from Hosseini et al. [15] and modified by authors.

may be questionable [9]. The dependence on these users and interviewed people can be minimized by applying a more independent revision of a qualitative assessment approach considered a second phase of assessment (proposed by Woolf et al. [30]). The proposed tool [30] is appropriate for project planning to encourage the multidimensional nature of resilience toward holistic environmental, economic, and social inclusivity. With limited interviews, such a framework can enable local stakeholders and community-based agencies to assess the projects' resilience.

Conceptual Frameworks
Most qualitative approaches to assess system resilience are based on conceptual frameworks. In light of that, the generic framework proposed by Resilience Alliance can be used to assess the resilience of socio-ecological systems [33]. The framework consists of the following seven steps:

1) Defining the system under investigation,
2) Determining the proper scale for assessing resilience,
3) Identifying the internal and external disturbances and drivers of the system,

Table 11.3 Comparison of the key definitions and characteristics of urban system resilience theory.

Urban system components	Urban system definition	Resilience attributes	Reference
Natural, human, and mechanical	Cities are dynamic and complex metasystems in which technical and social elements interact	Redundancy, diversity, efficiency, autonomy, strength, interdependency, adaptability, and collaboration	[8, 9]
Physical and social	-	Robustness, redundancy, and rapidity	[9, 27]
Physical (processes, resources) and social (activities, institutions, and people)	Cities are complicated adaptive systems that are constantly out of balance	-	[7, 9]
Natural, built, social, and economic	Urban systems comprise a complex web of interactions between people, the natural and built surroundings, and between the natural and constructed environments	-	[9, 28]
Leadership and strategy, health and well-being, the economy and society, infrastructure, and the environment	Individuals, communities, institutions, businesses, and systems that makeup cities	Reflectiveness, resourcefulness, robustness, redundancy, flexibility, inclusiveness, and integration	[9, 29]

Source: reproduced from [7–9, 28, 29] and modified by authors.

Table 11.4 Qualitative framework categories and indicators.

Categories	Number of Indicators	List of Indicators	Reference
Qualities	5	Robustness, redundant, well-located, diverse, and equitable	[9, 30, 31]
Capacities	4	Adaptivity, resourcefulness, flexibility, and learning	[9, 30, 32]
Assets	5	Environmental, physical, human, economic, and social assets	[9, 30]
External resources	3	Information and connections, natural resources, and services	[9, 30]

4) Finding the system's key players,
5) Defining recovery activities through the development of conceptual models,
6) Informing policymaking by using the results of conceptual models,
7) Incorporating the findings from the results of conceptual models.

Another example of conceptual resilience framework was proposed by Kahan et al. [27], which applies to homeland security. The framework is based on eight guiding principles: (1) assessing the potential hazards and threats; (2) robustness of the system; (3) mitigating the consequences of hazards; (4) adaptive capacity of the system; (5) planning based on risk-related information; (6) investing based on risk-related information; (7) harmonizing purposes; and (8) scope inclusivity. Similarly, Labaka et al. [34] proposed a holistic resilience framework consisting of two types of resilience: external and internal, with policies and sub-policies. A framework for survivability and resilience of communication networks assessment was developed by Sterbenz et al. [35], along with a survey of resilience disciplines. The framework, however, does not quantify the system's resilience; it only provides insights. Factors contributing to designing resilient networks include detection, defense, diagnosis, remediation, recovery, and refining.

The resilience of transportation systems can be improved through some technological solutions, as proposed by Bruyelle et al. [36]. A resilience framework for an earthquake-prone area at the community level was proposed by Ainuddin and Routary [37]. The framework comprises the identification of disaster characteristics, determining community vulnerability, preparedness, and improving physical (e.g., shelter systems), social (e.g., health care and education), and economic (e.g., employment and housing) resources. Similarly, the most influential properties of telecommunication network resilience have been identified, including availability, reliability, safety, maintainability, integrity, interactions, and performance [38].

Semiquantitative Indices

With a set of questions to measure several resilience-based system features (e.g., redundancy and resourcefulness), the semiquantitative index approach is usually performed using a percentage or a Likert scale, 0–100% or 0–10 points, respectively. The approach assesses resilience characteristics based on the collected opinions of experts, which are eventually combined to form a resilience index. In different processes, the semiquantitative approach is used to assess resilience engineering by defining relevant resilience variables (or disaster characteristics), including robustness, redundancy, and resourcefulness. The variables (or characteristics) are scored based on collected data over a scale of 0 to 100; this approach was conducted by Cutter et al. [39]. The variables or characteristics can be grouped into different sub-indices (i.e., community capital, infrastructure, economic, social, and institutional) based on a resilience assessment scope, where each score is calculated. The total score is then calculated by weighting or unweighting the scoring average of all sub-indices.

Quantitative Resilience Assessment

Quantitative assessment approaches for assessing the resilience of urban systems are popularly used. Many authors have presented the concept of fundamental resilience in performance curves that show the difference in a system's functionality over time [9, 16, 22, 40]. According to Marta Suárez et al. [41], quantitative methods are the most widely used, whereas scenario modeling is the most used methodology. At the same time, Geographic Information System GIS-based models are frequently used to simulate the performance and behavior of an urban system after disturbances [42]. Other popular methods used to assess urban resilience include indicators and indexes applied to different research fields and for general and specific resilience. Table 11.5 shows several quantitative approaches used to assess resilience.

Table 11.5 Example of quantitative resilience assessment approaches.

No.	Assessment category	Approaches	Features
1	General resilience approaches		General resilience approaches assess resilience by measuring system's performance and comparing measures across different contexts
1.1		Deterministic performance-based approach	Uncertainty is not incorporated into the metric (e.g., probability of disruption) [15] An example of a deterministic performance-based approach is the assessment conducted by Bruneau et al. [16] to measure a community's resilience to loss to an earthquake
1.2		Probabilistic performance-based approach	Stochasticity associated with system behavior is captured [15] The probabilistic performance-based approach was introduced by Chang and Shinozuka [43]. The approach is measured with two elements: (i) Loss of performance. (ii) Length of recovery.
2	Structural-based models		These models examine the impact of a system structure, on its resilience, by observing the system's behavior and modeling its characteristics This approach consists of three models: optimization, simulation, and fuzzy logic
		Optimization models	These are mathematical models used to assess and optimize the resilience of airports and maximize the resilience of runway and taxiway networks [44]. The models are based on quick restoration of post-event capacities to their pre-disruption levels. The model considers time, resource, space, operational, physical, and budget restrictions
		Simulation models	Simulation models, such as the discrete event simulation model produced by Albores and Shaw [45], are applied to assess the preparedness of relevant departments during terrorist attacks
		Fuzzy logic models	The fuzzy model is used to assess organizational resilience [46], and the fuzzy cognitive map (FCM) is used to assess factors of engineering resilience [47]. The main inputs of this model are redundancy and adaptability. Also, fuzzy architecture can be used to assess the resilience of critical infrastructure [48]

Source: Adapted from S. Hosseini et al. [15] and modified by authors.

General Resilience Approaches (Measures)

Irrespective of a system's structure, general resilience approaches assess resilience by measuring a system's performance and comparing measures across different contexts. Generic metrics calculate a system's resilience by comparing its performance before and after disturbance, neglecting the system's characteristics [15]. These general measures are classified as deterministic and stochastic, describing static and dynamic system behavior.

Deterministic Performance-based Approach

Uncertainty is not incorporated into the metric (e.g., probability of disruption) [15]. The following four resilience dimensions are defined for this type of assessment [16]:

- *Robustness*: The ability of a system to prevent damage due to disruption.
- *Redundancy*: The extent to which a system can minimize the impact of a disruption.
- *Rapidity*: The speed at which a system returns to an acceptable degree of functionality, or its original state, after being disturbed.
- *Resourcefulness*: The ability to apply resources and materials in responding to a potential disturbance.

An example of a deterministic performance-based approach is an assessment conducted by Bruneau et al. [16] that measured the loss of a community's resilience due to an earthquake.

Probabilistic Performance-based Approach

The probabilistic performance-based approach was introduced by Chang and Shinozuka [43]. The approach is measured with two elements:

- Loss of performance
- Length of recovery

The above-suggested metric is distinguished, in resilience quantification, by its acceptance of uncertainty. The metric is used to measure the resilience of infrastructures and communities. However, when performance loss and recovery time exceed the maximum allowable limits, the metric does not apply a penalty.

Structural-based Models

Structural-based models are models that examine the impact of system structure on its resilience by observing a system's behavior and modeling its characteristics. This approach consists of three models: optimization, simulation, and fuzzy logic.

Optimization Models

These are mathematical models used to assess and optimize the resilience of transportation systems and maximize the resilience of runway and taxiway networks [44]. The models are based on quickly restoring post-event capacities to their pre-disruption levels. The models consider time, resource, space, operational, physical, and budget restrictions. In these models, preparedness and recovery activities may also be considered.

The stochastic mathematical model can have three stages, as introduced by Faturechi and Miller-Hooks [49]: (1) mitigation (pre-event), (2) preparedness, and (3) response (post-event). The resilience of public transportation networks at a metropolitan level can be analyzed by applying a two-stage stochastic model proposed by Jin et al. [50]. The vulnerability and recoverability of waterway networks can be assessed using a similar model as in the research of Barker et al. [51] and Baroudetal. [52].

Simulation Models

Simulation models, such as discrete event simulation models produced by Albores and Shaw [45], are applied to assess the preparedness of relevant departments during terrorist attacks. This model examines the preparedness of rescue and fire services, which is considered a critical factor in pre-event disruption resilience [15]. The proposed two simulation models by Albores and Shaw [45] deal with the decontamination of a population and resource allocation harmonization. Similarly, simulation models can be applied to assess the resilience of other business attributes. For example, a supply chain's resilience can be measured by applying discrete event simulation, in which redundancy and flexibility are taken into consideration as primary elements [53]. Some other simulation models can be applied to assess the resilience of some urban systems. These models include the Monte Carlo simulation, which is used in studying water storage reservoir resilience [52], using the conditional probability proposed by Hashimoto et al. [54]. For critical infrastructure networks, time-dependent simulation models can be applied, as suggested by Adjetey-Bahun et al. [55], to evaluate the resilience indicators of railway transportation systems [15].

Fuzzy Logic Models
Fuzzy models, proposed by Aleksic et al. [46], are used to assess organizational resilience. They have also been used by Azadeh et al. [47], who used a fuzzy cognitive map (FCM) to assess engineering resilience factors. The main inputs of this model are redundancy and adaptability. On the other hand, fuzzy architecture is used to assess the resilience of critical infrastructure [48]. Furthermore, integrated fuzzy models have been applied to measure organizational resilience using a fuzzy technique for order preference by similarity to ideal solution (FTOPSIS) and analytic hierarchy process (AHP) [15, 56].

Frameworks and Tools for Measuring Resilience

In developing a generic technique to quantify resilience, the multi-scalar and multi-dimensional character of resilience presents various practical obstacles [30]. Although many assessment procedures are tailored to study a single home, community, or city, a few can scale across various systems. Indicators that examine resilience at one scale (e.g., national level) would mark over elements that affect resilience at other scales (e.g., community level), as well as make trade-offs between scales [57]. Yet, a framework must be generic enough to allow comparisons between different populations in various settings [58]. Furthermore, because resilience is frequently examined after a shock, such as a natural disaster, techniques for assessing resilience in terms of cost or asset depletion have been widely used.

According to Levine [60], five approaches are identified and currently used to measure resilience. The approaches are (1) quantitative analysis based on functionality, (2) quantitative analysis based on indicators and characteristics, (3) quantitative analysis based on food access, (4) quantitative analysis based on activities, and (5) quantitative analysis based on theoretical resilience frameworks. Yet, the frameworks of foreign assistance organizations for implementing resilience initiatives have made little or no attempt to measure the impact of their interventions [58, 60]. Instead, they serve as a set of best practice guidelines. Various indicators have been proposed to evaluate and quantify urban or community resilience. Table 11.6 outlines examples of community resilience to climate change frameworks. More resilience frameworks for assessing community and urban resilience are given in Table 11.A.1.

Table 11.6 Example of community resilience to climate change frameworks.

No.	Resilience framework	Framework dimensions / scope	Concept	Resilience indicator	Website
1	Arup's City Resilience Framework I Rockefeller Foundation [61–63]	Place: infrastructure and environment People: health and well-being Organization: economy and society Knowledge: leadership and strategy	The framework provides a globally comprehensive and technically robust basis for measuring and assessing resilience at a city scale The framework assesses the ability of individuals, systems, communities, and organizations to withstand, adapt, and thrive no matter what climate stresses or acute and chronic shocks they experience	There are 52 resilience indicators adding definitions to 12 goals. The indicators are assessed through 156 questions based on qualitative and quantitative data	(1) https://www.arup.com/projects/city-resilience-index (2) https://www.cityresilienceindex.org/#/
2	The UK Department for International Development Building Resilience and Adaptation to Climate Extremes and Disasters (BRACED) framework [64]	Adaptive capacity, including adaptability of livelihoods and basic services Anticipatory capacity, including preparedness, capacity, and risk information Absorptive capacity, including safety nets and diverse resources Transformation, including decision-making process, strategic planning, and innovative technology	The ability to predict, avoid, plan for, cope with, recover from, and adapt to climate change-related shocks and stresses	Qualitative and quantitative indicators span different data types related to climate change impacts, ecological, economic, institutional, livelihood, social planning, and decision-making processes	http://www.braced.org/resources/i/?id=cd95acf8-68dd-4f48-9b41-24543f69f9f1

(Continued)

Table 11.6 (Continued)

No.	Resilience framework	Framework dimensions / scope	Concept	Resilience indicator	Website
3	Assessments of Impacts and Adaptation of Climate Change (AIACC) sustainable livelihood approach [65]	It covers the five capitals (natural, physical, human, financial, and social)	Life quality improvement without compromising livelihood options for others	Qualitative and quantitative indicators measure the ability of communities to cope with and recover from stresses, shocks, ecological integrity, income stability, economic efficiency, and social equity	http://www.start.org/ Projects/AIACC_Project/ working_papers/ Working%20Papers/ AIACC_WP_No017.pdf
4	United States Agency for International Development (USAID) Measurement for Community Resilience	Ecosystem health, food security, health, nutrition, social capitals (bonding, bridging, and linking social capital), assets, and poverty	The community is expected to improve living standards, absorb change, and transform livelihood systems while maintaining the natural resource base	Health index, food security and nutrition index, social capital index, asset index, and economic and poverty index. Considering the potential shocks and stresses nature, the capacity of the community to measure resilience, and areas of collective action to analyze the baseline values. The values include disaster risk reduction, social protection, natural resource, conflict, and public services management	https://www.usaid.gov/ resilience/resources
5	Characteristics of a Disaster-Resilient Community	Four sectors: resilience and disaster risk reduction, disaster preparedness and mitigation, disaster response and recovery, and resilience assessments and measurement.	(a) Key concepts are disaster risk reduction, resilience, and the resilient community. (b) The capacity to achieve the following:	Three subdimensions: resilience components, characteristics of an enabling environment, and disaster-resilient community. The three subdimensions, along with the five thematic areas, guide multiple dimensions for analysis	https://www.fsnnetwork. org/resource/ characteristics-disaster-resilient-community-guidance-note

		Five Thematic Areas: risk assessment, risk management and vulnerability reduction, governance, knowledge and education, and disaster preparedness and response.	i) Anticipate and absorb potential stresses through resistance or adaptation; ii) maintain basic functions and structures during disastrous events; iii) recover after an event.		
6	United Nations Development Programme's (UNDP's) Community-Based Resilience Analysis (CoBRA) framework	It covers the five capitals (Natural, Physical, Human, Financial, and social)	Managing risks over time at different levels (individual, household, community, and societal) in ways that build capacity to manage and sustain development drive, minimize costs. Also, maximizing transformative potential and managing change by maintaining living standards during the shocks	Composite of qualitative and quantitative resilience indicators The tool allows communities to identify key resilience blocks. It also assess different interventions in achieving resilience characteristics	http://www.undp.org/content/undp/en/home/librarypage/environment-energy/sustainable_land_management/CoBRA/cobra-conceptual-framework.html

Table 11.A.1 Resilience tools and measurement efforts made by different institutes.

No.	Resilience tool	Developer	Application	Scope	Country/city	Released/updated	Dimensions/capacities	Indicators	Source
1.	Alliance for National and Community Resilience Benchmarking System	Resilience Alliance	Resilience benchmarking	Buildings, water, and energy	-	2018	-	49	https://www.resilientalliance.org/
2.	Baseline Resilience Indicators for Communities (BRIC)	Cutter, Burton, and Emrich	A quantitative index of pre-disaster community resilience	-	United States	2010	Economic, social, capital, ecosystems, infrastructure, and institutional capacity	-	http://artsandsciences.sc.edu/geog/hvri/baseline-resilience-indicators-communities-bric
3.	Characteristics of a Disaster-resilient Community	John Twigg	-	-	-	2007	Risk assessment, governance, risk management, etc.	-	https://www.preventionweb.net/files/2310_Characteristicsdisasterhighres.pdf
4.	City Resilience Index (CRI, also referred to as the City Resilience Framework or CRF)	Arup, with the support from the Rockefeller Foundation	Based on intensive site visits and consultation with resilience literature	-	-	2014	Four domains	52	https://www.rockefellerfoundation.org/report/city-%ADresilience-index
5.	Climate Resilience Screening Index (CRSI)	The Environmental Protection Agency		Five domains (built environment, risk, natural environment, governance, and society)	County level	2017	-	20 (117 matrices)	https://cfpub.epa.gov/si/si_public_record_Report.cfm?dirEntryId=350154&Lab=CEMM

6.	Climate Risk and Adaptation Framework and Taxonomy (CRAFT)	Arup, under guidance from Bloomberg Philanthropies and C40	The standard for participating C40 cities on adaptation actions and climate risk experiences	2015	-	https://www.c40.org/
7.	Coastal Resilience Decision Support System	The Nature Conservancy's coastal resilience program	Benchmarking process	Approach and a series of geospatial mapping tools that depict a variety of indicators	Has not yet been reported	-	https://coastalresilience.org/
8.	Coastal Resilience Index	Mississippi-Alabama Sea Grant Consortium and the National Oceanic and Atmospheric Administration's Coastal Storms Program	Questions across five physical and social categories	-	2010	-	https://toolkit.climate.gov/tool/coastal-resilience-index
9.	Community Assessment of Resilience Tool (CART)	The National Consortium for the Study of Terrorism and Responses to Terrorism (START),	Community survey begun by scholar	Community resilience	2007	-	https://www.start.umd.edu/research-projects/community-assessment-resilience-tool-cart

(Continued)

Table 11.A.1 (Continued)

No.	Resilience tool	Developer	Application	Scope	Country/city	Released/updated	Dimensions/capacities	Indicators	Source
10.	Community Disaster Resilience Index (CDRI)	Texas A&M Scholars—funded by National Oceanic and Atmospheric Administration			US Gulf Coast counties	2010	Overlaid multidimensional capitals onto disaster stages	Series of indicator	http://hrrc.arch.tamu.edu/_common/documents/10-02R.pdf
11.	Community Resilience Indicators and National-Level Measures	The Federal Emergency Management Agency	Community resilience	Local and national levels	United States	2020 update	10 core capacities	28	https://www.fema.gov/sites/default/files/documents/fema_2022-community-resilience-indicator-analysis.pdf
12.	Community Resilience Manual	Canadian Centre for Community Renewal	No specific applications have been noted	Community's social capital and cohesion	Canada	2000	Social capital	-	http://communityrenewal.ca/sites/all/files/resource/P200_0.pdf
13.	Community Resilience Planning Guide	The National Institute for Standards and Technology	Local resilience plans related to buildings and infrastructure systems	Six-step planning process	-	2015	Guidance regarding the subject dimensions	No specific recommendations for measurement	https://www.nist.gov/community-resilience/planning-guide
14.	Community Resilience System (CRS)	Community resilience to disasters	Reviews the knowledge base of community resilience with CART survey	Qualitative process	United States	2010	Seven community capacities and attributes	The Community and Regional Resilience Institute (CARRI) at Oak Ridge National Laboratory	https://floodsciencecenter.org/products/crs-community-resilience/

#	Name	Organization	Focus	Scale	Country	Year	Key dimensions	Indicators	Link
15.	Community Resilience: Conceptual Framework and Measurement Feed the Future Learning Agenda	The US Agency for International Development commissioned an exploratory	Community resilience and household resilience	Community resilience and household resilience	United States	2013	Social dimensions and institutional resilience activities	Indicators (assets, social dimensions, and areas of collective action)	https://www.fsnnetwork.org/resource/community-resilience-conceptual-framework-and-measurement-feed-future-learning-agenda
16.	Community-Based Resilience Analysis (CoBRA)	The United Nations Development Programme's Drylands Development Centre	Rural communities (food security and infrastructure)	Measuring resilience in the community and household levels	Ethiopia, Kenya, and Uganda	2012	Five capitals	-	https://www.undp.org/search?q=Community-Based+Resilience+Analysis+%28CoBRA%29&form_build_id=form-O_aqoHDv7OF_pf-fh_tqWkz07b_SZNn7xhbropvG0sc&form_id=undp_solr_search&op=Submit
17.	Conjoint Community Resilience Assessment Measure (CCRAM)	A scientific process by a group of content experts from seven academic institutions and governmental agencies	Community	Community resilience	-	-	Community resilience	Comparisons of community resiliency across time and place	https://www.researchgate.net/publication/278434223_The_Conjoint_Community_Resiliency_Assessment_Measure
18.	Disaster Resilience Scorecard	IBM and AECOM	Reduce risks of disasters	Communities	-	2016	Wide range of resilience dimensions	10 "essentials"	https://www.undrr.org/publication/disaster-resilience-scorecard-cities

(*Continued*)

Table 11.A.1 (Continued)

No.	Resilience tool	Developer	Application	Scope	Country/city	Released/updated	Dimensions/capacities	Indicators	Source
19.	Disaster Resilience Scorecard for Cities	United Nations International Strategy for Disaster Risk Reduction	Qualitative assessments	Disaster Resilient Scorecard for cities	Many cities outside the United States	2014	Capacity measures (organization, planning, response, and recovery)	85 criteria plus additional indicators	https://www.undrr.org/publication/disaster-resilience-scorecard-cities
20.	Earthquake Recovery Model	The San Francisco Planning and Urban Research Association (SPUR)	Earthquake scenario	Infrastructure facilities	San Francisco	2008	-	Not bound to any particular indicator	https://www.spur.org/
21.	Evaluating Urban Resilience to Climate Change	The Environmental Protection Agency	Urban resilience to climate change	Qualitative indicators	United States	2016	-	Dozens of indicators across eight sectors	https://cfpub.epa.gov/ncea/global/recordisplay.cfm?deid=322482
22.	Flood Resilience Measurement Framework	Zurich Insurance Group	(Qualitative): pre-disaster conditions in relation to flood events	Data sourced from community meetings, focus groups, household surveys, interviews, and third-party sources	Zurich (pilot-tested worldwide in 100 communities)	2014	Five capitals	88 indicators	https://www.zurich.com/en/sustainability/people-and-society/zurich-flood-resilience-alliance
23.	IFRC Framework for Community Resilience	The International Federation of the Red Cross	Community resilience	Community resilience	-	2014	Multiple dimensions of resilience	Sample indicators	https://www.ifrc.org/document/ifrc-framework-community-resilience

24.	Indicators of Disaster Risk and Risk Management	Inter-American Development Bank	Disaster economic risk, aggregate local risks, risk management, and vulnerability	Other resilience measurement efforts		2010	Four indexes	[59]
25.	National Health Security Preparedness Index	The Centers for Disease Control and Prevention	To prepare and improve the awareness of health security		United States	2013	Group of 18 measures	134 individual measures	https://nhspi.org/explore-the-index/
26.	PEOPLES Resilience Framework	The National Institute of Standards and Technology	Resilience of people	No clear use of the framework	-	2010	Multiple dimensions	Seven categories of community indicators	http://peoplesresilience.org/
27.	Resilience Capacity Index (RCI)	Foster	The resilience of metropolitan regions	-	US metropolitan regions	2011	-	12 social and economic indicators	
28.	Resilience Index Measurement and Analysis (RIMA)	The United Nations Food and Agriculture Organization	Resilience measurement framework	Social service, delivery, and access	-	2010	City level (range of communities)	-	https://www.fao.org/agrifood-economics/areas-of-work/rima/en/
29.	Resilience Inference Measurement (RIM)	UNISDR	Community resilience measurement	To quantify resilience across three elements.	-	2016	Multiple dimensions	Exposure, damage, and recovery indicators	https://www.unisdr.org/campaign/resilientcities/toolkit/article/community-resilience-inference-measurement.html

(Continued)

Table 11.A.1 (Continued)

No.	Resilience tool	Developer	Application	Scope	Country/city	Released/updated	Dimensions/capacities	Indicators	Source
30.	Resilience Measurement Index (RMI)	The Infrastructure Assurance Center at Argonne National Laboratory with the Protective Security Coordination Division of the U.S. Department of Homeland Security	To support decision making	To characterize the impact and response resilience of critical infrastructure	United States	2013	Risk management, disaster response, and maintenance of business continuity.	-	https://www.anl.gov/
31.	Resilience Scorecard	Urban planning scholars	Impacts on a community's physical and social vulnerability to hazards	Assessment tool for planning vulnerability gaps of districts	-	2015	-	Index derived from 11 social indicators	https://coastalresiliencecenter.unc.edu/page/2/?s=Resilience+Scorecard
32.	Resilience United States (ResilUS)	Miles and Chang	To project losses and recovery time in communities after disaster	Community resilience	San Francisco	2007	-	Social, economic, and infrastructure indicators	https://scarp.ubc.ca/publications/resilus-community-based-disaster-resilience-model
33.	Rural Resilience Index (RRI)	Canadian scholars (Cox and Hamlen)	Qualitative resilience self-assessment process and toolkit devised for rural communities	To address the social cohesion resilience dimension and emergency management	Canada	2014	Series of yes/no questions	-	https://www.researchgate.net/publication/322852244_The_rural_resiliency_index

There is a need for independent assessment techniques that consider location, context, and time frame [30]. The requirements for such techniques may include several elements identified by Béné [58]. First, the resilience indicators must be designed to measure subjective perceptions and objective changes (subjective and objective). Second, the indicators should not be limited to a specific aspect of resilience, such as individuals, communities, or cities (multiscale), and should capture change in resilience at different scales. Third, since resilience covers different strategies (coping, absorption, adaptation, and transformation), the strategies should also capture ex-post and ex-ant (anticipation) strategies (multidimensional) [30]. Fourth, the indicators should be measured independently from other factors affecting resilience (Independently built). Lastly, the indicators should allow for scaling up and replication and be built on specific agendas, contexts, and circumstances.

A resilient approach that may be easily deployed for localized community-based services is required. This means the measure must be relevant to the local context and simple to implement by local organizations. The Atlas of Social Protection Indicators of Resilience and Equity (ASPIRE) toolkit and framework have been widely used by organizations such as Habitat for Humanity [29] to conduct sustainability assessments for community-based projects in different parts of Asia. The framework was developed to integrate a poverty and sustainability agenda into infrastructure projects while defining project boundaries, scale, and timeframe. The good feature of the ASPIRE framework is that it meets the requirements set by Béné [58], briefly discussed above. As a result, NGOs, development agencies, and policymakers are expected to use these new methodologies to analyze the resilience-building benefits of initiatives (especially infrastructural in nature) in rural, urban, and peri-urban slum populations [30].

Conclusion

Although efforts have been made to assess the resilience of engineering systems, challenges still exist. This chapter summarizes the applicable methods and approaches to assessing urban and community resilience. The chapter provided insights into the different existing resilience frameworks and the two main categories of tools into which they are classified: qualitative and quantitative. Conceptual frameworks and semiquantitative methodologies are examples of qualitative assessment approaches. Conceptual frameworks provide information about the concept of resilience but do not provide a numerical number. Compiling expert opinion across various variables into a semiquantitative indicator is common. Quantitative assessment is divided into generic resilience measurements and structural models. Generally, general resilience measurements examine resilience by comparing a system's performance before and after a disruption. Some metrics are static, while others provide a time-dependent view of a system's performance. A recent trend in resilience measures has been to use stochastic techniques to account for aleatoric and epistemic uncertainty. Structural-based techniques focus on a system's structure or characteristics in order to calculate its resilience.

References

1 Donovan, A. and Oppenheimer, C. (2016). Resilient science: the civic epistemology of disaster risk reduction. *Sci. Public Policy* 43: 363–374. https://doi.org/10.1093/scipol/scv039.

2 Kovacic, Z. (2018). Conceptualizing numbers at the science–policy interface. *Sci. Technol. Hum. Values* 43: 1039–1065. https://doi.org/10.1177/0162243918770734.

3 Borie, M., Pelling, M., Ziervogel, G., and Hyams, K. (2019). Mapping narratives of urban resilience in the global south. *Glob. Environ. Chang.* 54: 203–213. https://doi.org/10.1016/j.gloenvcha.2019.01.001.

4 Burgess, J., Stirling, A., Clark, J. et al. (2007). Deliberative mapping: a novel analytic-deliberative methodology to support contested science-policy decisions. *Public Underst. Sci.* 16: 299–322. https://doi.org/10.1177/0963662507077510.

5 Brown, K., Eernstman, N., Huke, A.R., and Reding, N. (2017). The drama of resilience: learning, doing, and sharing for sustainability. *Ecol. Soc.* 22: https://doi.org/10.5751/ES-09145-220208.

6 Committee on Measuring Community Resilience (2019). *Building and Measuring Community Resilience*. Washington, DC: National Academies Press ISBN 978-0-309-48972-0.

7 Desouza, K.C. and Flanery, T.H. (2013). Designing, planning, and managing resilient cities: a conceptual framework. *Cities* 35: 89–99. https://doi.org/10.1016/j.cities.2013.06.003.

8 Godschalk, D.R. (2003). Urban hazard mitigation: creating resilient cities. *Nat. Hazards Rev.* 4: 136–143. https://doi.org/10.1061/(ASCE)1527-6988(2003)4:3(136).

9 Rus, K., Kilar, V., and Koren, D. (2018). Resilience assessment of complex urban systems to natural disasters: a new literature review. *Int. J. Disaster Risk Reduct.* 31: 311–330. https://doi.org/10.1016/j.ijdrr.2018.05.015.

10 Serdar, M.Z., Koç, M., and Al-Ghamdi, S.G. (2022). Urban transportation networks resilience: indicators, disturbances, and assessment methods. *Sustainable Cities Soc.* 76: 103452. https://doi.org/10.1016/j.scs.2021.103452.

11 Council, N.R = National Research Council (2012). *Disaster Resilience*. Washington, D.C: National Academies Press ISBN 978-0-309-26150-0.

12 Cutter, S.L. (2016). Resilience to what? Resilience for whom? *Geogr. J.* 182: 110–113. https://doi.org/10.1111/geoj.12174.

13 Tong, P. (2021). Characteristics, dimensions and methods of current assessment for urban resilience to climate-related disasters: a systematic review of the literature. *Int. J. Disaster Risk Reduct.* 60: 102276. https://doi.org/10.1016/j.ijdrr.2021.102276.

14 McClymont, K., Morrison, D., Beevers, L., and Carmen, E. (2020). Flood resilience: a systematic review. *J. Environ. Plan. Manag.* 63: 1151–1176. https://doi.org/10.1080/09640568.2019.1641474.

15 Hosseini, S., Barker, K., and Ramirez-Marquez, J.E. (2016). A review of definitions and measures of system resilience. *Reliab. Eng. Syst. Saf.* 145: 47–61. https://doi.org/10.1016/j.ress.2015.08.006.

16 Bruneau, M., Chang, S.E., Eguchi, R.T. et al. (2003). A framework to quantitatively assess and enhance the seismic resilience of communities. *Earthq. Spectra* 19: 733–752. https://doi.org/10.1193/1.1623497.
17 Collier, F., Hambling, J., Kernaghan, S. et al. (2014). Tomorrow's cities: a framework to assess urban resilience. *Proc. Inst. Civ. Eng. Urban Des. Plan.* 167: 79–91. https://doi.org/10.1680/udap.13.00019.
18 Bozza, A., Asprone, D., and Fabbrocino, F. (2017). Urban resilience: a civil engineering perspective. *Sustainability* 9: 103. https://doi.org/10.3390/su9010103.
19 Tahir, F., Ajjur, S.B., Serdar, M.Z. et al. (2021). Proceedings of the Qatar Climate Change Conference 2021. In: *Proceedings of the Qatar Climate Change Conference 2021*, 44.
20 Farhad, N., Garg, S., Huxley, R., and Pillay, K. (2019). *Understanding infrastructure interdependencies in cities. Multiple* .
21 Tepes, A. and Neumann, M.B. (2020). Multiple perspectives of resilience: a holistic approach to resilience assessment using cognitive maps in practitioner engagement. *Water Res.* 178: 115780. https://doi.org/10.1016/j.watres.2020.115780.
22 Cimellaro, G.P. (2016). *Urban Resilience for Emergency Response and Recovery: Fundamental Concepts and Applications*, Geotechnical, Geological and Earthquake Engineering, vol. 41. Cham: Springer International Publishing; ISBN 978-3-319-30655-1.
23 Courtois, L.A. (1976). Respiratory responses of Gillichthys mirabilis to changes in temperature, dissolved oxygen and salinity. *Comp. Biochem. Physiol. Part A Physiol.* 53: 7–10. https://doi.org/10.1016/S0300-9629(76)80002-3.
24 Cutter, S.L., Ash, K.D., and Emrich, C.T. (2014). The geographies of community disaster resilience. *Glob. Environ. Chang.* 29: 65–77. https://doi.org/10.1016/j.gloenvcha.2014.08.005.
25 Cimellaro, G.P., Tinebra, A., Renschler, C., and Fragiadakis, M. (2016). New resilience index for urban water distribution networks. *J. Struct. Eng.* 142: https://doi.org/10.1061/(ASCE)ST.1943-541X.0001433.
26 Lu, P. and Stead, D. (2013). Understanding the notion of resilience in spatial planning: a case study of Rotterdam, the Netherlands. *Cities* 35: 200–212. https://doi.org/10.1016/j.cities.2013.06.001.
27 Kahan, J.H., Allen, A.C., and George, J.K. (2009). An Operational Framework for Resilience. *J. Homel. Secur. Emerg. Manag.* 6, https://doi:10.2202/1547-7355.1675.
28 Harrison, C.G. and Williams, P.R. (2016). A systems approach to natural disaster resilience. *Simul. Model. Pract. Theory* 65: 11–31. https://doi.org/10.1016/j.simpat.2016.02.008.
29 Foundation, T.R. The Rockefeller Foundation, 100 Resilient Cities, https://www.rockefellerfoundation.org/100-resilient-cities/ (accessed on 25 December 2021).
30 Woolf, S., Twigg, J., Parikh, P. et al. (2016). Towards measurable resilience: a novel framework tool for the assessment of resilience levels in slums. *Int. J. Disaster Risk Reduct.* 19: 280–302. https://doi.org/10.1016/j.ijdrr.2016.08.003.
31 Al-Humaiqani, M.M. and Al-Ghamdi, S.G. (2022). The built environment resilience qualities to climate change impact: concepts, frameworks, and directions for future research. *Sustainable Cities Soc.* 80: 103797. https://doi.org/10.1016/j.scs.2022.103797.

32 Al-Humaiqani, M.M. and Al-Ghamdi, S.G. (2023). Assessing the built environment's reflectivity, flexibility, resourcefulness, and rapidity resilience qualities against climate change impacts from the perspective of different stakeholders. *Sustainability* 15: 5055. https://doi.org/10.3390/su15065055.

33 Resilience Alliance Assessing Resilience in Social-Ecological Systems (2010). Workbook for practitioners. *S. Afr. Med. J.* 2: 54.

34 Labaka, L., Hernantes, J., and Sarriegi, J.M. (2015). Resilience framework for critical infrastructures: an empirical study in a nuclear plant. *Reliab. Eng. Syst. Saf.* 141: 92–105. https://doi.org/10.1016/j.ress.2015.03.009.

35 Sterbenz, J.P.G., Cetinkaya, E.K., Hameed, M. et al. (2011). Modelling and analysis of network resilience. In: *Proceedings of the 2011 Third International Conference on Communication Systems and Networks (COMSNETS 2011)*, 1–10. Bangalore, India: IEEE.

36 Bruyelle, J.-L., O'Neill, C., El-Koursi, E.-M. et al. (2014). Improving the resilience of metro vehicle and passengers for an effective emergency response to terrorist attacks. *Saf. Sci.* 62: 37–45. https://doi.org/10.1016/j.ssci.2013.07.022.

37 Ainuddin, S. and Routray, J.K. (2012). Community resilience framework for an earthquake prone area in Baluchistan. *Int. J. Disaster Risk Reduct.* 2: 25–36. https://doi.org/10.1016/j.ijdrr.2012.07.003.

38 Vlacheas, P., Stavroulaki, V., Demestichas, P. et al. (2013). Towards end-to-end network resilience. *Int. J. Crit. Infrastruct. Prot.* 6: 159–178. https://doi.org/10.1016/j.ijcip.2013.08.004.

39 Cutter, S.L., Barnes, L., Berry, M. et al. (2008). A place-based model for understanding community resilience to natural disasters. *Glob. Environ. Chang.* 18: 598–606. https://doi.org/10.1016/j.gloenvcha.2008.07.013.

40 Henry, D. and Emmanuel Ramirez-Marquez, J. (2012). Generic metrics and quantitative approaches for system resilience as a function of time. *Reliab. Eng. Syst. Saf.* 99: 114–122. https://doi.org/10.1016/j.ress.2011.09.002.

41 Suárez, M., Baggethun, E.G., and Onaindia, M. (2020). Assessing socio-ecological resilience in cities. In: *The Routledge Handbook of Urban Resilience*, 197–216.

42 Schwind, N., Minami, K., Maruyama, H. et al. (2016). Computational framework of resilience. In: *Advanced Sciences and Technologies for Security Applications*, 239–257. Springer.

43 Chang, S.E. and Shinozuka, M. (2004). Measuring improvements in the disaster resilience of communities. *Earthq. Spectra* 20: 739–755. https://doi.org/10.1193/1.1775796.

44 Faturechi, R., Levenberg, E., and Miller-Hooks, E. (2014). Evaluating and optimizing resilience of airport pavement networks. *Comput. Oper. Res.* 43: 335–348. https://doi.org/10.1016/j.cor.2013.10.009.

45 Albores, P. and Shaw, D. (2008). Government preparedness: using simulation to prepare for a terrorist attack. *Comput. Oper. Res.* 35: 1924–1943. https://doi.org/10.1016/j.cor.2006.09.021.

46 Aleksić, A., Stefanović, M., Arsovski, S., and Tadić, D. (2013). An assessment of organizational resilience potential in SMEs of the process industry, a fuzzy approach. *J. Loss Prev. Process Ind.* 26: 1238–1245. https://doi.org/10.1016/j.jlp.2013.06.004.

47 Azadeh, A., Salehi, V., Arvan, M., and Dolatkhah, M. (2014). Assessment of resilience engineering factors in high-risk environments by fuzzy cognitive maps: a petrochemical plant. *Saf. Sci.* 68: 99–107. https://doi.org/10.1016/j.ssci.2014.03.004.

48 Muller, G. (2012). Fuzzy architecture assessment for critical infrastructure resilience. *Procedia Comput. Sci.* 12: 367–372. https://doi.org/10.1016/j.procs.2012.09.086.

49 Faturechi, R. and Miller-Hooks, E. (2014). Travel time resilience of roadway networks under disaster. *Transp. Res. Part B Methodol.* 70: 47–64. https://doi.org/10.1016/j.trb.2014.08.007.

50 Jin, J.G., Tang, L.C., Sun, L., and Lee, D.-H. (2014). Enhancing metro network resilience via localized integration with bus services. *Transp. Res. Part E Logist. Transp. Rev.* 63: 17–30. https://doi.org/10.1016/j.tre.2014.01.002.

51 Barker, K., Ramirez-Marquez, J.E., and Rocco, C.M. (2013). Resilience-based network component importance measures. *Reliab. Eng. Syst. Saf.* 117: 89–97. https://doi.org/10.1016/j.ress.2013.03.012.

52 Baroud, H., Barker, K., Ramirez-Marquez, J.E., and Claudio, M.R.S. (2014). Importance measures for inland waterway network resilience. *Transp. Res. Part E Logist. Transp. Rev.* 62: 55–67. https://doi.org/10.1016/j.tre.2013.11.010.

53 Carvalho, H., Barroso, A.P., Machado, V.H. et al. (2012). Supply chain redesign for resilience using simulation. *Comput. Ind. Eng.* 62: 329–341. https://doi.org/10.1016/j.cie.2011.10.003.

54 Hashimoto, T., Stedinger, J.R., and Loucks, D.P. (1982). Reliability, resiliency, and vulnerability criteria for water resource system performance evaluation. *Water Resour. Res.* 18: 14–20. https://doi.org/10.1029/WR018i001p00014.

55 Adjetey-Bahun, K., Birregah, B., Chatelet, E. et al. (2014). A simulation-based approach to quantifying resilience indicators in a mass transportation system. In: *Proceedings of the Proceedings of the 11th international ISCRAM conference*. Pennsylvania: University Park.

56 Tadić, D., Aleksić, A., Stefanović, M., and Arsovski, S. (2014). Evaluation and ranking of organizational resilience factors by using a two-step fuzzy AHP and fuzzy TOPSIS. *Math. Probl. Eng.* 2014: 1–13. https://doi.org/10.1155/2014/418085.

57 Engle, N.L., de Bremond, A., Malone, E.L., and Moss, R.H. (2014). Towards a resilience Indicator framework for making climate-change adaptation decisions. *Mitig. Adapt. Strateg. Glob. Chang.* 19: 1295–1312. https://doi.org/10.1007/s11027-013-9475-x.

58 Béné, C. (2013). Towards a quantifiable measure of resilience. *IDS Work. Pap.* 2013: 1–27. https://doi.org/10.1111/j.2040-0209.2013.00434.x.

59 Cardona, O.D. *Indicators of Disaster Risk and Risk Management- Program for Latin America and the Caribbean Summary Repor – 2t*; 2008; ISBN 9789584402196.

60 Levine, S. (2014). Assessing Resilience: Why Quantification Misses the Point, London.

61 Rockefeller & Arup (2018). City Resilience Index. https://www.cityresilienceindex.org#/ (accessed on 8 August 2021).

62 Arup (2014). The City Resilience Index supported by the Rockefeller Foundation. https://www.arup.com/projects/city-resilience-index (accessed on 8 August 2021).

63 U.S. EPA (2017). *Evaluating Urban Resilience to Climate Change: A Multi-Sector Approach (Final Report)*, Washington, DC, vol. 10.

64 BRACED (2015). The 3As: Tracking Resilience across BRACED. http://www.braced.org/resources/i/?id=cd95acf8-68dd-4f48-9b41-24543f69f9f1 (accessed on 9 August 2021).

65 Elasha, B.O., Elhassan, N.G., Ahmed, H., and Zakieldin, S. (2005). *Sustainable Livelihood Approach for Assessing Community Resilience to Climate Change: Case Studies from Sudan*; AIACC Working Paper. Sudan: Higher Council for Environment and Natural Resources.

Part III

Resilient Urban Systems: Navigating Climate Change and Enhancing Sustainability

Part 3 of the book delves into the critical domains of energy and healthcare systems within urban environments, focusing on developing and implementing resilience strategies to mitigate the impacts of climate change. As cities continue to face the challenges posed by climate change, it becomes imperative to understand and strengthen the resilience of key infrastructures such as energy and healthcare. These systems play vital roles in supporting urban life and are particularly vulnerable to the disruptions caused by extreme weather events and changing climate patterns.

Chapter 12 explores the pivotal role of urban planning in building climate resilience within the built environment. It emphasizes the wide-ranging impacts of climate change on urban areas and highlights the urgent need for resilience. The chapter delves into various urban planning strategies, such as land use planning, transportation planning, and green infrastructure development, which effectively address these impacts. It also discusses the significance of risk assessment and adaptation in urban planning and the policy and governance aspects involved.

Chapter 13 emphasizes the integration of green–blue–gray infrastructure (GBGI) for sustainable urban flood risk management. It introduces the concept of GBGI, which combines green infrastructure (GI), gray infrastructure (GRAI), and blue infrastructure (BI) to provide comprehensive solutions for adapting to and mitigating climate change impacts. The chapter highlights the benefits of GI in regulating temperature, managing floods, and enhancing urban resilience. It underscores the importance of integrating green and blue infrastructure elements into existing gray infrastructure systems to optimize flood risk management.

Chapter 14 focuses on enhancing energy system resilience in the face of climate change and security challenges. It explores the complex terminologies associated with energy system resilience, the influence of climate change and weather events, and approaches to enhancing resilience in energy systems. The chapter emphasizes the need for resilience planning and strategic investments to ensure reliable and secure energy supplies, particularly in countries heavily dependent on energy imports.

Chapter 15 addresses the inclusion of climate change impacts in health policy development to enhance the resilience of the healthcare sector. It explores strategies for incorporating climate change considerations into health policies, including surveillance and monitoring systems, risk assessment and management, adaptation measures, and public health interventions. The chapter emphasizes interdisciplinary collaboration, data-driven decision-making, and community engagement in addressing the impacts of climate change on human health.

Chapter 16 highlights the importance of surveillance strategies in monitoring the spread of vector-borne diseases as a measure of resilience. It explores key aspects of surveillance, including monitoring human cases, identifying pathogens and vectors, and considering climatic and environmental factors. The chapter emphasizes the need for proactive surveillance and response systems to mitigate the impact of vector-borne diseases and enhance community resilience.

Overall, Part 3 provides a comprehensive understanding of developing and implementing resilience strategies in energy and healthcare systems to mitigate climate change impacts. It emphasizes the urgent need to enhance the resilience of these critical systems, offers insights into effective strategies and approaches, and provides valuable knowledge for researchers, practitioners, and policymakers involved in urban planning, energy management, and healthcare system development.

12

Building Climate Resilience Through Urban Planning: Strategies, Challenges, and Opportunities

Nisreen Abuwaer[1,2], Safi Ullah[1,2] and Sami G. Al-Ghamdi[1,2]

[1] Environmental Science and Engineering Program, Biological and Environmental Science and Engineering Division, King Abdullah University of Science and Technology (KAUST), Thuwal, Saudi Arabia
[2] KAUST Climate and Livability Initiative, King Abdullah University of Science and Technology (KAUST), Thuwal, Saudi Arabia

Introduction

The field of urban planning is concerned with the different patterns in which urban areas have formed, developed, and are designed. The way in which they are functioning, being managed, and being maintained along with the desired goals and projected future of those urban areas. Urban planning is an interdisciplinary field that focuses on various aspects of a city. This permits urban planners to bring all the stakeholders of urban areas to the table, giving them the opportunity to learn, influence, and impact the trajectory of the city. This process involves multiple stakeholders, ranging from architects and engineers to government officials and policymakers to community members and anyone affiliated with the urban area.

Urban planners can be seen as doctors and urban areas as their patients; the responsibility of a planner is to identify the diseases and illnesses that impact the city, in other words, what is making the city unhealthy. Planners need to collect a variety of data and conduct various examinations to identify potential problems. Climate change is one of the main illnesses that make cities unhealthy, and vice versa, cities contribute to climate change, for instance, imagine someone smoking. Urban planners need to then see which areas of the patient are impacted and how to treat that illness. Treatments vary based on who is impacted and what measures are to be taken. Treatments for climate change include resilient and sustainable approaches for mitigation and adaptation. When treating an urban area, it is crucial to look at the different systems within a city, like transportation systems, how they are being designed, operated, and impact the rest of the area. It is also crucial to include everyone involved within the urban area by obtaining their insights and revealing the responsibilities and duties of everyone in committing to a healthier city.

The significance of urban planning in tackling climate change is pivotal as it has the capacity to redirect the course of current trends. Urban resilience is the immune system of urban areas, as it prepares and protects cities' systems from any threats and challenges that they

Sustainable Cities in a Changing Climate: Enhancing Urban Resilience. First Edition.
Edited by Sami G. Al-Ghamdi.
© 2024 John Wiley & Sons Ltd. Published 2024 by John Wiley & Sons Ltd.

might face. Urban resilience also smoothens and enhances the recovery process after a natural, economic, and/or social crisis. In terms of climate change, resilience in the built environment is extremely crucial. The built environment has a direct impact and a profound influence on the lives of people. Resilience in the built environment ensures that the city adopts strategies to withstand climate-related disasters such as green permeable pavements for flooding reduction. Resilience in the built environment also includes the functioning of water and electric systems during heat or cold waves or after a severe storm. Overall, resilience in the built environment determines the strength of a city and its ability to withstand challenges.

Understanding Climate Change Impacts on Urban Areas

The impacts of climate change are evident in urban areas due to their higher degree of exposure to climate extremes. For instance, rising temperatures, sea level rise, and extreme weather events are some of the many consequences that urban areas are experiencing as a result of human-induced disturbances to the environment and climate, as depicted in Figure 12.1, and further explained in this section. Urban areas are extremely vulnerable, as more than half of the world's population lives there and is impacted by the frequent occurrence of climate extremes. These extreme events lead to catoptric disturbances to the urban form and function, posing severe threats to the urban populations. Climate change can contribute to the escalation of built environment deterioration. The reoccurrence of threats to the built environment leads to destruction that often happens in the presence of individuals. Climate change events may also lead to disturbances in urban systems, including transportation, water, and power. For instance, power outages occurred as a result of Hurricane Katrina that caused significant disruptions to critical infrastructure systems like pipelines, impacting the prices of gasoline along with disturbance to the grain harvest exports through the Mississippi River [1]. The social and economic conditions will also be heavily impacted by such disturbances due to their dependency on these systems. The

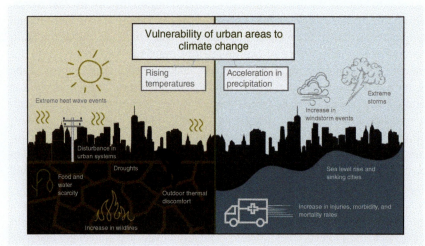

Figure 12.1 The various climate change impacts on urban areas, resulting in their vulnerability.

urban population will suffer severe consequences; the impacts can be economic, social, health-related and may even impact individuals' chances of survival. In other words, climate and urban environments mutually influence each other within a cyclical relationship, while cities contribute to the exacerbation of climate change impacts, these impacts, in turn, pose various threats to urban systems and outdoor livability. The interconnected impacts of climate change repel across multiple aspects of life, leading to a series of environmental impacts, human health concerns, disturbance in the built environment, economic and social challenges, food security, clean water access, droughts, and resource scarcity, and national security threats and conflicts.

Rising temperatures, extreme heat waves, droughts, and wildfires are climate extremes that are influenced by anthropogenic global warming and have significant impacts on urban areas. Rising temperatures alter the overall quality of life in cities by influencing the way individuals interact with outdoor areas; individuals will find themselves in a state of thermal discomfort, concerned about their health, and urged to avoid outdoor movement, limiting their social engagement and economic activities, and increasing their reliance on cars. The rise in temperatures increases the occurrence of heat waves, which pose serious threats to the urban population while imposing a heavier burden on the urban environment. During heat waves, cooling systems become more in demand, requiring more energy to be generated and consumed. Cities may not be prepared for such demands and may face shortages or disturbances in such systems; the failure of operation leads to serious health-related illness and may result in a high mortality rate. Longer periods of high temperatures, a lack of precipitation, and persistent dryness in the air lead to droughts and wildfires, which impact the food and water systems. The Arabian Peninsula is at extraordinary risk of rising temperatures and heat stress (the joint effect of extreme heat and humidity). If the ongoing rise in temperatures and humidity persist, there is a risk of surpassing the limits of human tolerance in the region, which could increase the likelihood of hyperthermia and ultimately, fatalities [2]. The impacts of climate change are not stationary as they can spread over multiple cities. A wildfire can start in one city, and its impacts (i.e., air quality) can be witnessed in neighboring cities.

Climate change-induced flooding and sea level rise have profound impacts on urban areas. Urban flooding, for instance, affects people and the built environment in urban areas as it causes severe destruction to homes and infrastructure, affecting different systems including water, energy, transportation, communication networks, and many other systems within a city. Flooding may leave people devastated and with no shelter along with a rise in homelessness levels. Mental health concerns will arise after the experience of such a traumatic event along with physical impacts, from the spread of waterborne diseases to injuries, morbidity, and mortality during and post-flooding. The occurrence of coastal floods includes the tragic event in Bangladesh in 1991, where a staggering death toll of 140,000 individuals and the displacement of 10 million people from their homes occurred [3]. Flooding also has negative impacts on the environment, including damage to natural ecosystems and biodiversity, soil erosion, and water contamination.

Cities may not be prepared to endure intense flooding events and may not have the capability to withstand the consequences. To comprehend the implications of future changes on the city's prevalent urban flooding issue, a detailed modeling study was conducted on Can Tho city's urban water system in Vietnam. In the scenario of a 100cm rise in sea levels and an escalation of the Mekong River runoff from rising emissions—described as the worst scenario—during periods of intense rainfall, water from roads, canals, and sewers converges into the canal. However, due to the limited drainage capacity of the canal, it is unable to handle the large volume of water effectively. Consequently, the nearby low-lying areas consistently suffer from flooding. The alarming concern is that the depth of the flooding in this area may rise by as much as 1.51 m, potentially posing a significant risk to human life [4].

Coastal cities and islands face a dire threat from climate change, as rising sea levels place them in imminent danger of sinking. The escalation of global temperatures leads to alarming increases in sea levels. With sea levels rising, coastal cities become very vulnerable as the shoreline approaches the city little by little, invading the spaces that people occupy. With enough time and persistence in following the same patterns, cities will find themselves drowning. The Maldives, for instance, is at great risk; if sea levels rise by 1 m, the country could be completely inundated by 2085. In the scenario of a 2-m rise, the Maldives may disappear around 2050 if protective measures are not taken [5].

Windstorms are another climate change-induced hazard that exacerbates the vulnerability of urban areas. The range of extreme windstorm-related events, from tornados to hurricanes, typhoons, and cyclones, has all brought devastation to the areas that they visit. Recent studies indicate that over a span of two decades, a total of 34 urban cyclonic events were documented in Soledad, a municipality in Colombia known for experiencing the majority of these destructive weather phenomena. The frequent occurrence of these cyclones has impacted 60 different neighborhoods, resulting in 7 fatalities, 14,552 injuries, and 5180 homes affected [6]. The catastrophic impacts are determined by the frequencies and intensities of these events. With the acceleration of climate change, the frequency and intensity of these events will cause more severe devastation to the urban environment.

Urban Planning Strategies for Mitigating Climate Change Impacts

To reduce the vulnerability of cities and avoid the catastrophic risks imposed by climate change, urban planners must implement strategic plans that prioritize combating climate change and achieving resilient and sustainable urban development. When an urban area aligns its vision and mission with climate change mitigation strategies, it establishes clear guidelines and prioritizes solutions related to climate challenges. This helps in the planning and development process by setting both long-term and short-term goals that focus on addressing climate issues. It ensures that climate considerations are at the forefront of decision-making and implementation, making it more feasible to plan for and tackle the impacts of climate change. Some of the urban planning strategies for climate change adaptation and mitigation that have been adopted across the globe are shown in Figure 12.2.

Transit-Oriented Development (TOD)

Transit-oriented development (TOD) serves the mitigation of climate change and contributes to the decrease of GHG emissions as it reroutes the direction of the city to foster pedestrian-friendly commuting options and seeks to replace the extensive use of individually driven private vehicles with a robust reliance on public transportation, all within a compact and

Figure 12.2 Urban planning strategies for mitigating climate change impacts.

centralized city. TOD provides various environmental benefits such as a decrease in air pollution, less congested cities with less travel time, and less GHG emissions.

The TOD framework depends on several concepts to be successfully implemented. This includes the availability, accessibility, and convenience of various transportation modes, such as trains and buses like the bus rapid transit (BRT) systems. Transit hubs allow for various transportation modes to be clustered in one area, providing a variety of options, allowing easy transitions from the different modes, and serving as a point of origin that provides a wider coverage of transit options to various parts of the city. TOD relies on infrastructure that serves human-powered transportation options such as walking and biking. This allows users to have better access to transit options. TOD aligns with the values of the *New Urbanism* movement and the *Traditional Neighborhood Development* that promote fundamental principles such as compactness and centralization where residents, services, and amenities are concentrated within the urban core. Centralization and compactness depend on various concepts such as density, land use diversity, and accessibility. TOD design and ordinance can facilitate achieving TOD. TOD offers cities an approach that establishes low carbon emissions.

Future cities should adopt three major paths when developing TODs including the examination of the built environment should encompass both the neighborhood and city levels. The effectiveness of TODs should be evaluated based on factors beyond public transit usage, including the prevalence of walking and the incorporation of urban planning and design elements such as green spaces and liveliness. It is neither feasible nor advantageous to enforce uniformly high density, diversity, and identical design in every TOD neighborhood [7].

The Comparative Study on Urban Transport and the Environment (CUTE) matrix is a classification tool for the different types of low-carbon transport measures, with three elements in the strategic axis, which include reducing unnecessary travel demand (AVOID), shifting travel to lower-carbon modes (SHIFT), and improving the intensity of transport-oriented emission (IMPROVE), with four instrumental elements that focus on how to implement the strategies. The four instruments include technology, regulation, information and economy, the implementation of land use control is under the AVOID strategy, and regulation instrument [8]. Promoting active transportation and reducing reliance on private vehicles will have a tremendous impact on the levels of CO_2 emissions. TODs are an excellent approach to redirecting the city to a path with less reliance on private vehicles by providing various transit options that allow for better alternatives. With fewer vehicles on the street, individuals are more likely to be encouraged to participate in active transportation as the level of safety increases.

Fifteen Minutes City (FMC)

Active transportation follows the path of TODs in decreasing GHG emissions; the *Fifteen Minutes City* (FMC) model is one approach to implementing active transportation within a city. It also serves the concept of TOD. Having cities that depend on active transportation is not purely new, as it was the norm before the invention of cars; however, it has been regaining interest in recent years, especially with the emergence of the FMC concept that has gained wide popularity, making it a buzzword in the literature and media. The FMC concept was first introduced in 2016 but gained greater attention after the Covid-19 pandemic as more individuals began to sense its importance [9]. The FMC promotes the idea of individuals having access to various everyday services and amenities either by walking or biking within a 15 minutes or less timeframe. This concept heavily depends on walkability, making it crucial to enhance the walkability experience for the FMC to be achieved, ensuring the usefulness, safety, comfort, and attractiveness of the walk. The safety of the road and enhancements to network connectivity, a high land-use mix, the urban character, and infrastructure development can all contribute to encouraging pedestrians to engage in the FMC experience [10]. To enhance walkability, a large variety of services and amenities need to be within walking distance, and the walking trips should be short and convenient. The safety of the trip is determined by various factors, including road safety, burglary, and crime safety but also includes safety and comfort during various weather conditions (i.e., hot or cold weather, rain or snow). Each city should be examined separately and designed based on its needs to ensure its safety and comfort. In order to encourage active transportation, designs and policies need to include all community members, including children, the elderly, and individuals with special abilities.

Compact Cities

Land use development and density are characteristics of compact cities, which serve to mitigate climate change impacts on the built environment. Compact cities are cities that focus on the clustering of people and diverse places in urban areas, i.e., New York. Population density and a highly diverse range of land use mix from residential to commercial to civil/institutional to open spaces, etc., contribute to strengthening compact cities. This clustering of people and services allows for putting limits to the stretching of cities through urban sprawl and allows for less reliance on cars and less time traveled. A resident of a city does not need to drive for miles to reach a supermarket to buy bread; instead, a neighborhood bakery minutes away within walking distance will provide this service. The proximity and diverse amenities make this possible. These need to be links that connect people and places, that is where access comes into play. Access, as discussed in the TOD Section, plays a crucial role in connecting people with places. The benefits of compact cities are not limited to the connectedness and vibrance of a community but also offer various other benefits. Besides encouraging physical activity, a compact neighborhood can reduce its vulnerability to climate change events. In the United States, 83 of the largest metropolitan regions were assessed for extreme heat events (EHEs) over a five-decade period, finding that the most sprawling metropolitan regions exhibited more than double the rate of increase in the annual number of EHEs compared to the most compact metropolitan regions [11].

Sustainable Land Use and Development Policies

Sustainable land use and development practices are of great importance when planning for climate change mitigation strategies, as they rigorously assess the different ways that land is being utilized and the patterns that it acquires. Ensuring that the utilization of land follows environment-friendly guidelines will heavily influence the city's contribution to climate change. In order to achieve such a goal, different measures and tools, such as land use laws, policies, and regulations along with zoning ordinances, are employed to regulate and influence land use management. This includes decision-making regarding how to proceed with redevelopment within vacant or brownfields, identifying, improving, and creating areas designated to ensure the availability of open and green spaces, ensuring the preservation of building structures and conservation of wilderness areas and natural habitats like forests and wetlands, along with identifying zones for each land use type and providing set policies for each zone. Another example is an industrial zone, providing guidelines on where factories are appropriate to be placed with clear buffer zones and ways of operation. It is crucial to wisely choose effective policies, as different policies direct the trajectory of cities in completely different directions. A hydrological model in an urbanized watershed in Taiwan was developed to evaluate diverse land use policies and climate change scenarios. Results indicated that the hydrological components displayed varying degrees of sensitivity to distinct demand and conversion policies across all land use scenarios [12]. Density and land use mix are also determined by land use policies, which drastically contribute to the impact of climate change on urban environment. They are also impacted by zoning. One way to monitor high and low density within an urban setting is through urban growth, which can be assessed through land cover and land use change (LCLUC) models [13]. Geographic information systems (GISs) and Urban Footprint are both tools that can be utilized to visualize different land uses and densities. Policies also play a role in influencing the design process concerned with land use as they set forth the design path for the urban fabric. Land use design is crucial as it determines when, where, and how green infrastructure is embedded along with identifying sufficient materials used like permeable pavements. The orientation of the urban design also plays a significant role in promoting TOD corridors, which allows for more public- and pedestrian-friendly options of transport.

Low-Impact Development (LID)

Low-impact development (LID) is a model that contributes to climate change mitigation and decreasing the risk of flooding in cities. The acceleration in precipitation and extreme storms requires innovative and sustainable solutions to mitigate such impacts. LID is a model that focuses mainly on stormwater runoff management and water quality in urban environments. It involves implementing green permeable infrastructure, bioswales, bioretention, and basins to manage storm runoff. The implementation of LID is often driven by regulations pertaining to stormwater quality improvement. It heavily relies on infiltration and evapotranspiration processes and aims to integrate natural elements into its design [14]. LID assists in reducing flood-related threats, and the heavy burden on urban systems and can manage stormwater efficiently, including filtering and storing it, establishing a more resilient drainage system. LID, in contrast to conventional urban stormwater management approaches, facilitates the restoration of runoff to the natural hydrologic cycle, including reducing runoff volume, improving infiltration, decreasing peak flow rates, prolonging lag time, lowering pollutant loads, and augmenting baseflow [14]. Certain limitations exist for the use of LID practices relying on infiltration, such as areas with high contaminant loading such as recycling centers or soil contamination rearing the risk of contaminating groundwater, steep slopes, shallow depth to bedrock, and seasonal high water tables. However, these conditions are rarely present throughout an entire site, allowing for the adoption of LID practices to minimize cumulative impacts on downstream water bodies [15]. Other stormwater management concepts aside from LID include the best management practices (BMPs), green infrastructure, water-sensitive urban design (WSUD), sustainable urban drainage systems (SUDSs), and sponge cities [16]. Sponge cities are further discussed in the following section.

Sponge Cities

Sponge cities have a huge contribution toward mitigating climate change impacts, especially those associated with flooding. The concept of a sponge city is devoted to ensuring water quality and management regarding hydrological cycle systems throughout the city. As of 2013, China has developed the strategy of sponge city. This innovative approach is guided by four fundamental principles: urban water resourcing, ecological water management, green infrastructures, and urban permeable pavement [17]. It has a diverse and broad range of values and strategies such as ensuring the quality of groundwater, restoration, and recovery of rivers; however, it also focuses on different strategies to mitigate flood-related impacts. The sponge city concept primarily focuses on nature-based solutions when implementing new projects. Green infrastructure is an example of a nature-based solution that mitigates urban flooding under the umbrella of sponge cities. Green roofs, open green spaces, trees, vegetation, and wetlands are examples of green infrastructure that function as permeable infrastructures which retain and manage stormwater, reducing the strain on urban drainage systems and helping to prevent flooding. The sponge city concept encompasses the following three objectives [18]:

1) *Implementation and development of LID concepts*: improve the management of urban peak runoff by effectively controlling stormwater by temporarily storing, recycling, and purifying it.
2) *Enhancement of traditional drainage systems*: efforts are focused on improving the resilience of conventional drainage systems by incorporating flood-resilient infrastructure, such as underground water storage tanks and tunnels. Additionally, LID systems are employed to mitigate peak discharges and reduce the surplus of stormwater.

3) *Integration of natural waterbodies*: the concept emphasizes the integration of natural waterbodies, including wetlands and lakes, into the urban environment. This integration serves multiple objectives within drainage design, such as enhancing ecosystem services.

There are various models used to forecast the impact of green infrastructure on addressing the quality and quantity of stormwater; the RECARGA model assesses the capacity and effectiveness of bioretention facilities, rain gardens, and infiltration practices. The Program for Predicting Polluting Particle Passage Through Pits, Puddles, and Ponds (P8 Urban Catchment Model) aims to achieve total suspended solid (TSS) removal through assessing design requirements for green infrastructure practices and, finally, stormwater management model (SWMM), which is one of the most common models that analyzes the performance of diverse green infrastructure practices and conducts runoff simulations across multiple spatial scales [19]. GIS can also be used to help understand where there is a lack of green infrastructure and where most areas are suffering either from floods or are impacted by the urban heat island (UHI) effect. In the smart city concept. Various technologies and sensors can be implemented to predict and assess the occurrence of storms and the capacities required for different systems to handle flooding.

Green Infrastructure and Urban Greening Initiatives for Cool Cities

Green infrastructure and urban greening initiatives are excellent approaches to mitigate the effects of rising temperatures and the impact of the UHI effect. Along with reducing the risks of urban flooding, as mentioned earlier, enhancing microclimate conditions, improving thermal comfort, reducing energy consumption, and the conserving of natural habitats are all positive outcomes of green infrastructure. Green infrastructure can be implemented by establishing tree canopies and shade, gardens with native plants, bioswales, vegetation, permeable pavements, green roofs, urban agriculture, and many uncountable approaches.

Green infrastructure has the potential to mitigate climate change by lowering local temperatures and contributing to cooling the outdoor environment of cities. The implementation of diverse forms of green infrastructure directly influences temperature levels, thereby affecting the UHI effect. Green infrastructure in it's many types along with open green spaces allow a reduction in the concrete surface area, which extensively absorbs heat. Green infrastructure, on the other hand, absorbs less heat and enhances the overall air quality. Green infrastructure, specifically tall trees that form tree canyons allows for a much cooler microclimate and provides shaded areas for pedestrians, enhancing the overall environment. This also enhances the thermal comfort of individuals, providing a more comfortable outdoor experience. To enhance urban forestry and tree coverage in neighborhoods, the 3-30-300 rule comes into effect. The 3 represents the availability of a minimum of 3 trees that can be visible from every home, the 30 represents the minimum percentage of tree canopy cover each neighborhood must implement to receive various benefits, and the 300 is the minimum distance in meters to reach the nearest park or green space. This rule is excellent for benchmarking, convenient for progress monitoring, easily conveyable, and can generate vast interest and support among different city members [20]. The 3-30-300 rule not only mitigates the UHI effect but also increases physical activity by encouraging outdoor recreation. Green infrastructure can also enhance nature-friendly cities that protect natural habitats, support biodiversity, and enhance a sustainable ecosystem.

Green infrastructure can also be implemented to decrease energy consumption. Energy-efficient and low-carbon urban design are main elements in decreasing cities contribution to greenhouse gas emissions. Green infrastructure can be classified under energy efficiency for its role in cooling down temperatures, decreasing the reliance on cooling systems, and reducing carbon in the atmosphere as a result of carbon sinks. Keeping in mind that buildings also have great potential for energy conservation and emitting carbon emissions. The British Building Research Establishment Environmental Assessment Method (BREEAM) assessment objectives center around the reduction of CO_2 emissions attributed to energy usage in buildings, while the American Leadership in the Energy and Environmental Design (LEED) system primarily strives to decrease annual energy costs in buildings [21]. These two systems are widely used and embedded by planners in various developments. Residents within a city are also incentivized to implement renewable energy sources like solar panels and are offered benefits like tax credits or points. This strategy is implemented within policies to motivate people to switch to renewable energy.

There are different approaches that urban planners use to implement such initiatives. For instance, land use policies are always an effective approach that set guidelines and requirements to consider before beginning any development. Besides policies, it is of vital importance to include mitigation strategies like green infrastructure as a part of the comprehensive plans of cities and is an aspect connected to the city's overall vision. Having a master plan specific for green infrastructure allows for more specific and detailed analysis, understanding, designing, and implementation procedures that solely focus on green infrastructure and their connection to better achieving the climate-related guiding criteria adopted by a city.

Waste Management and Recycling Systems, Public Participation, and Education

Enhancing waste management and recycling systems within urban areas plays a critical role in combating climate change. Effective waste management allows for a healthier environment as it reduces water and soil contamination, waste disposal in natural habitats, and air pollution. Recycling reduces the burden of extracting raw, virgin materials and reduces resource depletion. Proper reuse allows for a decrease in demand, leading to less production and less emissions. The planning process allows the facilitation of various programs to ensure the quality of waste management systems. This includes providing services like bins and a consistent waste pick-up schedule. Waste must be categorized and separated according to how it will be handled after it has been dispensed, including compost and recycling facilities. Recycling is one way to enhance waste management by reducing waste generation through reducing, reusing, and recycling to encourage the city to shift toward a circular economy. Cities must also provide recycling infrastructure to achieve that shift. It would be argued that community participation and awareness programs rank among the highest requirements; the city can provide all the infrastructure and programs needed, but without the willingness and participation of the public, no change will occur. The willingness of people to recycle can be expressed in different ways and for different purposes. In the city of Johannesburg, South Africa, informal waste pickers participate in the waste collection, sorting, and recycling processes, becoming a part of the municipal waste system [22]. In the state of Illinois, United States, four communities were studied to examine recycling behaviors and what factors motivate them; the results included convenience, voluntary recycling initiatives driven by environmental consciousness, and the persistence in motivational approaches [23]. It is crucial to understand the needs and motives of different communities and what motivates them to participate in activities like recycling. It also outlines the role governments have on influencing climate change and guiding people's lifestyle patterns. A city can have an entire efficient waste management system, but if there are no incentives, the system is of no use.

Risk Assessment and Adaptation in Urban Planning

The importance of risk assessment and vulnerability analysis in urban planning heavily influences the outcomes of the climate change impacts on urban areas. Risk assessment allows planners to identify any potential hazards and extremes that are projected to harm urban areas, their inhabitants, infrastructure, systems, social and economic aspects, etc. When assessing the risks, planners need to identify which areas are at the highest risk and are vulnerable to climate change impacts. Identifying the underlying risks and vulnerabilities allows urban planners to get a clear understanding of the problem/s that the city is facing, to be able to determine which areas demand the utmost priority and resource allocation, and to be able to set clear, impactful solutions. Table 12.1 highlights some of the methods that can be utilized by urban planners for risk assessment and vulnerability analysis.

Table 12.1 Approaches used in risk assessment and vulnerability analysis that could be utilized by urban planners.

Method	Purpose	Case study
The strengths, weaknesses, opportunities, and threats (SWOT)	Assesses the SWOT of a city	SWOT was conducted in the city of Faro, Portugal. Examples of strengths include, but are not limited to, investment in green facilities and wetland reserves. Weaknesses include abandoned areas, degradation within historical city centers and a lack of free space. Opportunities include the urban rehabilitation policy on green walls and roofs and urban agriculture. Threats include flooding, water scarcity, droughts, heat waves, and coastal erosion [24]
Early warning system (EWS)	Detect an approaching occurrence of a hazard	Emphasizes on the importance of EWS in achieving disaster resilience in Sri Lanka. It highlights the need for training, integration of disaster risk reduction (DRR) into development, and greater engagement from the government and private sector in ecosystem-based DRR. The report also emphasizes the evaluation of risk transfer measures, such as disaster risk financing and insurance, to ensure their applicability and effectiveness in the local context [25]
Multi-criteria analysis (MCA)	Guides decision making	MCA method is used to quantitatively evaluate climate change mitigation policy instruments [26]

(Continued)

Table 12.1 (Continued)

Method	Purpose	Case study
Transportation demand management (TDM)	Evaluate the performances of alternatives	Various TDM strategies have been identified in the Kingdom of Saudi Arabia as potential strategies to decrease vehicular emissions and congestion while promoting sustainable transportation options, including an increase in fuel prices, strong emission standards, and air quality management, incorporating a public participatory approach [27]
Geographic information system (GIS)	Map generation and analysis	Conducting drought/flood risk assessment through GIS either through "temperature information form" or through "precipitation information form" [28]
Participatory risk assessment	Allow participation of various stakeholders in assessing risks	Participatory risk assessment combined technical expertise with stakeholder engagement to understand and manage risks in complex systems [29]. It was used to develop a national policy on systemic water risks in a district of Vietnam where the piloting of national agricultural water reforms was taking place. The Risks and Options Assessment for Decision-Making (ROAD) process was used for the understanding and management of risks [29]
Participatory scenario planning (PSP)	Allow participation of various stakeholders in scenario planning	Examine the use of PSP to support adaptation planning for recreation sites in glacial mountain environments, specifically in the chosen study area of southeast Iceland. This involved conducting local stakeholder workshops where participants created maps depicting possible glacial land cover and land use in the near future. The process comprised several stages, including identifying potential drivers of land use change, developing multiple land use scenarios, and assessing the potential impacts of these scenarios while exploring adaptation options [30]
Climate change scenario modeling	Developing and modeling different climate change-related scenarios and assessing their impacts	Forecast the changes in surface temperature, relative humidity, and bioclimatic comfort zones in Kocaeli Province, considering climate comfort, under the scenarios of SSP 245 and SSP 585 as outlined by the Intergovernmental Panel on Climate Change (IPCC). Scenarios were modeled for the years 2040, 2060, 2080, and 2100. It analyzes the temporal and spatial variations in these climatic parameters to assess the potential impacts on human comfort [31]
Cost–benefit analysis	Assess the financial feasibility	Assessing the costs and benefits of technologies and strategies that address both local air pollution and global climate change. Integrated environmental policies that address both issues can generate net global welfare benefits [32]
Livelihood vulnerability analysis	Economic vulnerability to climate change impacts	Examined the level of vulnerability in the livelihoods of farmers residing in five communities within Dak Nong Province, Vietnam, who face regular exposure to drought conditions [33]
Social vulnerability index (SVI)	Social vulnerability to climate change impacts	Investigate associations between heat-related emergency department (ED) visits and mortality rates at the county level in Georgia from 2002 to 2008 in relation to SVI developed by the Centers for Disease Control and Prevention (CDC). The main aim is to analyze the impact of social vulnerability on heat-related health outcomes in different counties of Georgia [34]

Identifying and prioritizing climate change risks in urban areas must not be neglected. In today's modern world, cities are faced with numerous challenges and threats. Some planners may not adequately consider the influence of climate change or give it the significance it requires because they fail to grasp the consequences it poses for urban areas. Identifying and understanding how climate change impacts urban areas is crucial to help planners understand what causes some of the challenges that cities struggle with. After identifying the risks and the areas at risk, planners need to come up with innovative and effective approaches that will tackle the challenges of climate change by reducing or eliminating the risks and threats. One of the most common approaches planners utilize is climate change action plans. Action plans help identify the goals and targets that need to be achieved to reduce climate change's influence and impacts. Action plans have a great influence on guiding policy decisions. Climate plans of 29 major US cities were evaluated, providing insights into urban climate policy trends. These plans primarily focus on promoting high-quality buildings, mass transit, nonmotorized transportation, and reducing dependence on cars. However, the absence of policies related to dense development and parking restrictions in these cities' climate action plans limits their ability to effectively reduce building energy consumption and shift transportation modes. Notably, cities with high emissions in energy-intensive climates, where greenhouse gas reductions are most needed, tend to have inadequate climate action plans. In summary, the lack of coherence in many US cities' climate action plans undermines their potential for comprehensive success [35]. Preparedness plans are also another form of document used by urban planners to prepare the city for any type of threat that it might face. Various plans may be used to address climate change; however, planners need a deep understanding and a comprehensive approach when preparing such plans.

Developing adaptation strategies for urban infrastructure and buildings helps cities adapt to any of the challenges and threats that they might face. Adaptation plans are commonly used by urban planners to allow cities to adapt to various climate change impacts, decreasing their vulnerability. An adaptation strategy for rising temperatures and the UHI effects includes city canyons and/or green spaces, they help the city adapt to the rising temperatures, allowing life in urban areas to proceed with as little disturbance as possible. Planners must be aware of specific implementation processes to ensure that cities gain the most benefits. Carefully selecting and organizing different types of green spaces can improve their capacity to regulate air ventilation, humidity, and evapotranspiration, ultimately helping to mitigate the UHI effect [36]. In cases of flooding, permeable pavements and sewer systems play a crucial role in fluctuating the risks imposed by the city. After a city acknowledges that it is subjected to heavy rain, it takes improvement measures for its different systems. Planners can examine different flood-capacity-based scenarios that assess the level of floods a city will encounter and consult with engineers about which sewer model needs to be used to enhance the adaptation of the city.

Integrating nature-based solutions into urban planning can be used to enhance resilience in urban areas. The green infrastructure of various kinds is one of the most common strategies urban planners integrate into their plans for its countless benefits to nature and humans. Other scopes that planners focus on include embedding various laws to protect the soil and water resources from contaminants, restoring the natural habitats that have already been harmed, preserving the prairies, wetlands, forests, native species, and so on.

Specific habitat requirements and habitat fragmentation limit urban areas from becoming a substitute for the natural ecosystems; however, different urban habitat types, spanning the entire range of the four nature categories (pristine, agricultural, horticultural, and urban-industrial), can make substantial contributions to biodiversity conservation [37].

As emphasized many times earlier in this chapter, the importance of policy and governance for climate-responsive urban planning is essential. Policies are one of the most important strategies in urban planning as they have an influence on shaping every aspect related to urban areas from the behaviors of people to the design and functioning of a city. The *Green New Deal* is a framework that has gained substantial attention due to its policies and initiatives aimed at addressing climate change-related challenges. To ensure that cities are comprehensively addressing climate change and that plans are being implemented, every city should establish a governmental entity that is responsible for addressing climate change mitigation, adaptation, and resiliency. The US Environmental Protection Agency (EPA) is an agency formed to have a broader scope of protecting the environment. These agencies work closely with climate change challenges. On a smaller scale, different cities and local entities assign specific organizations, councils, and centers to address emerging climate-related challenges specific to their region.

In a globalized world where climate change is not confined to a certain area and its impacts spread across borders, adapting national and international policy frameworks for climate-responsive urban planning is needed. This ensures equitable measures between regions as it limits having some cities or countries contribute to most of the climate change impacts while other cities need to bear the burden created by other cities. The United Nations plays a key role in directing and establishing these frameworks between different countries. The most common international framework established by the United Nations is the Sustainable Development Goals (SDGs) that set various sustainability goals that countries are encouraged to achieve. Scheduled conferences by the United Nations, such as the United Nations Climate Change Conferences under the Conference of the Parties (COP), occur to bring countries to the table to evaluate progress, and further discuss and negotiate agreements between the different countries. Furthermore, the Paris Agreement Treaty is an international treaty that sets a specific goal of reducing greenhouse gas emissions to mitigate the impacts of climate change. The Green New Deal, covers the economic aspects such as jobs to the climate change context, while the Intergovernmental Panel on Climate Change (IPCC) influences the policy-making decisions in countries.

Urban planning allows for multilevel governance and stakeholder engagement when addressing climate change. Addressing climate change at an international level is of great importance but a bottoms-up approach is equally important. Working at the local, regional, and national levels allows the weight to be distributed into small portions, with each region focusing specifically on its challenges and strengths that are contributing to the overall impact. Each level of governance is under the same vision, scope, and guidelines, but each develops its own goals based on what each region needs. Stakeholder engagement means involving everyone in the climate change discussion, not just limiting it to scientists, architects, engineers, urban planners, city designers, researchers, governments, and private corporations, but also bringing everyone else, including citizens and local business owners, to the table to come to an agreement that benefits all with the least negative impacts on the environment. Residents, students, community organizations, businesses, etc., all have an impact. In planning, planners try to solve a problem by opening multiple doors, so if the policy door was not effective, opening the door of public awareness might achieve the desired outcomes. Residents and organizations can change their behavior to positively impact the environment. An example can be using reusable bags instead of single-use bags voluntarily or having a business not provide such an option. If it is a larger-scale issue like the contamination of a local river, residents and organizations can use their voices to demand changes in policies and developments. Organizations and educational institutions can have a big impact on spreading awareness and influencing the behaviors of residents by providing them with eye-opening information about the direct impacts climate change has on their lives. If residents are not responding to awareness and incentives, policies can be implemented to affect the behaviors of different stakeholders. The best practices for the framing of climate change within policies from a psychological perspective to enhance public engagement include [38]:

1) Highlight the urgency of climate change as a risk that is happening now, directly impacting local communities and individuals.
2) Foster emotional and experiential connections with climate change to enhance engagement and understanding.
3) Utilize social group norms that are relevant to specific communities to encourage climate action.
4) Present policy solutions in a way that emphasizes the immediate benefits and positive outcomes that can be achieved through proactive measures.
5) Emphasize the long-term environmental goals and outcomes that are inherently valuable to individuals and society as a whole [38].

It is crucial for stakeholders to understand the impact and power that they have but it is also essential for them to understand the duties that they have toward climate change.

Financial mechanisms for incentives regarding climate-resilient urban development are another door for planners to open to influence climate change impacts. In capitalist-driven societies, it is more likely to find individuals who are willing to change their behaviors if they are receiving benefits or rewards. Policies take advantage of such behavior to address climate-related challenges. Grants, subsidies, and loans may also be offered to individuals or businesses that are willing to take on an initiative to combat climate change. Tax incentives may also apply, an example includes receiving tax credits when implementing solar panels on houses. Other tax incentives include tax reductions. Usually, when incentives do not play an effective role, planners shift to recommending punishments and penalties such as fees, fines, suspending licenses, and more. For corporations that emit greenhouse gases above the national standards a carbon tax could be used. Germany has adopted various policies and policy types covering topics such as energy, transportation, and buildings. The policies range from tax, subsidies, information, regulatory, voluntary, and auction. Some initiatives include Eco-tax, Federal Energy Research Programme, Biofuel Tax Breaks and Mixing Obligation, and Energy Passport. [39].

Case Studies of Successful Climate-Responsive Urban Planning

This section highlights different cities that have implemented effective urban planning strategies to mitigate climate change impacts.

Case Study 1: Freiburg, Germany

Freiburg's success lies in history dating back to the 1970s of citizen action focused on rejecting nuclear power and promoting energy-efficient sources [40]. Such a campaign shifted the trajectory of the city and successfully navigated through the complex challenges to effectively incorporate and integrate adaptation solutions for climate change, facilitating a transformative process. This has led to significant changes, including [40]:

1) The accessibility of public transportation, which has been managed by:
   a) The techno-ecological solution, which included tram corridors that were designed with environmentally conscious features, such as grassed and pervious surfaces, to support WSUD strategies and enhance stormwater drainage efficiency.
   b) Subsidized public transport costs to reduce reliance on public vehicles.
   c) Mixed-use zoning to ensure the needs of individuals are met and services are accessible.
2) Prioritized the design and maintenance of extensive green spaces through the nature-based solution of active management of 600 hectares of parks and 5,000 hectares of forest, as nature-based solutions for:
   a) Climate protection.
   b) Human well-being.
   c) Reduced air temperatures during heat waves by 2–3 °C.
3) Reduce carbon emissions and increase energy savings by:
   a) A support program was established as a socio-technical solution to retrofit buildings, e.g., installing insulation, resulting in an impressive reduction of up to 38% in energy consumption per building across schools and offices in Freiburg.
   b) The adoption of renewable energy was encouraged through tax credits and subsidies provided by the federal government and regional utilities.

This successful case study portrays the power of grassroots movements and the influence people have on climate change. This case also outlines how successful climate-responsive urban planning takes time and dedication and should hold on to its values. Perseverance and a commitment to constant improvement are also visible in this case study, as improvement projects were constantly made. The community was not satisfied with just one solution but rather with various solutions within various aspects.

Case Study 2: Beddington, London, United Kingdom

Beddington Zero Energy Development (BedZED) in London is the UK's sole zero-carbon neighborhood, comprising 82 houses, 17 apartments, and around 1500 m^2 of workspace, designed to exclusively rely on on-site generated renewable energy sources [41]. The components of this neighborhood include [41]:

1) Design Elements:
   a) Compact urban development built on a brownfield site of former sewage works.

b) It promotes walkability with street designs that prioritize pedestrians and cyclists.
   c) Innovative features include a wood chip-based combined heat and power system, a "living machine" sewage system, shared electric cars, and photovoltaic panels on flat conservatories. It also includes individual small gardens and highly insulated three-story walk-up townhouses.
2) Energy Savings:
   a) Achieves major energy savings through insulation, airtightness, passive solar heating design, and solar photovoltaic arrays integrated into the buildings.
   b) The design of BedZED prioritized energy efficiency and renewable sources on-site, maximizing solar gain in south-facing blocks and utilizing occupants' equipment heat for heating. Other sustainable features include an on-site car club, ample secure cycle parking, excellent public transport links, sunspaces for solar gain in homes, and natural ventilation instead of electric fans.
   c) Monitoring showed significant energy and water use reductions: 88% for space heating, 57% for hot water, 25% for electricity (with 11% from solar panels), 50% for water consumption, and 65% for residents' car mileage.
3) Wind Cowls:
   a) In BedZED, a highly effective ventilation system was implemented, utilizing rooftop wind cowls to bring in fresh air and expel stale air. These brightly colored cowls, resembling traditional chimneys, have rotating vanes and collect fresh air at the front while expelling stale air at the back, warming the incoming air through a heat exchanger.
   b) The ventilation system in BedZED eliminates the need for electrically powered fans and promotes natural airflow, resulting in fresh and well-ventilated homes and workplaces with reduced energy consumption.
4) Heat Demand:
   a) While BedZED benefits from the warmth of its occupants and solar gain, additional heat is still required for some of its homes and workplaces, especially those on the northern sides without sunspace.
   b) To meet this demand, BedZED employs its own small-scale district heating system, circulating hot water through an underground network of insulated pipes. Each home has a large hot water tank that provides year-round hot water and supplemental warmth in winter, while conventional radiators heat the spaces on the northern sides using the district heating system.
5) Travel and Transport:
   a) Promotes sustainable transportation and significantly reduces greenhouse gas emissions from travel.
   b) Offers convenient access to public transport within a 10-minute walk, encourages bicycle use through incentives, and incorporates reduced car parking and road space.
   c) Partnered with City Car Club to implement a car-sharing scheme that includes electric vehicle.
   d) Monitored data from the first eight years of occupancy show a 60% reduction in private car usage compared to the local average.
6) Water Management:
   a) Water-efficient in response to the region's water stress, incorporating recycled water usage through a "Living Machine" system; however, the system is currently not operational.

b) The construction materials were sourced locally within 35 mi., reducing environmental impacts.
c) Most homes in BedZED have outdoor spaces like balconies and gardens, with some connected by bridges spanning the mews streets.

7) Houses:
   a) Offers a variety of housing options, including different sizes and market values, as well as shared home ownership and subsidized renting, resulting in a diverse social mix.

8) Lifestyle:
   a) BedZED authorities actively engage residents and the wider community through organized activities such as guided trips, exhibitions, seminars, and educational programs. These initiatives provide opportunities for people to learn about sustainable development.

BedZED is an excellent example of a holistic approach to developing a neighborhood in the face of climate change, as it addresses several challenges and provides various innovative solutions and initiatives. BedZED also contributes to raising awareness, promoting educatioal efforts, and fostering a sense of participation and collaboration among the residents and visitors.

Challenges and Opportunities

Various challenges and barriers are faced in the implementation of climate-responsive urban planning. The main issue that is faced is "placing the plans on the shelves" meaning that after plans have been crafted, outlining different approaches, strategies, goals, timeframes, etc., the municipality, city, or concerned entity does not implement the plans and the recommendations are left on paper instead of being transformed into reality. Implementation barriers include different reasons. It can simply be the irresponsibility of the person/entity assigned to implement the plan. The second and one of the most common barriers is financial feasibility; plans may provide unaffordable recommendations that are not within budget limits or do not specify any allocated resources for implementation. This is why some plans dedicate a portion to incorporating financial feasibility by providing different financial approaches that help manage various financial resources for implementation. Another important reason is political feasibility. Political feasibility has a wide range of aspects to consider, including current laws, regulations, and policies, and their formulation and implementation. Political feasibility is also closely tied to stakeholders, public support, and equity. Public support depends on many factors; one reason why the public might oppose is if it poses a threat to maintaining their jobs like shutting down a factory, for instance. People may also oppose it if it poses inequitable consequences such as gentrification; many developments, especially those associated with green infrastructure and accessibility improvements, become appealing, attracting more people, and this causes a rise in property values, leading to the displacement of many.

Another challenge faced is the deliverables of the plan after or during implementation. Expected or predicted results may not be reflected in real life as they were addressed on paper. There are several reasons, and the poor liaison among the relevant stakeholders is one of the major reasons. Another reason is "copying" and "pasting" different techniques, strategies, or best practices that have been sufficient in different cities without assessing if they fit the city being planned for. Each city has its own different unique characteristics and features, which require special innovations tailored just for it. Scalability is another example of how using a concept that was proposed for a neighborhood level and implementing it on a larger scale without carefully assessing it can differ the results. Using a concept that does not holistically fit within its surrounding environment, like implementing wind turbines in an area that barely has any wind or focusing on establishing a sponge city in an area that has limited precipitation instead of focusing on green infrastructure to combat the UHI. Other reasons include a lack of technical resources and human expertise. It is crucial to keep such challenges in mind when planning to overcome any shortcomings as climate change does not compensate for lost time.

With challenges, various opportunities come that can be carefully managed to enhance the role and effectiveness of urban planning. Urban planners can bring scientists, researchers, and engineers together with government officials, community members, business developers, and corporate owners to solve the challenges from different perspectives. Urban planners can also work on translating and finding common ground between all concerned parties like translating a complex proposed urban drainage system from an engineering perspective to a simplified and understandable version for community members. This allows the overcoming of many barriers and helps address challenges early on in the process. Planning is important in spreading awareness and shedding light on important topics related to climate change, broadening the scope of individuals and allowing them to see through different lenses. Community awareness programs are a great example of how to communicate to the public about climate change impacts affecting them and their city. This can also be done through various media forms. New emerging technologies are an excellent opportunity to optimize. Technologies such as digital twin cities, where different techniques, concepts, and strategies can be tested on the city through a digital model to assess their impact and effectiveness without implementing them in real life. This saves a lot of time, resources, and money and can even showcase how cities will be impacted in the future. One example is the digital urban climate twin (DUCT) in Singapore, which aims to create a decision support system that assists in prioritizing and selecting strategies across various environmental, economic, and social aspects [40].

Major Key Points

- Urban planning has a major role in mitigating, adapting, and enhancing resiliency to climate change.
- With climate change, cities are becoming more vulnerable.
- Urban planning connects various stakeholders, allowing them to holistically influence climate change.
- Urban planners identify areas at risk and their vulnerability.
- Urban planners use different strategies, methods, and concepts to address climate change-related challenges.

Conclusion

Urban planning plays a huge role in mitigating, adapting, and creating resilient cities that can withstand climate change. The knowledge urban planners have about the vulnerability of cities to extreme climate events allows them to put climate change as a top priority. Planners can use various tools, strategies, and skills to deal with climate change. Planners are also able to assess different urban areas with different scales and characteristics and suggest various concepts and theories that allow mitigation and adaptation to climate change. Urban planners assess which areas are at the most risk and what imposes such risks to be able to identify priorities, goals, and timeframes. Urban planning creates strong, durable built environments that allow cities to be more resilient. Overall, efficient urban planning can shift the trajectory of cities and have a positive impact on reducing climate change impacts.

References

1 Wilbanks, T.J. and Fernandez, S.J. (2014). *Climate Change and Infrastructure, Urban Systems, and Vulnerabilities* (ed. T.J. Wilbanks and S. Fernandez). Washington, DC: Island Press/Center for Resource Economics. http://link.springer.com/10.5822/978-1-61091-556-4.
2 Schär, C. (2016). *Climate Extremes: The Worst Heat Waves to Come*, vol. 6, 128–129. Nature Climate Change. Nature Publishing Group.
3 Fuchs, R., Conran, M., and Louis, E. (2011). Climate change and Asia's coastal urban cities: can they meet the challenge? *Environ. Urban. Asia.* 2 (1): 13–28.
4 Huong, H.T.L. and Pathirana, A. (2013). Urbanization and climate change impacts on future urban flooding in Can Tho city, Vietnam. *Hydrol. Earth Syst. Sci.* 17 (1): 379–394.
5 Anthoff, D., Nicholls, R.J., and Tol, R.S.J. (2010). The economic impact of substantial sea-level rise. *Mitig. Adapt. Strateg. Glob. Chang.* 15 (4): 321–335.
6 Pérez-Arévalo, R., Serrano-Montes, J.L., Jiménez-Caldera, J.E. et al. (2023). Facing climate change and improving emergency responses in Southern America by analysing urban cyclonic wind events. *Urban Clim.* 49: 101489.
7 Loo, B.P.Y. and du Verle, F. (2017). Transit-oriented development in future cities: towards a two-level sustainable mobility strategy. *Int. J. Urban Sci.* 21 (Sup 1): 54–67.
8 Nakamura, K. and Hayashi, Y. (2013). Strategies and instruments for low-carbon urban transport: an international review on trends and effects. *Transp. Policy (Oxf)* 29: 264–274. https://linkinghub.elsevier.com/retrieve/pii/S0967070X12001187.
9 Moreno, C., Allam, Z., Chabaud, D. et al. (2021). Introducing the "15-Minute City": sustainability, resilience and place identity in future post-pandemic cities. *Smart Cities.* 4 (1): 93–111.
10 Abdelfattah, L., Deponte, D., and Fossa, G. (2022). The 15-minute city: interpreting the model to bring out urban resiliencies. *Transp. Res. Proc.* 60: 330–337.
11 Stone, B., Hess, J.J., and Frumkin, H. (2010). Urban form and extreme heat events: are sprawling cities more vulnerable to climate change than compact cities? *Environ. Health Perspect.* 118 (10): 1425–1428.

12 Lin, Y.P., Hong, N.M., Wu, P.J., and Lin, C.J. (2007). Modeling and assessing land-use and hydrological processes to future land-use and climate change scenarios in watershed land-use planning. *Environ. Geol.* 53 (3): 623–634.

13 Martinuzzi, S., Gould, W.A., and Ramos González, O.M. (2007). Land development, land use, and urban sprawl in Puerto Rico integrating remote sensing and population census data. *Landsc. Urban Plan.* 79 (3, 4): 288–297.

14 Eckart, K., McPhee, Z., and Bolisetti, T. (2017). Performance and implementation of low impact development – a review. *Sci. Total Environ.* 607, 608: 413–432.

15 Dietz, M.E. (2007). Low impact development practices: a review of current research and recommendations for future directions. *Water Air Soil Pollut.* 186 (1–4): 351–363. http://link.springer.com/10.1007/s11270-007-9484-z.

16 Song, C. (2022). Application of nature-based measures in China's sponge city initiative: current trends and perspectives. *Natu. Based Solut.* 2: 100010.

17 Nguyen, T.T., Ngo, H.H., Guo, W. et al. (2019). Implementation of a specific urban water management – Sponge City. *Sci. Total Environ.* 652: 147–162.

18 Chan, F.K.S., Griffiths, J.A., Higgitt, D. et al. (2018). "Sponge City" in China – a breakthrough of planning and flood risk management in the urban context. *Land Use Policy.* 76: 772–778.

19 Jayasooriya, V.M. and Ng, A.W.M. (2014). Tools for modeling of stormwater management and economics of green infrastructure practices: a review. *Water Air Soil Pollut.* 225 (8): 2055.

20 Konijnendijk, C. (2021 [cited 2023 Jun 13]). The 3-30-300 rule for urban forestry and greener cities. *Biophilic Cities J.* 4 (2): https://www.biophiliccities.org/s/330300-Rule-Preprint_7-29-21.pdf.

21 Schwartz, Y. and Raslan, R. (2013). Variations in results of building energy simulation tools, and their impact on BREEAM and LEED ratings: a case study. *Energy Build.* 62: 350–359.

22 Simatele, D.M., Dlamini, S., and Kubanza, N.S. (2017). From informality to formality: perspectives on the challenges of integrating solid waste management into the urban development and planning policy in Johannesburg, South Africa. *Habitat. Int.* 63: 122–130.

23 Vining, J., Linn, N., and Burdge, R.J. (1992). Why recycle? A comparison of recycling motivations in four communities. *Environ. Manage.* 16 (6): 785–797.

24 Berte, E. and Panagopoulos, T. (2014). Enhancing city resilience to climate change by means of ecosystem services improvement: a SWOT analysis for the city of Faro, Portugal. *Int. J. Urban Sustain. Dev.* 6 (2): 241–253.

25 Hettiarachchi, S.S.L. and Weeresinghe, S. (2014). Achieving disaster resilience through the Sri Lankan early warning system: good practises of disaster risk reduction and management. *Procedia Econ. Financ.* 18: 789–794.

26 Konidari, P. and Mavrakis, D. (2007). A multi-criteria evaluation method for climate change mitigation policy instruments. *Energy Policy.* 35 (12): 6235–6257.

27 Masiur Rahman, S. and Al-Ahmadi, H.M. (2010). Evaluation of transportation demand management (TDM) strategies and its prospect in Saudi Arabia. *Jordan J. Civ. Eng.* 4: 170–182.

28 Belal, A.A., El-Ramady, H.R., Mohamed, E.S., and Saleh, A.M. (2014). Drought risk assessment using remote sensing and GIS techniques. *Arab. J. Geosci.* 7 (1): 35–53.

29 Wyrwoll, P.R., Grafton, R.Q., Daniell, K.A. et al. (2018). Decision-making for systemic water risks: insights from a participatory risk assessment process in Vietnam. *Earths Future.* 6 (3): 543–564.

30 Welling, J., Ólafsdóttir, R., Árnason, Þ., and Guðmundsson, S. (2019). Participatory planning under scenarios of glacier retreat and tourism growth in Southeast Iceland. *Mt. Res. Dev.* 39 (2): 1–13.

31 Isinkaralar, O. (2023). Bioclimatic comfort in urban planning and modeling spatial change during 2020–2100 according to climate change scenarios in Kocaeli, Türkiye. *Int. J. Environ. Sci. Technol.* 20 (7): 7775–7786.

32 Bollen, J., van der Zwaan, B., Brink, C., and Eerens, H. (2009). Local air pollution and global climate change: a combined cost-benefit analysis. *Resour. Energy Econ.* 31 (3): 161–181.

33 Thao, N.T.T., Khoi, D.N., Xuan, T.T., and Tychon, B. (2019). Assessment of livelihood vulnerability to drought: a case study in Dak Nong Province, Vietnam. *Int. J. Disaster Risk Sci.* 10 (4): 604–615.

34 Lehnert, E.A., Wilt, G., Flanagan, B., and Hallisey, E. (2020). Spatial exploration of the CDC's social vulnerability index and heat-related health outcomes in Georgia. *Int. J. Disaster Risk Reduct.* 46: 101517.

35 Deetjen, T.A., Conger, J.P., Leibowicz, B.D., and Webber, M.E. (2018). Review of climate action plans in 29 major U.S. cities: comparing current policies to research recommendations. *Sustain. Cities Soc.* 41: 711–727.

36 Lin, J., Qiu, S., Tan, X., and Zhuang, Y. (2023). Measuring the relationship between morphological spatial pattern of green space and urban heat island using machine learning methods. *Build. Environ.* 228: 109910.

37 Kowarik, I. (2011). Novel urban ecosystems, biodiversity, and conservation. *Environ. Pollut.* 159 (8, 9): 1974–1983.

38 van der Linden, S., Maibach, E., and Leiserowitz, A. (2015). Improving public engagement with climate change. *Perspect. Psychol. Sci.* 10 (6): 758–763.

39 Weidner, H. and Mez, L. (2008). German climate change policy: a success story with some flaws. *J. Environ. Dev.* 17 (4): 356–378.

40 Lin, B.B., Ossola, A., Alberti, M. et al. (2021). Integrating solutions to adapt cities for climate change. *Lancet Planet Health.* 5 (7): e479–e486.

41 Oktay, D. (2022). Promoting energy-efficient neighbourhoods: learning from BedZED. 841–847. https://link.springer.com/10.1007/978-3-030-76221-6_93.

13

Integrating Green–Blue–Gray Infrastructure for Sustainable Urban Flood Risk Management: Enhancing Resilience and Advantages

Mohammed M. Al-Humaiqani[1] and Sami G. Al-Ghamdi[1,2,3]

[1] Division of Sustainable Development, College of Science and Engineering, Hamad Bin Khalifa University, Qatar Foundation, Doha, Qatar
[2] Environmental Science and Engineering Program, Biological and Environmental Science and Engineering Division, King Abdullah University of Science and Technology (KAUST), Thuwal, Saudi Arabia
[3] KAUST Climate and Livability Initiative, King Abdullah University of Science and Technology (KAUST), Thuwal, Saudi Arabia

Introduction

Nowadays, cities and urban areas are expanding in rapidly to the rapid growth in the population [1]. They are made of complicated structures, essential infrastructure networks, and many complex systems with different functions. Consequently, the sustainability and safety of the different systems in the cities and urban areas need to be enhanced to withstand the impacts imposed by climate change, such as flood risks, pollution, and heat stress. [2]. Today, climate change impacts should be seriously considered in any decision-making related to the development of cities and urban areas, mainly those located in areas suspected to direct exposure to environmental threats and risks. The acceleration in public and economic growth in many countries put massive stress on the governments to accelerate in developing the plans toward delivering the infrastructure systems in a sustainable way that not only meets the traditional standards but also adheres to the resiliency requirements and ensures proper functioning under different environmental conditions [3, 4]. In particular, infrastructure systems are usually built to withstand frequent severe weather events [2, 5]; hence, they must be designed to ensure high performance and deliver high resilience capacity that copes with the requirements related to climate change [6–8].

The increasing resilience of cities to cope with potential environmental hazards is crucial for urban society [9]. The exposure of these systems to direct or indirect environmental hazards like floods, one of the climate change aspects, results in many environmental, social, and economic consequences [10–14]. Many countries have suffered from flood impacts; for example, in the United States, Fort Lauderdale in the state of Florida, and Houston in Texas cities have suffered from flood disruptions in critical infrastructure (CI) and sinks at a rate of 5 cm per year [9, 15]. Other cities worldwide have also suffered from

---

Sustainable Cities in a Changing Climate: Enhancing Urban Resilience. First Edition.
Edited by Sami G. Al-Ghamdi.
© 2024 John Wiley & Sons Ltd. Published 2024 by John Wiley & Sons Ltd.

floods, such as Jakarta, Lagos, Houston, Dhaka, Venice, and many more. The World Economic Forum reported that eleven cities worldwide are in danger of disappearing, and many of these cities could be underwater by 2050 [15]. In 2007, half of the city of Jakarta was submerged under nearly four meters of water in height, causing damages at $550 and displacing 500 thousand people [16]. Some of these cities have set flood risk management plans towards moving some miles from the current locations to the safest places to protect people from future extreme flood risks. Climate change forced other countries to reconsider their defense plans against floods [16]. Climate change consequences continue to increase as long as the temperature of the planet causes global warming is still happening [17].

Green–blue–gray infrastructure (GBGI) is a term used mostly in association with the adaptation to climate change. As per Fritz [18], each type of these infrastructures could be designed or driven in a way that results in saving the environment by playing its significant role properly; however, the integration of the three infrastructures would result in a better model contributing to the integrated solutions that eventually help in adaptation to mitigating the relevant climate change impacts. Applying green and blue infrastructure (GBI) can increase resilience and reduce vulnerability to multiple threats through its multifunctional approach [19].

Green Infrastructure (GI)

Green infrastructure (GI) replaces gray infrastructure (GRAI), especially in highly sustainable developments. In the United States, GI is a prevalent embodiment often used in the form of street trees, parks, and bioswales to serve stormwater management. As per John Parker [20], GI is preferred to be used either as a replacement or complement to GRAIs such as water-related networks and pumping systems, as well as the traditional engineered structure and systems. According to a study by Marissa, M.A. [21], GI has demonstrated its cost-effectiveness compared to gray ones and provided additional co-benefits to its designed primary function by mitigating the heat island effects and improving the air quality in urban areas. The study notes that nature is still not fully integrated as an infrastructure asset. Besides, nature can serve as an infrastructure or a legally mandated service provider; it is important to bring political shifts for implementation.

As a set of strategies in built environments, GI is gaining a reputation as a core sustainable urban system serving various ecosystem needs. It is found that GI effectively controls temperature, reduces energy consumption, saves public resources, improves air quality and urban systems, manages floods, and increases drought resilience [22–28]. Several studies have been conducted on the design of GI to explore its multiple environmental benefits [28, 29]. Although these studies confirm the usefulness of GI in many environmental aspects, such as mitigating floods and reducing the heat island effect, other studies state GI has not yet accomplished widespread uptake as a core flood management strategy [30–32]. As such, there have been numerous recommendations and suggestions for further research on the role of GI in better flood management [32–34]. Moreover, it has been found that most studies do not apply any theoretical frameworks [28, 35].

Gray Infrastructure (GRAI)

Gray infrastructures (GRAIs) are traditional infrastructures that have become essential in urban and rural communities [22]. They include pipelines, roads, railways, bridges, and other essential networks such as water and wastewater, telecommunications, oil, and gas [22, 28, 36]. GRAIs play a significant role in urbanizing areas and advancing the civilization of cities and, therefore, countries. The fast-paced development of land destroys the natural cover. It converts surfaces to impervious, eventually preventing rainwater from infiltrating the ground, leading to flood risks, especially in cities with less vegetation [22, 28, 36, 37]. The acceleration in bringing such gray systems into existence as a replacement for the natural systems results in many environmental impacts, including the degradation of natural land cover, such as flora, which is considered one of the most important natural systems. It also exposes vulnerabilities and risks to infrastructure systems in different ways due to climate change impacts, such as intense rainfall that we did not experience before, resulting in high flooding that increases city problems [5, 28, 38, 39]. GRAIs have demonstrated acceptable performance in reducing the risk of flooding; however, GIs demonstrate additional benefits that cannot be offered by GRAIs [40].

Green–Blue–Gray Infrastructure Combination

GBGIs in the urban area are three alternatives that complement each other and offer many benefits. The benefits include flood risk reduction, energy and water savings, carbon sequestration maximization, and air quality improvement, resulting in the best adaptation strategy [40]. So far, the GRAI is still the dominant system used to discharge stormwater in many cities. Thus, using GI as an option for climate adaptation is still ambiguous. This is because of the deficiency in evaluating the required technology needed to help integrate such infrastructures for sustainable behavior [41]. This section is divided into subsections, in which each subsection explains the benefits and co-benefits of GBGI in urban areas from different perspectives. Overall, the section describes the environmental impacts of green climate change adaptation in terms of flood management in areas where high-intensity rainfalls occur. The section also elaborates on the regional progress in the GBGI nexus research.

Benefits of Combining Green–Blue–Gray Infrastructure (GBGI) Systems

GBGIs are the three main infrastructure types that could exist simultaneously in a single urban development zone as a combined or hybrid system. The hybrid system has demonstrable benefits and co-benefits translated in terms of its effectiveness in reacting and adapting to climate change impacts [21, 27, 40]. The benefits of combining two infrastructure systems were found to be much higher compared to the benefits of a single infrastructure system. For example, it has been demonstrated that GBI has effectively reduced flood hazards [4, 21, 22, 27, 36, 40, 42–45]. However, the viability of GBI for flood mitigation can be improved if co-benefits are considered.

Complementary GRAI with GI results in highly effective stormwater management. The life cycle cost of the integrated system (i.e., during design, construction, operation, and maintenance stages) was saved by more than 90% compared to the traditional GRAI as well as resulted in a saving of approximately 13% of commercial loan interest for homebuyers in China [36]. Nevertheless, the GRAI is still a measure of disaster risk reduction due to its design against natural hazards, including floods. Furthermore, using GI as a replacement for GRAI is an argument still questioning the desirable circumstances to apply a combination of both infrastructures [21, 23, 46–48]. However, considering the optimization of coupled GI and GRAI systems under multiple criteria is recommended, especially for environmental, economic, and safety benefits.

The GI performance is found to be satisfactory in mitigating flood hazards. Small catchments can be implemented in urban areas to collect stormwater. It was found that applying GI Little Stringybark Creek (LSC) catchment reduced the downstream flooded area by 29% and 91% when GI is implemented at full capacity [47]. Besides, the full implementation of GI may lower the intensity of the flow by approximately 83%. The GI is preferable in some areas when the population is relatively small [46]. In contrast, the GRAI option appears less economically viable if co-benefits are considered [40]. GI has been applied as a sustainable way to protect the environment, reduce water footprints, and save potable water in several cities worldwide, such as Philadelphia, Singapore, Berlin, Melbourne, and Sino-Singapore Tianjin Eco-city [42].

Green–Blue–Gray Infrastructure (GBGI) for Flood Risk Management

Flooding is an increasing issue, particularly frequent floods due to heavy rains that directly hit buildings and civil infrastructure, causing serious damages. The fast-paced development of lands and rapid urbanization resulted in converting nature to impervious surfaces in the form of hardscapes, which require strong systems to manage stormwater runoff and discharge safely [37]. Although the initial methods applied to manage stormwater runoff are adequate for the design purpose, the increase in rainfall intensity due to climate change makes rainwater runoff management an urgent challenge [28, 38, 49]. GBI offers various complementary benefits regarding flood management to GIs [50], whereas the gray solutions approach, such as pipes, provides low sustainability. However, the traditional and sustainable flood management strategies and the benefits offered by GBI are not properly assessed from a holistic perspective [40]. Figure 13.1 shows a conceptual framework linking the domains of GBGI for flood management along with the characteristics and benefits that can influence the built environment system's performance.

Environmental Impacts of Floods and Green Climate Change Adaptation

The environment is gradually exposed to several fundamental threats and serious impacts due to climate change. The threads include an increased risk of floods due to heavy precipitation and sea level rise [51]. Therefore, the urban environment is concerned with water submergence, disruptions in CIs, and network failure [52]. Flooding is classified as one of the most common natural hazards and the third most damaging worldwide [53]. It causes damages to traditional GRAI systems and may impose huge disruption to the new systems,

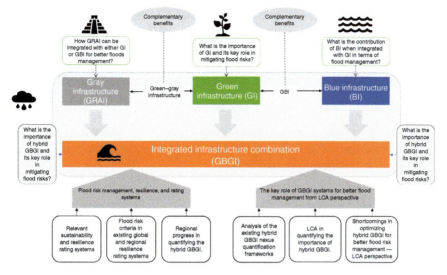

Figure 13.1 Conceptual framework linking the domains of GBGI for flood management along with the characteristics and benefits that can influence the built environment system's performance.

eventually resulting in economic losses [54–56]. Sustainable design for GRAI becomes a necessity to withstand such hazards as well as to show some resilience in combating such kinds of threads [57]. In this regard, interdisciplinary research has been growing to address the problem by proposing solutions and alternatives that could help integrate different options and combine multiple systems to combat possible natural hazards and threads [52, 53, 58–68].

GRAI is the main system for discharging rainwater after it is collected and directed to the drainage points. GRAI can be optimized during the design and construction stages for better management of the food risks; however, GRAIs alone are not necessarily to be always the only system responsible for managing rainwater, especially when rainfall intensity is higher than the designed value. Consequently, there is a need to change or update the strategies to manage flood risks better [57]. Among the needed updates and upgrades could be the greening of gray systems [48] and orienting the other systems in the urban area to play a role in absorbing the risk that emerges due to the excess rainwater runoff [20, 28, 42].

Regional Progress in GBGI Nexus Research

Today, hybrid green–blue–gray infrastructures (HGBGIs) are the most promising system in urban areas as these three infrastructures complement each other. The HGBGI combines the adaptability, sustainability, and multifunctionality of GBI and the reliability and resilience of the GRAI [69]. The hybrid system performs better in combating flood risks and shows a better adaptation strategy than other systems [4, 40]. In contrast, GBI is more effective in cost reduction [69, 70]; however, HGBGI economically competes with traditional GRAI systems [69]. However, the optimal combination of GI, traditional GRAI, and blue infrastructure (BI) that mimics natural solutions is still under debate by planners.

The HGBGI benefits and limitations of implementing and managing HGBGI solutions in an informal urban setting have been reported through a study of ten completed public space projects [71].

Recently, sponge city design has been introduced as a new way of managing urban water [72]. The idea of the sponge city is to develop an HGBGI system. The GI takes place at the source of the stormwater, which is connected to the GRAI that is provided midway to receive the water from GI and transfer it to the BI as a receiving waterbody. Figure 13.2 shows the suitable methodological framework for analyzing the combination of GBGI based on existing information. On the other hand, Table 13.1 outlines the methods and demographic details of the most recent studies.

Flood Risk Management Resilience

As a result of anthropogenic climate change, flood risks are expected to increase, causing damage to built environment systems and losses of lives and economy [53]. Over the past two decades, it has been reported that floods have been responsible for about 11% and 23% losses of life and economy, respectively [82]. The flood, heavy rainfalls, rising sea level rise, and increasing witness of the catchments. Table 13.2 outlines the research development, methods, and demographic details of the most recent studies on flood resilience.

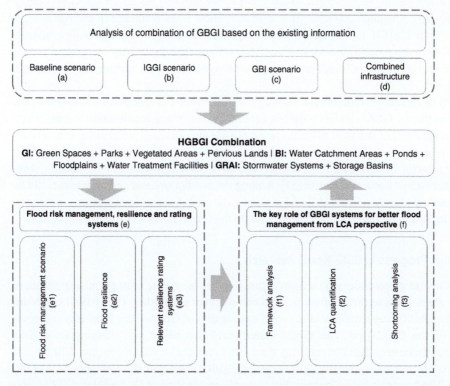

Figure 13.2 The suitable methodological framework for analyzing the combination of GBGI based on existing information.

Table 13.1 Methods, demographic details, methodologies, and findings of the most recent studies.

Region	Nexus type	Level of study (regional/country/city/system)	Scope	Methodology	Findings	Reference
Europe and Central Asia	GI	Country level (United Kingdom)	Presenting a new framework for the delivery of high-quality green infrastructure.	• Review of current guidelines and standards, academic literature, and national and local policies related to green infrastructure from across the United Kingdom	The resulting framework is presented as 23 principles for delivering green infrastructure. The framework is organized across four areas: • Core principles, • Principles related to health and well-being, • Principles related to sustainable water management, and • Principles related to nature conservation outcomes. The framework that defines the characteristics of high-quality green infrastructure at each stage of the planning and development process is set out as a series of 23 principles	[73]
	GI	Waterway Corridors (WWC) in Manchester	To explore green infrastructure enhancement possibilities along waterways and derelict vacant plots	• Study area defined. • Existing waterway, green infrastructure network, and hub identification. • Training and test site selection. • Potential GI intervention sites identification/prediction sites. • Selecting input variables, data extraction, and normalization. • Modeling the likelihood of sites transformed to green or GRAI.	This research found that multi-criteria-based machine-learning models such as ANN and ANFIS provided superior accuracy in understanding green and gray development spatial patterns. The results also demonstrated that future vacant or unused intervention sites along waterway corridors in Manchester have a higher likelihood of transforming into green infrastructure, whereas derelict plots are more likely to become gray	[74]

(Continued)

Table 13.1 (Continued)

Region	Nexus type	Level of study (regional/country/city/system)	Scope	Methodology	Findings	Reference
	GGI	City Level (Delft – Netherland)	The work presents a novel methodology to select, evaluate, and place different green-gray practices (or measures) for retrofitting urban drainage systems	• Hydrodynamic model and multi-objective optimization to design solutions at a watershed level	The application of multi-objective optimization processes for drainage configuration design may become a very promising tool, allowing the reduction of investments without compromising the efficiency of systems	[75]
	GGI	City Level (Delft – Netherland)	The aim of this method is to assist decision-makers in selecting and planning measures, which afterward can be part of either high-level scoping analysis or more complex studies, such as model-based assessment	• Multi-criteria Decision Analysis (MCDA) methods allow to structure of complex problems and help a better understanding of the trade-offs implied	A novel method for general assessment and selection of green and gray measures to reduce flood risk is presented in this work. This new method is based on a multi-criteria assessment of infrastructures oriented to reduce different types of flood risk. Three goals are considered to assess performance measures holistically: flood risk reduction, cost minimization, and co-benefits enhancement. The main objective is to help decision-makers, assisting in selecting adequate combinations of measures	[76]
	GBGI	City-level (the Dutch side of Sint Maarten Island, located in the Caribbean region).	• Presenting a method to include the monetary analysis of these co-benefits into a cost-benefit analysis of flood risk mitigation measures	• Expected annual damage (EAD) calculation. • Co-benefits calculation. • Costs calculation	• The mix of GBGI will likely result in the best adaptation strategy as these three alternatives complement each other. • GRAI performs well at reducing the risk of flooding, while GI brings in multiple additional benefits that GRAI cannot offer.	[40]

Region	Type	Scale/Location	Objective	Methods	Findings	Ref
			Assessing the performance of different green-blue-gray measures and their combinations in achieving flood risk reduction and improving other benefits. This study aims to compare green-blue, gray, and hybrid strategies for flood mitigation from an economic point of view and show how this comparison changes when co-benefits are considered.	Pre-processing method using a multi-criteria analysis considering local characteristics and needs. A hydrodynamic model with evaluating other options to improve the performance.	• Optimization is a helpful decision-making tool for stormwater management when several strategies are considered. • The combination of green-blue-gray measures is the best strategy in the case of the selected study. • There are inevitable trade-offs among different benefits obtained from different green-blue-gray measures. • It is proved that combining green-blue-gray measures is particularly important in urban spaces when several benefits are considered simultaneously.	[4]
East Asia and Pacific	GI	System-level (the Little Stringybark Creek [LSC] watershed located on the eastern fringe of Melbourne, Australia).	To demonstrate a modeling framework to translate the flood hazard benefits of GI into maps that end users have found useful for decision-making and to test the hypothesis that the small-scale flood risk mitigation benefits afforded by GI can propagate to benefits downstream in the catchment	Study site and modeling overview. Hydrologic modeling.	Full implementation of GI in the watershed can attenuate numerous metrics of flood impacts, including flood extent (91%), low intensity (83%), and flood duration (79%).	[47]
	GI	City Level: Sponge City Program (SCP)	Undertake integrated assessments of the development of GI for flood mitigation to support robust decision-making regarding sponge city construction in urbanized watersheds	An evaluation framework based on the Storm Water Management Model (SWMM) and life cycle cost analysis (LCCA)	• Results confirmed the effectiveness of GI implementation in the study area in mitigating urban flooding. • The bio-retention cell (BC) plus vegetated swale (VS) combination was found to be the most cost-effective GI option for unit investment under all rainfall events.	[70]

(Continued)

Table 13.1 (Continued)

Region	Nexus type	Level of study (regional/country/city/system)	Scope	Methodology	Findings	Reference
	GI	China	Hydrological aspect and knowledge about GI in managing stormwater	Review of the past research performed on GI	• GI has been popular in many countries and regions. • GI is one of the five strategic areas that significantly enhance the sustainability of the communities. • GIs have a profound impact on mitigating urban flooding. • Regulatory, institutional, financial, and technological are the four barriers to implementing the GIs. • GI can achieve great environmental benefits.	[77]
	GGI	Case study (A typical residential area in Nanjing city in China)	Analytic hierarchy process (AHP) and life cycle costing (LCC) to evaluate the environmental and economic benefits of various types of coupled GI and GRAI systems.	• Stormwater management model (SWMM). • Life cycle costing (LCC).	• The scenario that includes green space, permeable pavement, green roof, and stormwater detention cell can achieve better performance in urban residential drainage and flood control than other scenarios. • The total cost of green infrastructure can save 94% compared with the original gray scenario.	[36]
	GGI	Sponge city in China	To address the concept and principle of the Integration of Green and GRAI for water sustainability.	• Ecological environment factor. • Economic factors. • Ecological service. • Environmental consistency. • Evaluation method.	It introduces the concepts of gray infrastructure, green infrastructure, sponge city, etc., understands its core connotation, proposes different water problems brought about by urbanization, ecological problems, and management issues of building sponge cities and integrating gray–green infrastructure.	[22]

Northern America	GI	City level (western Washington State, and the City of Portland OR)	Advancing GI in the regional Pacific Northwest (PNW).	Summit and Survey	• Finding the most effective means to foster new, novel, and innovative partnerships to advance GI from site to national scales. • Quantifying the designs, uses, costs, and values of GI and seeking and ensuring equity in GI at all levels	[78]
	GI	City Level: Five "green leaders": Milwaukee, WI; Philadelphia, PA; Syracuse, NY; New York City, NY; and Buffalo, NY.	Influence of governance structure on green stormwater infrastructure investment	This analysis uses the existing literature and municipal reports as the primary data sources	• Results indicate that transitions to green approaches can and did occur in localities with complex organizational structures. • Results suggest that issuance of the 2007 EPA memorandum allowing green infrastructure technologies to meet CSO permitting requirements opened a policy window, which, coupled with the right conditions, such as those in Philadelphia and Milwaukee, created momentum for the adoption of green infrastructure at the city scale.	[79]
	GGI	System-level (Case study: The Cross-Bayou Watershed of Pinellas County - Florida)	The goal of this study was to develop a multi-scale modeling platform for drainage infrastructure resilience assessment in a coastal watershed. The model employs scale-dependent informatics, including hydro informatics, climate informatics, and geoinformatics, to support comprehensive hydrodynamic stormwater and hydrologic model, called the Interconnected Channel and Pond Routing Model.	An informatics-based multi-scale modeling approach	• Results indicate that the effectiveness of LID depends on the rainfall type being considered, such as convective storm versus frontal rain and sub-daily rainfall patterns, as well as a groundwater table analysis.	[80]

(Continued)

Table 13.1 (Continued)

Region	Nexus type	Level of study (regional/country/city/system)	Scope	Methodology	Findings	Reference
	GGI	City level (Portland, United States)	To explore the institutional knowledge system challenges of valuing urban nature as infrastructural assets.	Interviews were conducted with municipal staff at departments that manage GI in Portland.	There are a variety of conceptual challenges that have stalled this integration, including epistemological mismatches between financial accounting and ecological knowledge systems that have been reviewed	[21]
	GBGI	Community (Mariners Cove community - Cross Bayou watershed near Tampa Bay - Florida)	Identifying components of flood risk and quantifying them by an integrative approach	Formula Calculations. • Risk formulation • Resilience formulation • The proposed risk formulation and framework • Hazard variables. • Copula functions. • Vulnerability. • Exposure. • Adaptive measures.	Findings indicate that coupling flood risk and infrastructure resilience is achievable through carefully formulating flood risk associated with a resilience metric, which is a function of the predicted hazards, vulnerability, and adaptive capacity. Insights into improving existing methodologies for municipalities in flood management practices, such as incorporating a multi-criteria flood impact assessment that couples risk and resilience in a common evaluation framework.	[81]
The Middle East and North Africa	GBGI	Case study/City level (Ahvaz Iran)	Investigate the performance of HGBGIs considering different degrees of centralization	Presenting a simulation-optimization framework to optimize urban drainage systems considering HGBGI alternatives and different degrees of centralization	Site selection and software analysis (EPA's SWMM version 5.1 software)	[69]

Region	Type	Scale/Location	Objectives	Methods	Results	Ref
Eastern, Western, and Central Africa	GBGI	Case study/City level (Nairobi, Kenya)	• Benefits and limitations of implementing nature-based drainage solutions at the "local community scale" in an informal urban setting to navigate the inherent constraints of space, land tenure, maintenance, and participation? • Implications of "community-managed" and nature-based urban drainage regarding interaction with municipal governance systems and for the larger-scale adoption, integration, and management of GBGI?	Case study of sustainable management of storm and wastewater drainage. Web-based survey.	Results show that involvement in the co-development of small-scale green infrastructure changed people's valuation, perception, and stewardship of nature-based systems and ecosystem services. These results have implications for the larger-scale adoption, integration, and management of urban drainage infrastructure. They also suggest that hybrid infrastructure and governance systems constitute a resilient approach to incremental and inclusive upgrading. There are limitations, including the need to develop technical capacity among community managers and the costs of maintenance.	[71]
Australia, New Zealand	GI	– Five cities from around the world (Singapore, Berlin, Melbourne, Philadelphia, and Tianjin Eco-city). – Data were collected from open sources.	Share best practices for the transition to sustainable urban water management and to gain insight into the role, if any, of GI in urban water management.	Potential case cities were listed from open sources, recommendations from experts at international conferences, and personal networks.	Profiles are described for five cities in sustainable urban water management.	[42]

Table 13.2 Research development, methods, and demographic details of the most recent studies.

No.	Indicator/Index	Method/Tool	Author	Purpose
1	D-Depth, DF-Duration Factor, DD-Dwelling Density, In-Income per Capita, IS-Inadequate Sanitation	Hydrodynamic modeling – MODCEL (a pseudo-2D hydrodynamic model). Design storm. Flood Risk Index. Flood Resilience Index.	[83]	Quantitatively measuring and assessing city resilience to flooding
2	Flood resilience indicators through extensive literature review	Flood Vulnerability Index (FVI) method. Abstraction Hierarchy (AH) systems method	[82]	Derive measures of susceptibility, resilience, and exposure
3	Areas for the individual indicator of social, economic, institutional, and physical resilience	Subjective method	[10]	—
4	Indicator-based approach consisting of 88 measures of potential sources of resilience	Quantitative data analysis methods	[13]	A framework and tool for measuring community-level resilience to flooding
5	15 key retailers' resilience thinking. Maladaptation, interactive visualization models, raising risk perceptions of the community, and placing attachment	Semi-structured interviews	[84]	Adapting to floods in terms of risk communication.
6	28 indicators have been selected for the sub-catchment Flood Vulnerability Index (FVI) equations	Distinguishing different characteristics at each identified spatial scale for analyzing local indicators	[14]	Computing a Flood Vulnerability Index (FVI)
7	A composite index using a principal components analysis (PCA) integrates 24 resilience indicators of floods and relevant social, ecological, infrastructural, and economic aspects	Weights for the variables are generated for a principal component analysis (PCA) statistical method	[85]	Measuring and mapping the spatial distribution of the levels of flood resilience across a landscape
9	Indicators cover the three aspects of resilience: the amplitude of the reaction, graduality of the increase of reaction with increasingly severe flood waves, and recovery rate	Unit-loss method	[12]	Defining and testing indicators for the resilience of flood risk management systems

Conclusion

GBGI is a term that is used mostly in association with the adaptation to climate change. GRAIs are traditional infrastructures essential to urban and rural communities. GI replaces GRAI, especially in highly sustainable developments. GI is preferred to be used as a replacement or complement to GRAI. As a set of strategies in built environments, GI is gaining a reputation as a core sustainable urban system serving various ecosystem needs. The GI performance is satisfactory in mitigating flood hazards. GBI offers various complementary benefits regarding flood management to GIs. GBG in the urban area are three alternatives that complement each other and offer many benefits. The benefits include flood risk reduction, energy and water savings, carbon sequestration maximization, and air quality improvement, resulting in the best adaptation strategy. Complementary GRAI with GI results in highly effective stormwater management. Today, HGBGIs are the most promising system in urban areas, as these three infrastructures complement each other. The HGBGI combines the adaptability, sustainability, and multifunctionality of GBI with the reliability and resilience of the GRAI.

References

1 United Nations (2014). World urbanization prospects. https://population.un.org/wup/ (accessed on 25 December 2021).
2 IPCC (2012). Managing the risks of extreme events and disasters to advance climate change adaptation [cited 2020 Jul 4]. https://www.ipcc.ch/report/managing-the-risks-of-extreme-events-and-disasters-to-advance-climate-change-adaptation/.
3 UNDP (2012). Flood Risk Reduction: Innovation and technology in risk mitigation and development planning in Small Island Developing States: towards floor risk reduction in Sint Maarten. United Nations Development Programme (UNDP).
4 Alves, A., Vojinovic, Z., Kapelan, Z. et al. (2020). Exploring trade-offs among the multiple benefits of green-blue-grey infrastructure for urban flood mitigation. *Sci. Total Environ.* 703: 134980.
5 Jha, A.K., Bloch, R., and Lamond, J. (2012). *Cities and Flooding: A Guide to Integrated Urban Flood Risk Management for the 21st Century*. The World Bank: Cities and Flooding.
6 Tahir, F., Ajjur, S.B., Serdar, M.Z. et al. (2021). Proceedings of the Qatar Climate Change Conference 2021. In: *Qatar Climate Change Conference 2021*, Hamad Bin Khalifa University Press, p. 44.
7 Al-Humaiqani, M.M. and Al-Ghamdi, S.G. (2022). The built environment resilience qualities to climate change impact: Concepts, frameworks, and directions for future research. *Sustain Cities Soc.* 80: 103797.
8 Al-Humaiqani, M.M. and Al-Ghamdi, S.G. (2023). Assessing the built environment's reflectivity, flexibility, resourcefulness, and rapidity resilience qualities against climate change impacts from the perspective of different stakeholders. *Sustainability.* 15 (6): 5055.

9 de Bruijn, K.M., Maran, C., Zygnerski, M. et al. (2019). Flood resilience of critical infrastructure: approach and method applied to Fort Lauderdale, Florida. *Water (Switzerland).* 11 (3): 1–21.
10 Qasim, S., Qasim, M., Shrestha, R.P. et al. (2016). Community resilience to flood hazards in Khyber Pukhthunkhwa province of Pakistan. *Int. J. Disaster Risk Reduct.* 18: 100–106.
11 Garvin, S, Hunter, K, McNally, D et al. (2016). Property flood resilience database: an innovative response for the insurance market. *E3S Web of Conferences, 3rd European Conference on Flood Risk Management (FLOODrisk 2016)*, vol. 7, p. 22002.
12 Batica, J. and Gourbesville, P. (2016). Resilience in flood risk management – a new communication tool. *Procedia Eng.* 154: 811–817.
13 Keating, A., Campbell, K., Szoenyi, M. et al. (2017). Development and testing of a community flood resilience measurement tool. *Nat. Hazards Earth Syst. Sci.* 17 (1): 77–101.
14 Balica, S.F., Douben, N., and Wright, N.G. (2009). Flood vulnerability indices at varying spatial scales. *Water Sci. Technol.* 60 (10): 2571–2580.
15 Forum, W.E. (2017). These 11 sinking cities are in danger of disappearing | World Economic Forum [cited 12 April 2020]. https://www.weforum.org/agenda/2019/09/11-sinking-cities-that-could-soon-be-underwater/.
16 The Economist (2019). In deep trouble – Climate change is forcing Asian cities to rethink their flood defences | Asia | The Economist [cited 12 April 2020]. https://www.economist.com/asia/2019/09/21/climate-change-is-forcing-asian-cities-to-rethink-their-flood-defences.
17 National Geographic (2020). Global warming and climate change effects: information and facts [cited 12 April 2020]. https://www.nationalgeographic.com/environment/global-warming/global-warming-effects/.
18 Fritz, M. (2017). Nature-based solutions to climate change adaptation in urban areas: linkages between science, policy and practice – Theory and Practice of Urban Sustainability Transitions, Nadja Kabisch, Horst Korn, Jutta Stadler, Aletta Bonn, Springer pp. 1–342. doi: https://doi.org/10.1007/978-3-319-56091-5.
19 Dipeolu, A.A., Onoja Matthew Akpa, D., and Fadamiro D., A.J. (2019). Mitigating environmental sustainability challenges and enhancing health in urban communities: the multi-functionality of green infrastructure. *J. Contemp. Urban Aff.* 4 (1): 33–46.
20 Parker, J. (2015). Green vs. grey infrastructure – autocase. Autocase [cited 1 July 2020]. https://autocase.com/lawrencepark/.
21 Marissa, M.A. (2019). Making 'green' fit in a 'grey' accounting system: the institutional knowledge system challenges of valuing urban nature as infrastructural assets. *Environ. Sci. Policy.* 99 (May): 160–168.
22 Sun, Y., Deng, L., Pan, S.Y. et al. (2020). Integration of green and gray infrastructures for sponge city: Water and energy nexus. *Water Energy Nexus.* 3: 29–40.
23 Rafael, S., Vicente, B., Rodrigues, V. et al. (2018). Impacts of green infrastructures on aerodynamic flow and air quality in Porto's urban area. *Atmos. Environ.* 190: 317–330.
24 Block, A.H., Livesley, S.J., and Williams, N.S.G. (2012). *Responding to the Urban Heat Island: A Review of the Potential of Green Infrastructure Literature Review*, 1–55. Australia: Victorian Centre for Climate Change Adaptation Research (VCCCAR).

25 Bowler, D.E., Buyung-Ali, L., Knight, T.M., and Pullin, A.S. (2010). Urban greening to cool towns and cities: a systematic review of the empirical evidence. *Landsc. Urban Plan.*, Elsevier 97: 147–155.

26 Kloos, J. and Renaud, F.G. (2016). Overview of ecosystem-based approaches to drought risk reduction targeting small- scale farmers in Sub-Saharan Africa. In: *Advances in Natural and Technological Hazards Research* (ed. F. Renaud, K. Sudmeier-Rieux, M. Estrella, and U. Nehren), 199–226. Netherlands: Springer.

27 Pugh, T.A.M., MacKenzie, A.R., Whyatt, J.D., and Hewitt, C.N. (2012). Effectiveness of green infrastructure for improvement of air quality in urban street canyons. *Environ. Sci. Technol.* 46 (14): 7692–7699.

28 Venkataramanan, V., Lopez, D., McCuskey, D.J. et al. (2020). Knowledge, attitudes, intentions, and behavior related to green infrastructure for flood management: a systematic literature review. *Sci. Total Environ.* 720 (February): 137606.

29 US EPA (2017). What is Green Infrastructure? [cited 4 July 2020]. https://www.epa.gov/green-infrastructure/what-green-infrastructure.

30 Zölch, T., Henze, L., Keilholz, P., and Pauleit, S. (2017). Regulating urban surface runoff through nature-based solutions – an assessment at the micro-scale. *Environ. Res.* 157: 135–144.

31 Eckart, K., McPhee, Z., and Bolisetti, T. (2017). Performance and implementation of low impact development – a review. In: *Science of the Total Environment*, vol. 607, 608, 413–432. Netherlands: Elsevier B.V.

32 Ahiablame, L. and Shakya, R. (2016). Modeling flood reduction effects of low impact development at a watershed scale. *J. Environ. Manage.* 171: 81–91.

33 Le Gentil, E. and Mongruel, R. (2015). A systematic review of socio-economic assessments in support of coastal zone management (1992-2011). *J. Environ. Manage.* 149: 85–96.

34 Brink, E., Aalders, T., Ádám, D. et al. (2016). Cascades of green: a review of ecosystem-based adaptation in urban areas. *Glob. Environ. Chang.* 36: 111–123.

35 Kellens, W., Terpstra, T., and De Maeyer, P. (2013). Perception and communication of flood risks: a systematic review of empirical research. *Risk Anal.* 33 (1): 24–49.

36 Xu, C., Tang, T., Jia, H. et al. (2019). Benefits of coupled green and grey infrastructure systems: evidence based on analytic hierarchy process and life cycle costing. *Resour. Conserv. Recycl.* 151 (September): 1–10.

37 McGrane, S.J. (2016). Impacts of urbanisation on hydrological and water quality dynamics, and urban water management: a review. *Hydrol. Sci. J.* 61 (13): 2295–2311.

38 Ahmed, F., Moors, E., Khan, M.S.A. et al. (2018). Tipping points in adaptation to urban flooding under climate change and urban growth: the case of the Dhaka megacity. *Land Use Policy.* 79: 496–506.

39 Hallegatte, S. and Corfee-Morlot, J. (2011). Understanding climate change impacts, vulnerability and adaptation at city scale: an introduction. *Clim. Change.* 104 (1): 1–12.

40 Alves, A., Gersonius, B., Kapelan, Z. et al. (2019). Assessing the co-benefits of green-blue-grey infrastructure for sustainable urban flood risk management. *J. Environ. Manage.* 239 (February): 244–254.

41 Engström, R., Howells, M., Mörtberg, U., and Destouni, G. (2018). Multi-functionality of nature-based and other urban sustainability solutions: New York City study. *Land Degrad. Dev.* 29 (10): 3653–3662.

42 Liu, L. and Jensen, M.B. (2018). Green infrastructure for sustainable urban water management: practices of five forerunner cities. *Cities.* 74 (January): 126–133.

43 Denjean, B., Denjean, B., Altamirano, M.A. et al. (2017). Natural assurance scheme: a level playing field framework for green-grey infrastructure development. *Environ. Res.* 159 (July): 24–38.

44 Lindley, S., Pauleit, S., Yeshitela, K. et al. (2018). Rethinking urban green infrastructure and ecosystem services from the perspective of sub-Saharan African cities. *Landsc. Urban Plan.* 180 (September): 328–338.

45 Wouters, P., Dreiseitl, H., Wanschura, B., Wörlen, M., Moldaschl, M., Wescoat, J., and Noiva, K. (2016). Blue-green infrastructures as tools for the management of urban development and the effects of climate change. Ramboll Environ, Madrid.

46 Onuma, A. and Tsuge, T. (2018). Comparing green infrastructure as ecosystem-based disaster risk reduction with gray infrastructure in terms of costs and benefits under uncertainty: a theoretical approach. *Int. J. Disaster Risk Reduct.* 32 (May 2017): 22–28.

47 Schubert, J.E., Burns, M.J., Fletcher, T.D., and Sanders, B.F. (2017). A framework for the case-specific assessment of Green Infrastructure in mitigating urban flood hazards. *Adv. Water Resour.* 108: 55–68.

48 Naylor, L.A., Kippen, H., Coombes MA et al. (2017). Greening the grey: a framework for integrated green grey infrastructure (IGGI). University of Glasgow report, pp. 1–26. http://eprints.gla.ac.uk/150672/.

49 Miller, J.D. and Hutchins, M. (2017). The impacts of urbanisation and climate change on urban flooding and urban water quality: a review of the evidence concerning the United Kingdom. *J. Hydrol. Reg. Stud.* 12: 345–362.

50 Vojinovic, Z. (2015). Flood risk: the holistic perspective. IWA Publishing [cited 4 July 2020]. https://books.google.com.qa/books?hl=en&lr=&id=GmIbCgAAQBAJ&oi=fnd&pg=PP1&ots=QgKXPQX8HM&sig=ZYqdgJb12s7WwCewDSTVxALqiEQ&redir_esc=y#v=onepage&q&f=false.

51 World Wild Life Organization (2020). Effects of climate change | threats | WWF [cited 11 July 2020]. https://www.worldwildlife.org/threats/effects-of-climate-change.

52 Serre, D. and Heinzlef, C. (2018). Assessing and mapping urban resilience to floods with respect to cascading effects through critical infrastructure networks. *Int. J. Disaster Risk Reduct.* 30 (October 2017): 235–243.

53 Wilby, R.L. and Keenan, R. (2012). Adapting to flood risk under climate change. *Prog. Phys. Geogr.* 36 (3): 348–378.

54 Solomon, S. (2007). The Physical Science Basis: Contribution of Working Group I to the Fourth Assessment Report of the Intergovernmental Panel on Climate Change. Intergovernmental Panel on Climate Change (IPCC), Climate Change 2007, Intergovernmental Panel on Climate Change (IPCC), p. 996.

55 Masson, V., Marchadier, C., Adolphe, L. et al. (2014). Adapting cities to climate change: a systemic modelling approach. *Urban Clim.* 10 (P2): 407–429.

56 CWT (2007). Contribution of Working Groups I, II and III to the Fourth Assessment Report of the Intergovernmental Panel on Climate Change. IPCC 2007 Climate Change 2007 Synth Rep, p. 104.

57 Moura Rezende, O., da Cruz, R., de Franco, A.B. et al. (2019). A framework to introduce urban flood resilience into the design of flood control alternatives. *J. Hydrol.* 576 (June): 478–493.

58 Leandro, J., Chen, K.F., Wood, R.R., and Ludwig, R. (2020). A scalable flood-resilience-index for measuring climate change adaptation: Munich city. *Water Res.* 173: 115502.

59 Perrone, A., Inam, A., Albano, R. et al. (2020). A participatory system dynamics modeling approach to facilitate collaborative flood risk management: a case study in the Bradano River (Italy). *J. Hydrol.* 580 (November 2019): 124354.

60 Percival, S. and Teeuw, R. (2019). A methodology for urban micro-scale coastal flood vulnerability and risk assessment and mapping. *Nat. Hazards* 97: 355–377.

61 Pearson, J., Punzo, G., Mayfield, M. et al. (2018). Flood resilience: consolidating knowledge between and within critical infrastructure sectors. *Environ. Syst. Decis.* 38 (3): 318–329.

62 Khunwishit, S., Choosuk, C., and Webb, G. (2018). Flood resilience building in Thailand: assessing progress and the effect of leadership. *Int. J. Disaster Risk Sci.* 9 (1): 44–54.

63 McClymont, K., Morrison, D., Beevers, L., and Carmen, E. (2020). Flood resilience: a systematic review. *J. Environ. Plan. Manag.* 63 (7): 1151–1176.

64 Fekete, A. (2019). Critical infrastructure and flood resilience: Cascading effects beyond water. *Wiley Interdiscip. Rev. Water.* 6 (5): 1–13.

65 Murdock, H.J., de Bruijn, K.M., and Gersonius, B. (2018). Assessment of critical infrastructure resilience to flooding using a response curve approach. *Sustain.* 10 (10): 3470.

66 Wang, Y., Meng, F., Liu, H. et al. (2019). Assessing catchment scale flood resilience of urban areas using a grid cell based metric. *Water Res.* 163: 114852.

67 Vamvakeridou-Lyroudia, L.S., Chen, A.S., Khoury, M. et al. (2020). Assessing and visualising hazard impacts to enhance the resilience of critical infrastructures to urban flooding. *Sci. Total Environ.* 707: 136078.

68 Sweya, L.N. and Wilkinson, S. (2020). A tool for measuring environmental resilience to floods in Tanzania water supply systems. *Ecol. Indic.* 112 (5): 106165.

69 Bakhshipour, A.E., Dittmer, U., Haghighi, A., and Nowak, W. (2019). Hybrid green-blue-gray decentralized urban drainage systems design, a simulation-optimization framework. *J. Environ. Manage.* 249 (August): 109364.

70 Mei, C., Liu, J., Wang, H. et al. (2018). Integrated assessments of green infrastructure for flood mitigation to support robust decision-making for sponge city construction in an urbanized watershed. *Sci. Total Environ.* 639: 1394–1407.

71 Mulligan, J., Bukachi, V., Clause, J.C. et al. (2020). Hybrid infrastructures, hybrid governance: new evidence from Nairobi (Kenya) on green-blue-grey infrastructure in informal settlements: "Urban hydroclimatic risks in the 21st century: Integrating engineering, natural, physical and social sciences to build". *Anthropocene* 29: 100227.

72 Leng, L., Mao, X., Jia, H. et al. (2020). Performance assessment of coupled green-grey-blue systems for Sponge City construction. *Sci. Total Environ.* 728: 138608.

73 Jerome, G., Sinnett, D., Burgess, S. et al. (2019). A framework for assessing the quality of green infrastructure in the built environment in the UK. *Urban For. Urban Green.* 40 (October 2018): 174–182.

74 Labib, S.M. (2019). Investigation of the likelihood of green infrastructure (GI) enhancement along linear waterways or on derelict sites (DS)using machine learning. *Environ. Model Softw.* 118 (March): 146–165.

75 Alves, A., Sanchez, A., Vojinovic, Z. et al. (2016). Evolutionary and holistic assessment of green-grey infrastructure for CSO reduction. *Water (Switzerland).* 8 (9): 402.

76 Alves, A., Gersonius, B., Sanchez, A. et al. (2018). Multi-criteria approach for selection of green and grey infrastructure to reduce flood risk and increase co-benefits. *Water Resour. Manag.* 32 (7): 2505–2522.

77 Li, C., Peng, C., Chiang, P.C. et al. (2019). Mechanisms and applications of green infrastructure practices for stormwater control: a review. *J. Hydrol.* 568 (October 2018): 626–637.

78 Jayakaran, A.D., Moffett, K.B., Padowski, J.C. et al. (2020). Green infrastructure in western Washington and Oregon: perspectives from a regional summit. *Urban For. Urban Green* 50 (February): 126654.

79 Hopkins, K.G., Grimm, N.B., and York, A.M. (2018). Influence of governance structure on green stormwater infrastructure investment. *Environ. Sci. Policy.* 84 (November 2017): 124–133.

80 Joyce, J., Bin, C.N., Harji, R. et al. (2017). Developing a multi-scale modeling system for resilience assessment of green-grey drainage infrastructures under climate change and sea level rise impact. *Environ. Model Softw.* 90: 1–26.

81 Joyce, J., Bin, C.N., Harji, R., and Ruppert, T. (2018). Coupling infrastructure resilience and flood risk assessment via copulas analyses for a coastal green-grey-blue drainage system under extreme weather events. *Environ. Model Softw.* 100: 82–103.

82 Beevers, L., Walker, G., and Strathie, A. (2016). A systems approach to flood vulnerability. *Civ. Eng. Environ. Syst.* 33 (3): 199–213.

83 Miguez, M.G. and Veról, A.P. (2017). A catchment scale Integrated Flood Resilience Index to support decision making in urban flood control design. *Environ. Plan. B Urban Anal. City Sci.* 44 (5): 925–946.

84 Chiang, Y.C. and Ling, T.Y. (2017). Exploring flood resilience thinking in the retail sector under climate change: a case study of an estuarine region of Taipei City. *Sustainability (Switzerland)* 9: 1650.

85 Kotzee, I. and Reyers, B. (2016). Piloting a social-ecological index for measuring flood resilience: a composite index approach. *Ecol. Indic.* 60: 45–53.

14

Enhancing Energy System Resilience: Navigating Climate Change and Security Challenges

Muhammad Imran Khan[1,2] *and Sami G. Al-Ghamdi*[1,3,4]

[1] *Division of Sustainable Development, College of Science and Engineering, Hamad Bin Khalifa University, Qatar Foundation, Doha, Qatar*
[2] *Department of Mechanical Engineering, College of Engineering, Prince Mohammad Bin Fahd University, Al-Khobar, Saudi Arabia*
[3] *Environmental Science and Engineering Program, Biological and Environmental Science and Engineering Division, King Abdullah University of Science and Technology (KAUST), Thuwal, Saudi Arabia*
[4] *KAUST Climate and Livability Initiative, King Abdullah University of Science and Technology (KAUST), Thuwal, Saudi Arabia*

Introduction

Energy is essential to our daily life and its sustainability is one of the most important and challenging issues in the transition to a low-carbon economy. Energy is provided from the point of generation to end users through a system referred to as an energy system. An energy system is an interconnected network of physical infrastructure, or organizational arrangements, related to the production, conversion, delivery, management, and utilization of energy. The physical infrastructure of an energy system can take many forms and can be interconnected, depending upon the form of primary or secondary energy. For example, Figure 14.1 illustrates the main components of an energy system.

The overall energy supply in urban areas is diverse and comprises a number of sources and modes, ranging from centralized power plants, oil and gas for heating, or local power generation and heat networks. Evidence shows that these diverse energy systems can be very vulnerable to the diverse and evolving threats caused by climate change, natural disasters (e.g., hurricanes, earthquakes, floods, snowstorms, lightning, and solar flares), or human-made threats such as physical or cyberattacks (Figure 14.2). All these risks can cause temporary or permanent damage to an energy system and result in power outages, which can bring economic activity to a halt. Conventional energy systems are not designed to withstand these extreme conditions [2]. Article 7 of the Paris Agreement explicitly asks the committed parties to put in place policies and frameworks to improve their "adaptation" and "resilience" to climate change [3]. Therefore, in addition

Sustainable Cities in a Changing Climate: Enhancing Urban Resilience. First Edition.
Edited by Sami G. Al-Ghamdi.
© 2024 John Wiley & Sons Ltd. Published 2024 by John Wiley & Sons Ltd.

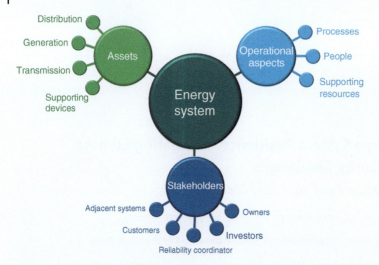

Figure 14.1 Components of an energy system. *Source:* Bukowski et al. [1]/U.S. Department of Energy National Laboratory/Public Domain.

Figure 14.2 Damages to energy systems by natural disasters. *Source:* (Bottom left) dpa picture alliance/Alamy Images; (Bottom right) Associated Press/Alamy Images.

to providing sustainable solutions that enhance the share of green energy, in the overall energy basket, to decarbonize the energy supply, it is also essential to make these energy systems resilient to the aforementioned threats.

Adapting the Theory of Resilience to Energy Systems

While resilience has become a commonly used term, there is no widely agreed-upon definition within the energy sector. The concept of resilience within the energy system domain is complex and multifaceted [4] and can be used to address extraordinary conditions of different natures. There is no standard definition for the resilience of energy systems [5], and it varies depending on the context and objective [6]. Several definitions for energy system resilience have been offered by various sources, e.g., governments, industries, regulators, professional organizations, and academia. Although the definitions share some common attributes, there is no formal, widely agreed-upon definition of energy system resilience. A general understanding of these available definitions of energy system resilience implies the ability of an energy system to return to its normal state after being exposed to a low-probability, high-impact disruptive event that affects its original state. Some of the most notable definitions of energy system resilience are listed below:

- The US Department of Energy (DOE) defines resilience of the energy system as "the ability of a system or its components to adapt to changing conditions and withstand and rapidly recover from disruptions" [7].
- The International Energy Agency (IEA), defines resilience of the energy system as "the capacity of the energy system or its components to cope with a hazardous event or trend, responding in ways that maintain their essential function, identity and structure while also maintaining the capacity for adaptation, learning and transformation" [8].
- The UK Energy Research Centre defines resilience of the energy system as "The capacity of an energy system to tolerate disturbance and to continue to deliver affordable energy services to consumers. A resilient energy system can speedily recover from shocks and can provide alternative means of satisfying energy service needs in the event of changed external circumstances" [9].

In addition to the above definitions, there are numerous definitions of resilience that are pertinent to energy systems proposed in academic literature. The review of these resilience definitions indicates that there is no unique insight on how to define the resilience of energy systems. As shown in Figure 14.3, several common themes from these many definitions can be overserved in academic literature, e.g., Prepare (P), Detect (D), Mitigate (M), Adapt (A), Recover (R), and Normalcy (N). Figure 14.4 illustrates the relationship between these different themes of resilience and the occurrence of a disruptive event in a typical energy system.

The descriptions found for energy system resilience in academic literature are included in Table 14.1, where the second column indicates which key themes are present in each definition.

In light of the above literature, the resilience of an energy system can be categorized by assessing the five Rs. These include three preventative attributes, i.e., Robustness, Resourcefulness, and Redundancy and two performance attributes, i.e., Recoverability and Response (see Figure 14.5). Table 14.2 illustrates the qualities and elements for enhancing the resilience of a typical electrical power system associated with the five Rs.

Figure 14.3 Key themes in resiliency definitions available in academic literature pertinent to energy systems.

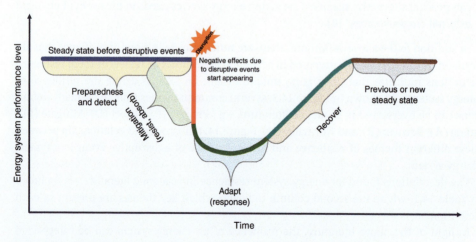

Figure 14.4 Relationships between key elements in resiliency definitions and a disruptive event.

Table 14.1 Key elements in resiliency definitions for engineering systems that are available in academic literature.

				Key elements			
Reference	P	D	M	A	R	N	Additional aspects
Dehghani et al. [10]			✓		✓	✓	
Plotnek and Slay [11]	✓	✓			✓	✓	
Elsayed [12]			✓	✓		✓	
Raoufi et al. [13]			✓		✓	✓	Sustaining critical social services
Hulse et al. [14]			✓	✓	✓	✓	
Kandaperumal and Srivastava [15]			✓		✓		Ensuring supply of critical loads
Hickford et al. [16]	✓	✓	✓	✓			
Azzuni and Breyer [17]			✓	✓		✓	Retaining essential function and identity of the system
Baros et al. [18]			✓		✓		
Li et al. [19]	✓			✓	✓		
Liu et al. [20]			✓		✓		
Erker et al. [21]			✓				Retaining essential function and identity of the system
Bie et al. [22]		✓	✓	✓	✓		
Panteli et al. [23]		✓	✓	✓	✓	✓	
Xu et al. [24]	✓		✓	✓	✓		
Friedberg et al. [25]			✓	✓	✓		
Arghandeh et al. [26]			✓	✓			
Espinoza et al. [27]			✓	✓	✓		Lessons learned
Bahramirad et al. [28]			✓		✓		
Alderson et al. [29]			✓	✓		✓	
Panteli and Mancarella [4]			✓	✓	✓		Lessons learned
Linkov et al. [30]	✓	✓	✓	✓	✓	✓	
Khodaei [31]			✓		✓	✓	Ensuring the least possible interruption in service supply
Cutter et al. [32]	✓	✓	✓	✓	✓		
Ouyang et al. [33]			✓	✓	✓	✓	
Vugrin et al. [34]			✓			✓	

(Continued)

Table 14.1 (Continued)

Reference	P	D	M	A	R	N	Additional aspects
Rudnick [35]			✓	✓	✓	✓	With limited performance degradation and constricted costs
Hollnagel [36]	✓	✓	✓	✓		✓	
López-Marrero and Tschakert [37]			✓	✓		✓	
Mili and Center [38]				✓	✓		
Aven [39]				✓		✓	Within acceptable degradation parameters and costs
Longstaff et al. [40]				✓	✓		Retaining essential function and identity of the system
O'Brien and Hope [40]				✓	✓		Minimizing vulnerabilities and exploiting beneficial opportunities through socio-technical co-evolution

Figure 14.5 Five Rs of resilience energy system.

Table 14.2 Resilience characteristics and examples of mitigations.

	Attributes What does it need to be?	Qualities How does it make it resilient?	Elements What about the quality makes it resilient?
Preventative	Robustness	• Performance monitoring • Hardened infrastructure • Physically secure • Cyber secure	• Risk management framework-compliant control systems • Active vs. passive performance monitoring • Maintenance schedule and checklist • Power quality voltage/frequency/phase match • Site access protocols (physical security) • Seismic design in earthquake zones • Elevated platforms in floodplains • Cyber-secure access to controls and networks
	Resourcefulness	• Community coordination • Available power generation • Energy storage • Recurring and relevant training exercises	• Community planning and resource integration • Uninterruptible power supply (UPS) • Diversified generation sources including generators, renewable energy, and storage • Nearby generation • Load shedding to prioritize more critical loads • Reduced operations and maintenance planning window
	Redundancy	• Eliminate single points of failure • Distributed generation topology	• Mesh and ring with bidirectional flow • Modular units of accounting for maintenance and downtime • Redundant lines (power and communication) and equipment • Backup staff (microgrid operator)

(Continued)

Table 14.2 (Continued)

	Attributes What does it need to be?	Qualities How does it make it resilient?	Elements What about the quality makes it resilient?
Performance	Recoverability	• Standardized components • Spare parts inventory • Damage assessment • Prioritization of repowering	• Centralized management of spares • Open architecture software • Commercial off-the-shelf parts • Portfolio and equipment consolidation • Utility coordination and agreements • Distributed generation systems • Black start sequence
	Response	• Automated • Self-healing • Forecasting/threat assessment • Performance indicators • Training and exercises	• Maintenance staff training and exercises • Energy consumption data collection and predictive analysis • Fault tolerance (failover or failsafe) • Inclement weather response plans • "Smart" control systems with built-in response protocol • Documented procedures available during an emergency • Condition-based maintenance • Data collection and predictive analytics • Smart control systems

Source: Adapted from Rits [41]; Mishra et al. [42].

Why Incorporate Resilience into Energy Systems?

The energy sector facilitates economic growth and supports key service sectors that drive the development of a country. Disruptions to energy systems can create widespread economic and social impacts, including losses in productivity, health and safety issues, and, in the most extreme case, loss of life [43]. The broader economic impact of disruption to the energy system puts greater emphasis on the need to strengthen the sector's resilience and keep the system operating. The longer there is disruption to the energy system, the more financial and economic loss there is to the economy [44]. For example, on 14 August 2003, contact between an Ohio power line and an overgrown tree, complicated by human error, software issues, and equipment failures, resulted in a widespread power outage, with an estimated $ 6 billion impact on the US and Canadian economy [45]. In recent years, the resilience of energy systems, their vulnerability, and the risks stemming from their failure have been received increasing attention in the scientific literature. The development of resilient energy systems is essential for the control of climate change and, in many parts of the world, a necessary adaptation strategy. Recent studies [7, 46–48] indicate that, in most modeled scenarios accounting for the growing impacts of climate change, the benefits conferred by resilient electricity systems considerably exceed the associated costs [49]. It is estimated that for every dollar invested in climate-resilient infrastructure, six dollars can be saved. According to the World Bank, if the actions needed for resilience are delayed by 10 years, the cost will almost double [50]. For instance, underground transmission and distribution cables, which require a higher upfront outlay than above-ground systems, can significantly reduce potential damage from climate impacts and save recovery costs [49]. Transmission and distribution lines above ground tend to be more vulnerable to climate hazards such as high-speed winds, wildfires, floods, and landslides than underground systems [51]. When storm Gudrun hit Sweden in January 2005, outages in rural areas lasted up to 20 days due to the damaged distribution lines, while those in urban areas, with underground cabling, lasted only a few hours [52]. Due to the lengthy outages, the Swedish network operators lost around EUR €250 million. Moreover the socioeconomic losses due to the interruptions were estimated at EUR €3 billion. Enhancing energy sector resilience protects not only energy companies but also the economies and populations that rely upon the energy services these companies provide [53]. Therefore, governments have a compelling interest in enhancing the resilience of their energy sectors. Building resilience into energy systems can also support mitigation efforts by lowering carbon emissions while decreasing vulnerability [54]. For example, investing in renewable energy projects can address resilience and mitigation objectives by diversifying the energy supply while also reducing carbon emissions from fossil fuel combustion.

The advantage of focusing on resilience as a key concept is that it can be seen as an intrinsic characteristic of the energy system itself. It does not require one to think about the underlying causes of a particular shock, only that a particular kind of shock is possible. For example, a prolonged interruption of a gas supply [9]. Other opportunities could include the replacement of aging energy infrastructure, using decentralized energy systems with renewable energy sources, or supporting resilience and mitigation by reducing widespread power outages and GHG emissions [55]. In summary, an energy system with high resilience is of utmost importance in today's modern society, which is highly dependent on continued access to energy services.

What are the Threats to the Energy System?

Energy systems are exposed to several types of threats that require different and sometimes contradictory resilience measures. System changes that improve resilience against one threat may be completely ineffective, or even decrease resilience, against another threat [56]. As illustrated in Figure 14.6, based on their nature, the numerous types of

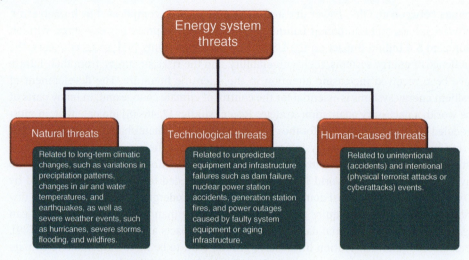

Figure 14.6 Different failure scenarios in energy systems.

threats can be classified into three main categories: (1) natural threats, (2) technological threats, and (3) human-caused threats

Among these various types of hazards, the most common are those associated with weather/climate change, thus requiring close attention when designing energy system resilience. The energy sector is exposed to several climate-related risks that arise from increasing temperatures, variability in rainfall patterns, and increased frequency of extreme weather events. A particular concern is that energy systems, with a higher penetration level of renewable energy technologies that cater to multiple energy services, may create new vulnerabilities given the high weather dependency of renewable energy sources, such as electricity, heating, and cooling, [57–59]. As a result of extreme weather events, such as heat waves, the impacts of climate change at periods of peak demand can reach well beyond simple changes in net annual demand, harming the operation of the energy infrastructures and undermining their reliability [60]. Climate change can induce intensified climate variations and consequently, stronger and more frequent extreme events [61]. Consequently, all parts of an energy system, from energy production and transformation to its transmission, distribution, storage, and demand, can be affected by climate and weather (see Figure 14.7). For example, gradual changes in the climate can affect the availability of important resources such as water for hydropower as well as for cooling thermal power plants. They can also affect energy demand, in particular in relation to heating, cooling, and water supply. Weather extremes such as floods and storms can lead to blackouts due to flooding electric substations or windfall on power lines. Sea level rise can threaten coastal and offshore energy infrastructure [49].

Other possible impacts of climate change on the various components of the energy sector (e.g., production, transformation, transportation and storage, and demand) are described in Table 14.3.

Therefore, when designing energy systems for the future, both short-term and long-term climatic variations should be considered, as both can influence the choice of energy

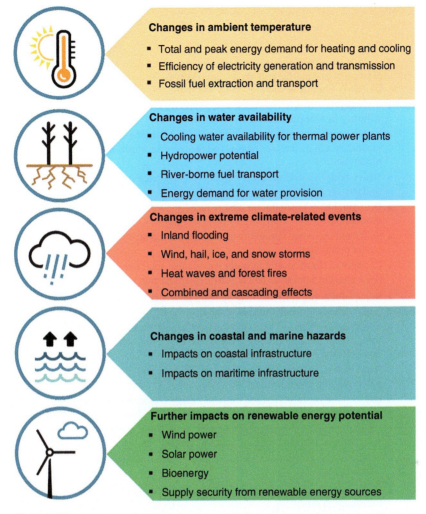

Figure 14.7 Important climatic changes and their impacts on the energy system. *Source:* Adapted from EEA [52].

technologies and the sizing of components [57]. In this regard, the development of a framework for assessing climate resilience can be useful as a means of addressing both short-term and long-term variations associated with climate change and uncertainties [6].

Domains of Resilience Approaches to Energy Systems

There are three major domains of resilience approaches applicable to energy systems, including (1) engineering resilience, (2) ecological resilience, and (3) adaptive resilience. Table 14.4 sums up the differences between these domains of resilience.

Table 14.3 Climate change affects all parts of the energy system.

	Impact
Primary energy production	• Melting of permafrost and sea ice improves access to oil and gas reserves, but also compromises land stability and damages infrastructure. • Increased risk of wildfires affects oil production (e.g., Fort McMurray wildfires in Alberta, Canada). • Water scarcity poses constraints on shale gas or tight gas developments, secondary and tertiary (enhanced) oil recovery approaches, and biofuel production. • Heavy rains increase moisture content (and decrease quality) of stockpiled coal surface mines. • Drought, heavy precipitation, and reduced snowpack affect hydropower production. • Shifts and increased variability of wind speed and direction affect wind power production. • Changes in cloud cover and water vapor affect solar energy (photovoltaic [PV], concentrated solar power [CSP], solar heating).
Energy transformation	• Sea level rise and storm activity increase flood risk for coastal infrastructure (e.g., refineries and nuclear power plants). • Wind, hail, and extreme precipitation increase damage to solar PV, flat plate collectors in solar thermal systems, on- and off-shore wind turbines, and hydroelectric dams. • Extreme heat reduces efficiency of solar PV cells and thermal conversion processes, and cooling efficiency in thermal power plants. • Lower reservoir levels reduce water-to-energy conversion in hydropower production. • Increased water temperatures constrain thermal power generation by reducing plant cooling efficiency and increasing cooling water demand. • Water scarcity constrains CSP and carbon capture and storage (CCS) technologies.
Transportation, transmission, storage, and distribution	• Higher temperatures increase transmission losses and reduce overall transmission efficiency. • Higher temperatures reduce viscosity of transported fuels. • Extreme events (e.g., flooding and landslides), erosion, and melting permafrost cause pipeline damage. • Melting sea ice opens up new shipping routes (e.g., Bering Strait and Northwest Passage). • Freeze/thaw cycles and extreme weather cause damage to paved roads; extreme precipitation increases washouts for unpaved roads and low-lying coastal routes.
Energy demand	• Rising air temperatures increase cooling demand (mostly electricity) in summer months and reduce heating demand (heating fuels and electricity) in winter months. • Net changes occur in energy demand depending on geographic location and access to energy technologies such as air conditioning. • Warming trends change attractiveness of tourist destinations and tourism-related energy demand.

Source: International Energy Agency [53]/Public Domain CC BY 4.0.

Table 14.4 Paradigms of resilience applicable to energy system.

Framework	Focus on	Context	Characteristics	System's behavior	Equilibrium state	Type of system
Engineering resilience	Recovery, constancy, stability, robustness, rapidity	Maintaining the static equilibrium state	Return time, efficiency, resistance, optimization	Close to stable point	Quick return to stable point	Predictable quasi-stationary system
Ecological resilience	Persistence, resourcefulness, redundancy	May shift to a new equilibrium state	Buffer capacity, withstand shock, maintain function	Moves out of basin of attraction. Adjusts the system. Transforms the system. Resists disturbance without much alteration of the system	Returns to one of multiple equilibrium states. Toward new equilibrium state. New system and state. Return to old equilibrium state	Complex adaptive system
Adaptive resilience	Adaptive capacity, transformability, learning capacity, innovation	Maintaining a dynamic equilibrium state	Interplay disturbance and reorganization, sustaining and developing	System learns and self-organizes	Adjustment to a new stable state	Complex adaptive system

Source: References [62–66].

Resilience Enhancement Approaches for Energy Systems

Achieving resilient energy systems within a community will not happen overnight [67]. A core resilience challenge for energy system owners and operators is to translate the definitions, objectives, and approaches for resilience into identifiable and implementable actions at the component and engineering levels [68]. Building resilience can be achieved only by bringing together a number of different mitigating measures. Broadly, these involve bearing down on energy demand, ensuring adequate capacity, diversifying supply, and making greater investments in infrastructure [9]. Depending upon the type of disaster, various strategies have been developed in the literature to improve the resilience of energy systems [69–71].

System Hardening

System hardening is a combined system of actions that create a strong infrastructure to better protect utility customers from weather-related disruptions, e.g., windstorms and floods. From physical structures to communications to effective documentation, energy system hardening entails an end-to-end approach that better protects infrastructure and improves levels of service in the event of extreme weather [72]. Hardening measures usually require substantial investment. Some common hardening practices are summarized as follows [22, 69, 73]:

- Undergrounding the distribution/ transmission lines;
- Upgrading the distribution system poles with stronger, more robust materials;
- Elevating substations and relocating facilities;
- Building flood defense systems for new or existing ground-level infrastructure;
- Redundancy in transmission and distribution system;
- Tree trimming/vegetation management.

Distributed Generation

Resilient energy systems often incorporate a diverse portfolio of energy generation technologies to increase reliability and allow for service disruptions to be mitigated and resolved quickly [74]. If the energy system of a city is set up on a single energy network, an outage in one area of the city will have a ripple effect on the rest of the energy system. Distributed generation can help as it expands the number and type of generation units that can be utilized at any time, so that if part of a network fails, the damage will not be as widespread [75]. There are a range of options for energy and electricity supply. For heat, natural gas plays the largest role, but biomass could also be a substitute to some extent. However, substantial use of natural gas for power generation may present issues of diversity and security of supply [9]. In this context, the increasing penetration of distributed (decentralizing) energy resources such as microturbines, wind turbines, photovoltaic (PV) panels, and energy storage systems can play a key role in providing resilience to external shocks. On-site renewable energy generation (or other generators) can reduce the disruption of a grid outage at a site, mitigating the impacts of a long-term outage. Microgrids, where on-site renewable energy is combined with energy storage to support resilience, are often implemented [76].

When integrated into a microgrid, distributed generation can increase survival time, when fuel supplies are limited during a grid outage [77]. In fact, generating, storing, and controlling energy locally without the need of long transmission lines can make a network less vulnerable to disasters as well as make the response to an emergency much faster and more efficient [78]. It is worth noting that distributed energy technologies can address specific resilience challenges but are only a part of the solution.

Energy Storage

Integrating storage into electricity grid systems and district heating networks can help smooth variation in renewable energy sources, such as wind and solar generation, while adding redundancy to the grid in the event of power outages [76]. While the most common energy storage technologies are batteries, other storage methods may include flywheels, compressed air energy storage, and pumped hydro power storage [76]. In countries with high direct normal irradiance ($\geq 2000\,kWh/m^2$), solar thermal energy storage systems are preferred methods of addressing the intermittent nature of solar power. While distributed or microgrid renewable energy systems can be incorporated without energy storage, adding the flexibility of energy storage technologies may allow microgrids or critical loads to operate for longer periods of time without grid connection [76].

Smart Grid Technology

Smart grid technologies can improve the overall efficiency of a power system's operation, increase its visibility, and improve its response to faults and outages [22]. A smart grid is an electricity network (Figure 14.8) that uses digital and other advanced technologies to monitor and manage the transport of electricity from all generation sources to meet the varying electricity demands of end users [79]. A smart grid provides two-way communication and

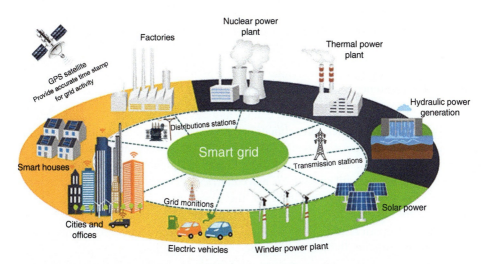

Figure 14.8 Layout of a typical smart-grid-based energy system.

flexible control within an electricity network in order to support flexibility, efficiency, and resilience. Automation technologies, or devices within a smart grid, collect data in real time that can be analyzed by utilities and system operators [76]. In the context of a climate-resilient energy system, smart grids can support several positive outcomes [71, 80]. For example, because of real-time communication on changing needs and circumstances associated with extreme climate-related events (e.g., storms and other disasters), responses and decisions regarding new power needs and power diversion can be made quickly. If a power generation source is disrupted by a storm, other generation sources can be accessed more quickly, and power can also be diverted to critical facilities, such as hospitals. Smart meters can allow for this type of responsiveness by sending data directly from electricity meters to energy providers and can enable islandable systems to support resilience during disasters or grid outages [76].

Enhancing Energy Efficiency

The connection between an efficient energy system and a resilient energy system has been identified as an opportunity to better integrate efficiency and resilience prior to disasters and during post-disaster rebuilds [81]. Energy efficiency improvements can be defined as reductions in the amount of energy needed to provide an energy service, like heating a home or driving a car. More energy-efficient technologies and measures allow less energy to be used in providing the same level of service [73]. Examples include energy-efficient household appliances, energy-efficient building designs and materials, more fuel-efficient vehicles, and switching from fossil fuels to electricity for energy end uses. Energy efficiency can reduce vulnerabilities to energy disruptions by lessening demand pressures on energy systems and enabling households to endure extreme events such as heat waves at a lower cost [73]. As shown in Table 14.5, the numerous benefits of energy efficiency can make it an effective strategy for improving the resilience of energy systems. In the latest UN Climate Change Conference, COP26 that took place on November 2021 in Glasgow, Scotland, the signatories were encouraged to double the energy efficiency of four key energy-use equipment by 2030 (motors, air conditioners, refrigerators, and lighting), which together currently account for over 40% of energy consumption [83].

Make Climate Resilience a Central Part of Energy System Planning

There is a clear linkage between climate change and more intense heat, damaging hurricanes, and wildfires [84]. Therefore, climate resilience should be integral to government policymaking as well as energy system planning by utilities and relevant industries [84, 85]. However, in many countries, the level of commitment and progress toward climate resilience in the energy sector still lag [49]. Even in countries where national strategies or plans for climate change adaptation are in place, the urgent need for climate resilience in the energy sector has often been overlooked. Mainstreaming climate resilience in energy and climate policies can send a strong signal to the private sector, inspiring businesses to consider climate resilience in their planning and operations [49].

Table 14.6 illustrates how the six components (shown in Figure 14.3) that define resilience are connected to the actions that enhance the capacity of an entity to be resilient.

Table 14.5 Resilience benefits of energy efficiency.

Benefit type	Energy efficiency outcome	Resilience benefit
Emergency response and recovery	Reduced electric demand	Increased reliability during times of stress on electric system and increased ability to respond to system emergencies
	Backup power supply from combined heat and power (CHP) and microgrids	Ability to maintain energy supply during emergencies or disruptions
	Efficient buildings that maintain temperatures	Residents can shelter in place as long as the buildings' structural integrity is maintained
	Multiple modes of transportation and efficient vehicles	Several travel options that can be used during evacuations and disruptions
Social and economic	Local economic resources may stay in the community	Stronger local economy that is less susceptible to hazards and disruptions
	Reduced exposure to energy price volatility	Economy is better positioned to manage energy price increases, and households and businesses are better able to plan for the future
	Reduced spending on energy	Ability to spend income on other needs, increasing disposable income (especially important for low- income families)
	Improved indoor air quality and emission of fewer local pollutants	Fewer public health stressors
Climate mitigation and adaptation	Reduced greenhouse gas emissions from power sector	Mitigation of climate change
	Cost-effective efficiency investments	More leeway to maximize investment in resilient redundancy measures, including adaptation measures

Source: Ribeiro et al. [82]/American Council for an Energy-Efficient Economy.

Further details of energy system hardening techniques can be found in the Electric Power Research Institute (EPRI) guide [86], which describes innovative technologies, strategies, tools, and systems that EPRI and electricity sector stakeholders are developing and applying to address the challenge of resiliency.

Conclusion

Modern society relies heavily on high energy consumption; thus, a reliable, secure, and affordable energy supply is essential to spur economic growth and development. However, energy security and the physical impacts of climate change are posing increasingly formidable challenges to energy systems; addressing this requires more resilient energy systems that can anticipate, prepare for, and adapt to these changing conditions. The resilience aspect of the energy system is relatively new, and so far, no consensus has been reached on

Table 14.6 Relationship between components of resilience and resilience-enhancing measures.

Resilience enhancing measures	Components of resilience	Task/activities required	Example
Preparedness	Anticipate	Preparedness refers to activities undertaken by an entity to define the hazard environment to which it is subjected	Creation of energy assurance plans and strategic energy plans Coordinating communications between responders Development of continuity, contingency, and strategic plans Training and exercising of plans
Mitigation	Resist Absorb	Mitigation refers to activities taken prior to an event to reduce the risk by reducing consequences, vulnerabilities, and threats/hazard	Fences Intrusion detection systems Closed circuit television Hardening, strengthening, and retrofitting Automation and smart monitoring investments Backup generators
Response	Respond Adapt	Response refers to immediate and ongoing activities, tasks, programs, and systems that have been undertaken or developed to manage the adverse effects of an event. These capabilities are typically associated with actions taken immediately following the event	Mobile incident management and command center Mutual aid agreements Coordinating agreements between energy system assets and emergency response entities
Recovery	Recover	Recovery refers to activities and programs designed to effectively and efficiently return conditions to a level that is acceptable to the entity. Recovery measures usually consist of longer-term remediation measures	Critical material provider priority plans Access to critical equipment Memorandum of understanding/memorandum of agreement activation, e.g., with material providers or outside contractors After-action reporting and lessons learned

Source: Adapted from: [68].

the universally accepted definition of a resilient energy system. Unlike reliability, which has well-established metrics, energy system resilience is not absolute, which means that incremental steps can be taken to develop a resilient energy system, and that total resilience is not possible in the face of all situations. In the context of energy systems, resilience is not a static concept, so a dynamic and continuous approach is required to adapt the frameworks and operations for superior readiness against unexpected external disruptive events. There is no one-size-fits-all solution for enhancing energy system resilience, and it needs to be tailored to each specific energy system costs, fuel supply chain, lifestyle of local communities, and weather risks. In developing a successful energy system approach, a combination of hardening and resiliency measures that address low-probability, high-impact, internal, and external disruptive events should be considered.

References

1 Bukowski, S.A., Culler, M.J., and Gentle, J.P. (2021). Distributed wind resilience metrics for electric energy delivery systems: comprehensive literature review. Idaho National Laboratory (INL), Idaho Falls, United States, INL/EXT-21-62149. https://inldigitallibrary.inl.gov/sites/sti/sti/Sort_37381.pdf.
2 Shandiz, S.C., Foliente, G., Rismanchi, B. et al. (2020). Resilience framework and metrics for energy master planning of communities. *Energy* 203: 117856.
3 The Paris Agreement (2015). United Nations framework convention on climate change. https://unfccc.int/sites/default/files/english_paris_agreement.pdf.
4 Panteli, M. and Mancarella, P. (2015). Influence of extreme weather and climate change on the resilience of power systems: impacts and possible mitigation strategies. *Electr. Power Syst. Res.* 127: 259–270. https://doi.org/10.1016/j.epsr.2015.06.012.
5 Panteli, M. and Mancarella, P. (2015). Modeling and evaluating the resilience of critical electrical power infrastructure to extreme weather events. *IEEE Syst. J.* 11 (3): 1733–1742.
6 Nik, V.M., Perera, A.T.D., and Chen, D. (2021). Towards climate resilient urban energy systems: a review. *Natl. Sci. Rev.* 8 (3): nwaa134.
7 U.S. Department of Energy, Office of Electricity Delivery and Energy Reliability (2013). Economic benefits of increasing electric grid resilience to weather outages. https://www.energy.gov/sites/prod/files/2013/08/f2/Grid%20Resiliency%20Report_FINAL.pdf.
8 International Energy Agency (IEA) (2015). Making the energy sector more resilient to climate change, Paris, France. https://www.iea.org/reports/making-the-energy-sector-more-resilient-to-climate-change.
9 Chaudry, M., Ekins P., Ramachandran, K. et al. (2011). Building a Resilient UK Energy System, UK Energy Research Centre, Research Report REF UKERC/RR/HQ/2011/001. https://nora.nerc.ac.uk/id/eprint/16648/1/UKERC_energy_2050_resilience_Res_Report_2011.pdf.
10 Dehghani, A., Sedighizadeh, M., and Haghjoo, F. (2021). An overview of the assessment metrics of the concept of resilience in electrical grids. *Int. Trans. Electr. Energy Syst.* 31 (12): e13159.
11 Plotnek, J.J. and Slay, J. (2021). Power systems resilience: definition and taxonomy with a view towards metrics. *Int. J. Crit. Infrastruct. Prot.* 33: 100411.
12 Elsayed, E.A. (2021). *Reliability Engineering*, 3e. John Wiley & Sons, Inc.
13 Raoufi, H., Vahidinasab, V., and Mehran, K. (2020). Power systems resilience metrics: a comprehensive review of challenges and outlook. *Sustainability* 12 (22): 9698.
14 Hulse, D., Hoyle, C., Tumer, I.Y. et al. (2020). Temporal fault injection considerations in resilience quantification. In: *International Design Engineering Technical Conferences and Computers and Information in Engineering Conference*, American Society of Mechanical Engineers, vol. 84003, p. V11AT11A040.

15 Kandaperumal, G. and Srivastava, A.K. (2020). Resilience of the electric distribution systems: concepts, classification, assessment, challenges, and research needs. *IET Smart Grid* 3 (2): 133–143.

16 Hickford, A.J., Blainey, S.P., Ortega Hortelano, A., and Pant, R. (2018). Resilience engineering: theory and practice in interdependent infrastructure systems. *Environ. Syst. Decis.* 38 (3): 278–291.

17 Azzuni, A. and Breyer, C. (2018). Definitions and dimensions of energy security: a literature review. *Wiley Interdiscip. Rev. Energy Environ.* 7 (1): e268.

18 Baros, S., Shiltz, D., Jaipuria, P. et al. (2017). *Towards Resilient Cyber-Physical Energy Systems*. United States: DSpace, Massachusetts Institute of Technology (MIT).

19 Li, Z., Shahidehpour, M., Aminifar, F. et al. (2017). Networked microgrids for enhancing the power system resilience. *Proc. IEEE* 105 (7): 1289–1310.

20 Liu, X., Ferrario, E., and Zio, E. (2017). Resilience analysis framework for interconnected critical infrastructures. *ASCE-ASME J. Risk Uncert. Eng. Syst. B: Mech. Eng.* 3 (2): 021001.

21 Erker, S., Stangl, R., and Stoeglehner, G. (2017). Resilience in the light of energy crises– Part I: a framework to conceptualise regional energy resilience. *J. Cleaner Prod.* 164: 420–433.

22 Bie, Z., Lin, Y., Li, G., and Li, F. (2017). Battling the extreme: a study on the power system resilience. *Proc. IEEE* 105 (7): 1253–1266.

23 Panteli, M., Mancarella, P., Trakas, D.N. et al. (2017). Metrics and quantification of operational and infrastructure resilience in power systems. *IEEE Trans. Power Syst.* 32 (6): 4732–4742.

24 Xu, Y., Liu, C.-C., Schneider, K.P. et al. (2016). Microgrids for service restoration to critical load in a resilient distribution system. *IEEE Trans. Smart Grid* 9 (1): 426–437.

25 Friedberg, I., McLaughlin, K., Smith, P., and Wurzenberger, M. (2016). Towards a resilience metric framework for cyber-physical systems. In: *4th International Symposium for ICS & SCADA Cyber Security Research 2016*, Association for Computing Machinery, 4, pp. 19–22.

26 Arghandeh, R., von Meier, A., Mehrmanesh, L., and Mili, L. (2016). On the definition of cyber-physical resilience in power systems. *Renew. Sustain. Energy Rev.* 58: 1060–1069. https://doi.org/10.1016/j.rser.2015.12.193.

27 Espinoza, S., Panteli, M., Mancarella, P., and Rudnick, H. (2016). Multi-phase assessment and adaptation of power systems resilience to natural hazards. *Electr. Power Syst. Res.* 136: 352–361.

28 Bahramirad, S., Khodaei, A., Svachula, J., and Aguero, J.R. (2015). Building resilient integrated grids: one neighborhood at a time. *IEEE Electrif. Mag.* 3 (1): 48–55.

29 Alderson, D.L., Brown, G.G., and Carlyle, W.M. (2015). Operational models of infrastructure resilience. *Risk Anal.* 35 (4): 562–586.

30 Linkov, I., Bridges, T., Creutzig, F. et al. (2014). Changing the resilience paradigm. *Nat. Clim. Change* 4 (6): 407–409.

31 Khodaei, A. (2014). Resiliency-oriented microgrid optimal scheduling. *IEEE Trans. Smart Grid* 5 (4): 1584–1591.

32 Cutter, S.L., Ahearn, J.A., Amadei, B. et al. (2013). Disaster resilience: a national imperative. *Environ. Sci. Policy Sustain. Dev.* 55 (2): 25–29.

33 Ouyang, M., Dueñas-Osorio, L., and Min, X. (2012). A three-stage resilience analysis framework for urban infrastructure systems. *Struct. Saf.* 36: 23–31.

34 Vugrin, E.D., Warren, D.E., and Ehlen, M.A. (2011). A resilience assessment framework for infrastructure and economic systems: quantitative and qualitative resilience analysis of petrochemical supply chains to a hurricane. *Process Saf. Prog.* 30 (3): 280–290.

35 Rudnick, H. (2011). IEEE-INST Electrical Electronics Engineers INC 445 Hoes Lane, Piscataway, NJ. *Nat. Disasters* 9 (2): 22–26.

36 Hollnagel, E. (2011). Prolog-the scope of resilience engineering. In: *Resilience Engineering in Practice* (ed. H. Erik). Japan: CiNii Research.

37 López-Marrero, T. and Tschakert, P. (2011). From theory to practice: building more resilient communities in flood-prone areas. *Environ. Urban.* 23 (1): 229–249.

38 Mili, L., Taxonomy of the characteristics of power system operating states. *2nd NSF-VT Resilient and Sustainable Critical Infrastructures (RESIN) Workshop*, 2011, pp. 13–15.

39 Aven, T. (2011). On some recent definitions and analysis frameworks for risk, vulnerability, and resilience. *Risk Anal. Int. J.* 31 (4): 515–522.

40 Longstaff, P.H., Armstrong, N.J., Perrin, K. et al. (2010). Building resilient communities: a preliminary framework for assessment. *Homel. Secur. Aff.* 6 (3): 1–23.

41 Rits, M. (2019). Energy resilience and the 5 R's. Energy Express. https://www.afcec.af.mil/Portals/17/documents/Energy/Energy%20Express/07%20Energy%20Express%20Aug%2019%20Final.pdf?ver=2019-08-12-160727-127.

42 Mishra, S., Anderson, K., Miller, B. et al. (2020). Microgrid resilience: a holistic approach for assessing threats, identifying vulnerabilities, and designing corresponding mitigation strategies. *Appl. Energy* 264: 114726. https://doi.org/10.1016/j.apenergy.2020.114726.

43 American Gas Association (AGA) and American Public Gas Association (APGA) (2021). Comments of AGA and APGA on the Department of Energy's 2021 Climate Adaptation and Resilience Plan. American Gas Association (AGA) and American Public Gas Association (APGA), USA. https://www.apga.org/HigherLogic/System/DownloadDocumentFile.ashx?DocumentFileKey=7d32ffc3-eb13-9b62-5dd6-73f49447451b&forceDialog=0.

44 Wijayatunga, P., Jayawardena, M., Kiatgrajai, M., and Jin, X. (2021). Building sustainability and resilience in the energy sector. Asian Development Bank (ADB), Webinar. https://www.preventionweb.net/news/building-sustainability-and-resilience-energy-sector.

45 Minkel, J. (2008). The 2003 Northeast blackout–five years later. *Sci. Am.* 13: 1–3.

46 Gluckman, P. and Bardsley, A. (2022). Policy and political perceptions of risk: the challenges to building resilient energy systems. *Philos. Trans. R. Soc. A* 380 (2221): 20210146.

47 Hallegatte, S., Rentschler, J., and Rozenberg, J. (2019). *Lifelines: The Resilient Infrastructure Opportunity*. World Bank Publications.

48 Anderson, K., Li, X., Dalvi, S. et al. (2020). Integrating the value of electricity resilience in energy planning and operations decisions. *IEEE Syst. J.* 15 (1): 204–214.

49 International Energy Agency (IEA) (2020). Power systems in transition: challenges and opportunities ahead for electricity security. International Energy Agency (IEA), Paris, France. https://iea.blob.core.windows.net/assets/cd69028a-da78-4b47-b1bf-7520cdb20d70/Power_systems_in_transition.pdf.

50 Hallegatte, S., Rozenberg, J., Maruyama Rentschler, J.E. et al. (2019). Strengthening new infrastructure assets: a cost-benefit analysis. World Bank Policy Research Working Paper, no. 8896.

51 Nicolas, C., Jun Rentschler; Sam Oguah, et al. (2019). Stronger power: improving power sector resilience to natural hazards. World Bank Group. https://openknowledge.worldbank.org/bitstream/handle/10986/31910/Stronger-Power-Improving-Power-Sector-Resilience-to-Natural-Hazards.pdf.

52 European Environment Agency, European Union (EU) (2019). Adaptation challenges and opportunities for the European energy system: building a climate-resilient low-carbon energy system. European Environment Agency, European Union (EU), EEA Report No 01/2019. https://www.eea.europa.eu/publications/adaptation-in-energy-system/download.

53 International Energy Agency (IEA) (2016). Energy, climate change and environment: 2016 Insights. International Energy Agency (IEA), Paris, France. https://iea.blob.core.windows.net/assets/6b2eaf11-d479-4ab2-b92f-a8832cda61e8/ECCE2016.pdf.

54 Rooban, A. and Terton, A. (2018). Safeguarding NDC Implementation: Building resilience into energy systems. International Institute for Sustainable Development (IISD), Lombard Ave, Canada. https://www.iisd.org/system/files/publications/safeguarding-ndc-implementation.pdf.

55 Ebinger, J.O. (2011). *Climate Impacts on Energy Systems: Key Issues for Energy Sector Adaptation*. World Bank Publications.

56 Bhusal, N., Abdelmalak, M., Kamruzzaman, M., and Benidris, M. (2020). Power system resilience: current practices, challenges, and future directions. *IEEE Access* 8: 18064–18086.

57 Nik, V.M. and Perera, A. (2020). The importance of developing climate-resilient pathways for energy transition and climate change adaptation. *One Earth* 3 (4): 423–424.

58 Perera, A., Nik, V.M., Chen, D. et al. (2020). Quantifying the impacts of climate change and extreme climate events on energy systems. *Nat. Energy* 5 (2): 150–159.

59 Nik, V.M. (2016). Making energy simulation easier for future climate–synthesizing typical and extreme weather data sets out of regional climate models (RCMs). *Appl. Energy* 177: 204–226.

60 Auffhammer, M., Baylis, P., and Hausman, C.H. (2017). Climate change is projected to have severe impacts on the frequency and intensity of peak electricity demand across the United States. *Proc. Natl. Acad. Sci.* 114 (8): 1886–1891.

61 Field, C.B., Barros, V., Stocker, T.F., and Dahe, Q. (2012). *Managing the Risks of Extreme Events and Disasters to Advance Climate Change Adaptation: Special Report of the Intergovernmental Panel On Climate Change*. Cambridge University Press.

62 Jesse, B.-J., Heinrichs, H.U., and Kuckshinrichs, W. (2019). Adapting the theory of resilience to energy systems: a review and outlook. *Energy, Sustain. Soc.* 9 (1): 1–19.

63 Wang, C. and Blackmore, J.M. (2009). Resilience concepts for water resource systems. *J. Water Resour. Plan. Manag.* 135 (6): 528–536. https://doi.org/10.1061/(ASCE)0733-9496(2009)135:6(528).

64 Liao, K.-H. (2012). A theory on urban resilience to floods—a basis for alternative planning practices. *Ecol. Soci.* 17 (4): 1–15.

65 Folke, C. (2006). Resilience: the emergence of a perspective for social–ecological systems analyses. *Glob. Environ. Change* 16 (3): 253–267.

66 Laboy, M. and Fannon, D. (2016). Resilience theory and praxis: a critical framework for architecture. *Enquiry ARCC J. Archit. Res.* 13 (1): 39–53.

67 National Institute of Standards and Technology (2016). Community resilience economic decision guide for buildings and infrastructure systems – Volume 2. National Institute of Standards and Technology, US Department of Commerce, United Staes, NIST special publication 197. https://nvlpubs.nist.gov/nistpubs/SpecialPublications/NIST.SP.1190v2.pdf.

68 Phillips, J., Finster, M., Pillon, J. et al. (2016). State energy resilience framework. Argonne National Lab.(ANL), Argonne, IL (United States).

69 Lin, Y., Bie, Z., and Qiu, A. (2018). A review of key strategies in realizing power system resilience. *Glob. Energy Interconnect.* 1 (1): 70–78. https://doi.org/10.14171/j.2096-5117.gei.2018.01.009.

70 Birnie, D.P. (2014). Optimal battery sizing for storm-resilient photovoltaic power island systems. *Solar Energy* 109: 165–173. https://doi.org/10.1016/j.solener.2014.08.016.

71 Nikkhah, S., Jalilpoor, K., Kianmehr, E., and Gharehpetian, G.B. (2018). Optimal wind turbine allocation and network reconfiguration for enhancing resiliency of system after major faults caused by natural disaster considering uncertainty. *IET Renew. Power Gener.* 12 (12): 1413–1423.

72 Richard, J. (2017). The keys to grid hardening: how to implement an effective solution. *Leidos* https://www.leidos.com/insights/keys-grid-hardening-how-implement-effective-solution (accessed Jun. 27, 2022).

73 Clark, D. and Kanduth, A. (2022). Enhancing the resilience of Canadian electricity systems for a net zero future. *Can. Inst. Clim. Choices* https://climateinstitute.ca/wp-content/uploads/2022/02/Resiliency-scoping-paper-ENGLISH-Final.pdf.

74 Cox, S., Gagnon, P., Stout, S. et al. (2016). Distributed generation to support development-focused climate action. National Renewable Energy Lab (NREL), Golden, CO (United States).

75 (2016). Large Corp. Combat Climate Change with RSI Solutions. Amigo Energy. https://amigoenergy.com/blog/building-sustainable-and-resilient-energy-systems/ (accessed 27 June 2022).

76 Hotchkiss, E.L. and Cox, S. (2019). Resilient energy platform: power sector resilience technical solutions," National Renewable Energy Lab (NREL), Golden, CO (United States).

77 Elgqvist, E.M., Becker, W.L., Gagne, D.A. et al. (2019). Energy exchange pre-conference workshop: distributed energy technologies for resilience and cost savings. National Renewable Energy Lab (NREL), U.S. Department of Energy, Office of Scientific and Technical Information, Golden, CO (United States).

78 Panteli, M. and Mancarella, P. (2015). A stronger, bigger or smarter grid? conceptualizing the resilience of future power infrastructure. *IEEE Power Energy Mag. (May)* 1–16.

79 Sendin, A., Matanza, J., and Ferrus, R. (2021). *Smart Grid Telecommunications: Fundamentals and Technologies in the 5G Era*. John Wiley & Sons.

80 Cox, S.L., Hotchkiss, E.L., Bilello, D.E. et al. (2017). Bridging climate change resilience and mitigation in the electricity sector through renewable energy and energy efficiency: Emerging climate change and development topics for energy sector transformation. National Renewable Energy Lab (NREL), U.S. Department of Energy, Office of Scientific and Technical Information, Golden, CO (United States).

81 Baechler, M.C. and Gilbride, T.L. (2018). Residential energy efficiency and the three Rs: resistance, resilience, and recovery. Pacific Northwest National Lab (PNNL), U.S. Department of Energy, Office of Scientific and Technical Information, Richland, WA (United States).

82 Ribeiro, D., Mackres, E., Baatz, B., Cluett, R., Jarrett, M., Kelly, M. and Vaidyanathan, S. (2015). Enhancing community resilience through energy efficiency. American Council for an Energy-Efficient Economy. Report U1508, American Council for an Energy-Efficient Economy, Washington, DC 20045.

83 (2021). Joint statement in support of the UK-IEA Product Efficiency Call to Action to raise global ambition through the SEAD initiative. UN Climate Change Conference (COP26) at the SEC – Glasgow. 04 November 2021. https://ukcop26.org/joint-statement-in-support-of-the-uk-iea-product-efficiency-call-to-action-to-raise-global-ambition-through-the-sead-initiative/ (accessed 27 June 2022).

84 R. E. L. and Fleming, J. (2021). The climate wolf at the door: why and how climate resilience should be central to building back better. *Brookings* 18: https://www.brookings.edu/research/the-climate-wolf-at-the-door-why-and-how-climate-resilience-should-be-central-to-building-back-better/ (accessed 27 June 2022).

85 Birol, F. (2021). How to make electricity systems resilient to climate risks. World Economic Forum. https://www.weforum.org/agenda/2021/07/climate-change-electricity-energy-security-extreme-weather/ (accessed 27 June 2022).

86 Electric Power System Resiliency (2016). Challenges and opportunities. Electric Power Research Institute, United States. https://www.epri.com/#/pages/product/000000003002007376/.

15

Building Resilient Health Policies: Incorporating Climate Change Impacts for Sustainable Adaptation

Furqan Tahir[1,2], Fama N. Dieng[3] and Sami G. Al-Ghamdi[1,2,3]

[1] Environmental Science and Engineering Program, Biological and Environmental Science and Engineering Division, King Abdullah University of Science and Technology (KAUST), Thuwal, Saudi Arabia
[2] KAUST Climate and Livability Initiative, King Abdullah University of Science and Technology (KAUST), Thuwal, Saudi Arabia
[3] Division of Sustainable Development, College of Science and Engineering, Hamad Bin Khalifa University, Qatar Foundation, Doha, Qatar

Introduction

The evidence supporting the observed increase in global air and ocean temperatures, extensive snow and ice melting, and rising sea levels relies on the climate change hypothesis. Extensive documentation demonstrates that all components of the Earth's climate system have experienced long-term effects of climate change [1]. The continuous emission of greenhouse gases contributes to ongoing planetary warming, exerting long-term impacts on the entire climate system [2, 3]. As a result, the likelihood and frequency of extreme climate events, including floods and heat waves, are expected to rise, necessitating immediate responses and risk reduction efforts. However, gradual shifts in average sea levels and air temperatures over time also allow urban areas to adapt [4, 5].

The impacts of climate change on different societal sectors are intricately linked [6]. For instance, drought can adversely affect food production and human health. Flooding has the potential to propagate diseases and cause damage to infrastructure and ecosystems. Additionally, health-related issues can decrease worker productivity, increase mortality rates, and influence food availability. Figure 15.1 provides a comprehensive overview of climate change impacts across various sectors. The pervasive effects of climate change can be observed in every facet of our world. Nevertheless, it is important to note that the impacts of climate change do not exhibit uniformity at national or global scales; even within a single town, disparities can exist among individuals or neighborhoods.

In addition to having social and economic implications, climate change impacts the environment. The world's surface is warming because of an increase in greenhouse gases in the atmosphere brought on by human activity. Temperatures will quickly rise, endangering

---

Sustainable Cities in a Changing Climate: Enhancing Urban Resilience. First Edition.
Edited by Sami G. Al-Ghamdi.
© 2024 John Wiley & Sons Ltd. Published 2024 by John Wiley & Sons Ltd.

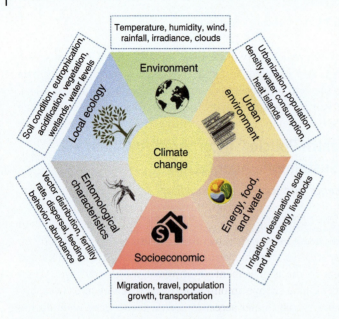

Figure 15.1 Examination of impacts of climate change on various sectors.

human health, along with thermal stress brought on by the climate, infectious diseases, and floods [7]. Such calamities increase the death rate because they overwhelm people's capacity to adjust [8]. Death rates rise as urban residents are more vulnerable than nonurban residents due to a lack of adequate housing equipped to handle thermal stress. Additionally, rising temperatures lead to more precipitation each year, and heavy rainfall, which causes human and animal waste to contaminate drinking water supplies and streams, gives rise to water-borne illnesses [9].

Climatic conditions also impact the burden of infectious diseases through their interactions with vectors, nonhuman reservoir species, and the pace at which infections grow [10]. An illustration of this is when the climate shifts in a region with low temperatures, tipping the ecological balance and sparking an epidemic [11]. In addition, it is projected that climate change will affect human migration patterns and the seasonal transmission of vector-borne diseases (VBDs). The spread of infectious diseases is accelerated by population mobility, particularly when temperatures and sea levels rise [12]. Human health and well-being are at risk because of the consequences of heat stress, the prevalence of infectious diseases, and extreme weather conditions [13]. Research and global health initiatives that tackle these issues must be prioritized to lessen their negative consequences on human health. This study explores the detrimental impacts of climate change on public health, such as infectious diseases, malnutrition, occupational health, respiratory diseases, and mental health. Furthermore, this chapter discusses how climate change impacts can be embedded in policy development in order to adapt to and mitigate climate change impacts.

Climate Change Impacts on Public Health

The increasing frequency and unpredictability of extreme weather events, such as severe droughts, rising temperatures, and elevated humidity, are clear indications of climate change. These climate shifts present multifaceted threats to human health and survival. Even with a mere 0.85 °C of warming, numerous threats have already manifested in the real world. Scientific evidence increasingly demonstrates that climate change substantially influences the likelihood of extreme weather events, often leading to severe health consequences [14]. Adverse health outcomes and conditions sensitive to climate include heat-related mortality or morbidity, respiratory illnesses associated with air pollution, infectious diseases, and health challenges displaced populations face due to forced migration. Figure 15.2 comprehensively depicts the health risks linked to climate change.

The progress made in recent years in areas such as development, global health, and poverty reduction is at risk of being undermined by climate-related disasters, which can potentially worsen existing health disparities between and within communities. This situation poses a significant threat to the achievement of universal health coverage (UHC) through various mechanisms. Climate change amplifies the disease burden already experienced and exacerbates the barriers that hinder access to healthcare services, often during critical periods. A considerable proportion, approximately 12%, of the global population—equivalent to over 930 million individuals—currently spends at least 10% of their household

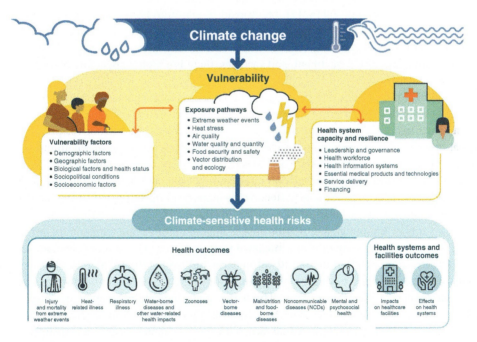

Figure 15.2 An outline of how climate change affects the associated health risks.
Source: WHO [15]/with permission of WHO.

income on healthcare expenses [15]. Furthermore, health shocks and pressures have already pushed an estimated 100 million people into poverty annually, and the effects of climate change further aggravate this trend, disproportionately affecting the most economically disadvantaged individuals who lack insurance coverage. As per the World Health Organization (WHO) fact sheet [15], it is stated:

- Climate change affects health's social and environmental determinants, including clean air, safe drinking water, sufficient food, and appropriate shelter.
- It is predicted that climate change will result in an additional 250,000 deaths per annum between 2030 and 2050, mostly from starvation, malaria, diarrhea, and heat stress.
- By the year 2030, it is estimated that the direct health damage expenses will range between USD $2 and 4 billion annually.
- Better dietary, transportation, and energy choices can reduce greenhouse gas emissions, enhancing health, especially by improving air quality.

Infectious Diseases

Even though efforts to eradicate infectious diseases succeeded in the twentieth century, there has been a reversal in recent years due to climate change, the deterioration of natural ecosystems, and increasing urbanization. The strain on healthcare systems in endemic areas was made worse by the cocirculation of mosquito-borne diseases along with the COVID-19 pandemic in 2020 [16, 17]. The spread of mosquito-borne diseases, especially those like dengue and chikungunya that are becoming increasingly widespread in recent years, may put many regions in danger [18–20]. Additionally, with insufficient infrastructure for testing, diagnosis, treatment, and vaccines, there is a higher possibility of shifting the burden to diseases.

Human health outcomes rely significantly on environmental changes, particularly climatic changes. The shift in temperature, humidity, and other factors resulting from climate change produces an environment that empowers health risks. In recent years, the presence of well-known diseases such as malaria and Lyme disease has expanded in the number of cases and geographical range [21]. For VBDs, three main factors are altered by climate change: vector distribution, transmission rates, and parasite development [22]. Although climate change may not be the sole reason for the increase in the burden of VBDs, there is sufficient global data to make predictions regarding the range of vectors [23].

In one study by Chapungu and Nhamo [24], changes in rainfall and temperature trends on VBDs, in Masvingo Province in Zimbabwe were analyzed. The authors concluded a positive correlation between the burden of malaria and temperature. There has been an increase in the average temperature of approximately 0.33 °C in Masvingo Province over the past 40 years. Furthermore, results from the study indicate a correlation between increases in malaria and a decrease in mean monthly precipitation. Another survey, taken by Roiz et al. [22] in the Mediterranean, examined 10 years of data biweekly to indicate that the effect of climatic conditions varied depending on the species of mosquitoes. For example, the number of *Culex pipiens* was higher during the winter when there was rainfall. While the aforementioned studies do not claim that climate change acts alone in the fluctuation of the burden of VBDs, they provide adequate evidence that climate change has a significant impact on the burden of VBDs and must be given special attention within the process of vector control and disease reduction.

Ren et al. [25] predicted a significant increase in the burden of malaria between the years 2030 and 2050. This study uses species distribution models for the years 2030–2040 and 2050–2060 and simulation-based approximations to evaluate the future impact of climate change on the four main malaria vectors. Climate change may not be the sole reason for an increase in the burden of VBD; however, it severely impacts the burden. Changes in temperature influence transmission rates as well as the state of the vectors themselves. An increase in temperature also affects environmental factors such as rainfall and relative humidity. Hence, in addition to vector-control methods, studying climate change and its impact on VBDs is crucial to preserving human health.

Diarrhea can arise from inadequate hygiene practices associated with water scarcity, particularly prevalent in impoverished nations. Alternatively, floods can lead to the contamination of drinking water sources across entire watersheds, facilitated by the runoff from sewage systems, containment lagoons, or general pollution. *Cryptosporidium* parasites, primarily linked to domestic cattle, can contaminate water intended for human consumption, particularly during periods of intense precipitation. A notable example occurred in 1993, when heavy spring rains and snowmelt runoff resulted in a cryptosporidiosis outbreak in Milwaukee, United States, causing over 50 fatalities and potentially exposing more than 400,000 individuals to *Cryptosporidium* [26].

Air Pollution

Climate change significantly impacts air pollution, exacerbating existing environmental challenges. The relationship between air temperature and ground-level ozone (a harmful pollutant) tends to be nonlinear, with a strong association observed only at temperatures above 32 °C. With people encouraged to grow and reproduce plants to help take in rising CO_2 levels, pollen levels in the air may also rise due to global warming. For instance, when ragweed (*Ambrosia artemisiifolia*) plants were experimentally subjected to high levels of CO_2, they multiplied the amount of pollen they produced, which may contribute to the rising quantities of ragweed pollen seen in recent years [27]. According to Ziska and Caulfield [28], ragweed grew more quickly, bloomed earlier, and generated more pollen in urban areas than in rural ones. This is likely due to urban regions' comparatively high CO_2 and air temperatures.

Moreover, the escalation of extreme weather events, including heat waves and wildfires, attributable to climate change worsens air pollution levels as these events release substantial amounts of pollutants into the atmosphere. The alterations in precipitation patterns resulting from climate change can disrupt the dispersion and deposition of contaminants, further impacting air quality. Consequently, climate change acts as a catalyst for the progressive deterioration of air quality, posing dire consequences for human health and the overall well-being of the environment. Thus, addressing the complex interplay between climate change and air pollution becomes imperative to mitigate the far-reaching impacts on our planet.

Extreme Events

Even though cultures worldwide have adapted to their local climates, vulnerable populations are still in danger during heat waves, and all climate change models predict considerable increases in heat waves [29]. Individuals with limited capacity for temperature regulation, such as those at extremes of age or those who are dehydrated, exhibit reduced tolerance to any changes in temperature. In the context of climate change, where rising temperatures are anticipated, these vulnerable populations are particularly susceptible to the adverse impacts of heat waves and other extreme weather events. The physiological limitations they face in maintaining temperature homeostasis render them less resilient in the face of temperature fluctuations brought about by climate change. Hence, it is crucial to recognize and address the heightened vulnerability of such individuals to ensure their well-being amid a changing climate.

Extremely high temperatures are known to be correlated with human sickness and mortality. Additionally, there is compelling evidence that, in various locations, the impacts of climate change are driving an increase in heat-related mortality [14]. The lives of numerous people have been harmed by recent floods, droughts, and violent storms, and killed millions worldwide. Globally, disasters claimed the lives of 123,000 individuals, on average, between 1972 and 1996. Although 80% of persons affected by natural disasters reside in Asia, Africa has the highest number of disaster-related fatalities [30]. The population's overall well-being can be significantly affected by post-traumatic stress disorder (PTSD) and other disaster-related mental disorders. Factors such as the unanticipated nature of the impact, the intensity of the experience, the extent of personal and community disruption, and prolonged exposure to the visual aftermath of the disaster can contribute to the development of mental health conditions. These conditions have a substantial impact on the well-being of affected populations.

Considerations in Health Policy Development

The results of health outcomes due to climate change significantly impact socioeconomic trends. Hence, community leaders must initiate and execute policy developments to reduce such negative health consequences. Because of climate change, health risks are more significant to vulnerable populations with inadequate healthcare services. One example of an effective policy that addresses climate change is emission abatement. Negative impacts of climate change can be reduced by emission reduction because it alters global trade and investments via changes in three areas: fossil fuel demands, competitiveness, and economic growth rates. Furthermore, climate change impacts are best addressed in an interdisciplinary manner that, instead of creating new initiatives, incorporates effective adaptation methods to healthcare initiatives that are already functioning [31]. The mitigation and adaptation measures that are thought to prevent adverse impacts on human health include:

Reducing Carbon Emissions

Nations must expedite the coal phaseout from the global energy mix to safeguard cardiovascular and respiratory health. Failing to replace coal with cleaner energy sources could lead to a proliferation of coal-fired facilities, thereby exacerbating the risks associated with human health [32]. Extensive scientific evidence demonstrates the detrimental impacts of coal combustion on air quality, including the emission of hazardous pollutants such as fly ash, particulate matter, sulfur dioxide, and nitrogen oxides [33]. These pollutants contribute to the development and exacerbation of cardiovascular and respiratory diseases, posing significant threats to public health. Therefore, a decisive shift from coal toward cleaner energy alternatives is essential to mitigate the health risks of coal combustion and secure a healthier future for communities worldwide.

Medical Interventions

In order to empower medical practitioners and guarantee that climate and health issues are fully incorporated into government-wide initiatives, new procedures should be devised to promote collaboration between health ministries and other government agencies [34]. It will be ineffective to combat climate change with fragmented strategies. It is crucial to acknowledge and address the potential impact of worldwide environmental changes, including deforestation, ocean acidification, biodiversity loss, and on human health and resilience to climate change.

Healthy Lifestyle

Governments are responsible for ensuring that urban areas not only foster healthy environments for people but also promote active lifestyles that benefit both individuals and the planet. An effective strategy involves developing energy-efficient buildings and creating accessible green spaces within urban settings. These measures enhance residents' capacity to adapt to a changing climate and significantly reduce the risk of various health conditions, including respiratory, mental, and cardiovascular diseases, as well as cancer, obesity, and diabetes. Simultaneously, by implementing such initiatives, urban areas can achieve the dual benefits of curbing carbon emissions and mitigating urban pollutants. By embracing these approaches, governments can create sustainable and livable cities that prioritize the well-being of their inhabitants, improve public health outcomes, and contribute to a healthier, greener future for generations to come.

Monitoring

The enhancement of national capacities to efficiently monitor and respond to disease outbreaks is of utmost significance. The collection and analysis of robust data play a vital role in climate change adaptation within human health. This includes gathering crucial information on health climate risks, vulnerabilities, and diseases linked to the impacts of climate change. By establishing comprehensive public health surveillance and monitoring systems, governments can acquire valuable data and information necessary to assess disease burdens, identify at-risk individuals and vulnerable groups, comprehend patterns of illness, and develop targeted public health interventions and response strategies [35]. Such data-driven approaches empower policymakers and public health practitioners to proactively address emerging health threats, allocate resources efficiently, and prioritize effective interventions. Strengthening the infrastructure for disease surveillance and monitoring is indispensable for safeguarding public health, building resilience to climate change, and effectively mitigating the adverse health consequences of a rapidly evolving world.

Proactive Approaches

The health sector plays a critical role in identifying and implementing solutions that address the root causes of vulnerability and effectively respond to the impacts of natural disasters. The health sector can significantly reduce the adverse effects of current and future disasters by proactively planning for such occurrences. For instance, implementing sound land management practices that prioritize reducing deforestation, preserving watersheds and coastal zones, and mitigating physical hazards like mudslides and storm surges can significantly diminish the health consequences associated with floods. Additionally, proactive measures by the health sector, such as implementing strategies to prevent the spread of vector-borne and water-borne diseases, can effectively mitigate the harmful health effects that often follow in the wake of flooding events [36]. By taking a comprehensive approach that combines environmental management and targeted public health responses, the health sector can play a pivotal role in minimizing the health impacts of natural disasters. Such proactive and coordinated efforts are essential for building resilient communities and safeguarding public health in the face of increasingly frequent and severe natural disasters.

Strengthening Institutions

The involvement of diverse entities and organizations is essential in developing and implementing health adaptation policies and activities, encompassing government agencies, nongovernmental organizations (NGOs), informal associations, family networks, and traditional institutions [37]. Effective health adaptation to climate change necessitates collaboration and strengthening of institutional capacities across various levels while assessing institutional needs [14]. Networks that foster connections and partnerships among enterprises and stakeholders are crucial to overcoming fragmentation, promoting cooperation, facilitating information flows, and enabling collective learning [38].

Specific types of studies must be conducted to create effective climate change policies that will aid in reducing associated health risks, such as:

1) Detection and attribution studies
2) Scenario modeling
3) Studies on resilience and risk management approaches.

The first type of studies that should be used are detection and attribution, which identify transformations in health risks and determine whether the transformations are due to anthropogenic climate change. The second type of study is scenario modeling. With scenario modeling, researchers can make predictions regarding the future impacts of climate change by using models based on transmission processes, vector distribution, and meteorological variables. The third type of study crucial to the foundation of climate change is healthcare policymaking, which is used to develop resilience and risk management approaches.

Rather than specifically focusing on climate change, these studies focus on producing more resilience toward climate change by expanding public health and disease-control interventions [39]. Furthermore, effects on health stemming directly from developmental and environmental policies heavily rely on the ability of ecosystem services to replace health risks within varying types of landscapes. Studying socioeconomic policy scenarios allows researchers to examine how factors, such as land use, interact with climate change to increase health risks [40].

Conclusion

This chapter examines the impacts of climate change on public health, specifically focusing on various adverse effects, such as infectious diseases, malnutrition, occupational health risks, respiratory disorders, and mental health implications. Additionally, it has been discussed how to incorporate the effects of climate change into creating policies that will adapt to and reduce those effects. Some of the key highlights are as follows:

- Governments should ensure a swift phaseout of fossil fuels from the world's energy mix to protect cardiovascular and respiratory health.
- Between 2030 and 2050, an estimated 250,000 deaths per year are expected due to climate change, most of which will be caused by malnutrition, malaria, diarrhea, and heat stress.
- Detection and attribution studies can identify transformations in health risks and determine whether the transformations are due to anthropogenic climate change.
- To empower medical practitioners and guarantee climate and health issues, new procedures should be devised to promote collaboration between health ministries and other government agencies.
- The health sector should work to develop solutions that address the underlying causes of vulnerability and plan for efficient responses to such occurrences to lessen the effects of current and upcoming natural disasters.
- The data and information collected through public health surveillance or monitoring systems serve multiple purposes, including assessing disease burdens and trends, identifying at-risk individuals and groups, understanding illness patterns, and informing the development of public health interventions and response strategies.
- With scenario modeling, researchers can make predictions regarding the future impacts of climate change by using models based on transmission processes, vector distribution, and meteorological variables.
- Studies on resilience and risk management approaches are crucial to the foundation of climate change healthcare policy.

References

1 Salimi, M. and Al-Ghamdi, S.G. (2020). Climate change impacts on critical urban infrastructure and urban resiliency strategies for the Middle East. *Sustain. Cities Soc.* 54: 101948. https://linkinghub.elsevier.com/retrieve/pii/S2210670719315872.

2 Tahir, F. and Al-Ghamdi, S.G. (2023). Climatic change impacts on the energy requirements for the built environment sector. *Energy Rep.* 9: 670–676. https://linkinghub.elsevier.com/retrieve/pii/S2352484722024192.

3 Zhang, J., You, Q., Ren, G. et al. (2023). Inequality of global thermal comfort conditions changes in a warmer world. *Earth's Future* 11 (2): https://onlinelibrary.wiley.com/doi/10.1029/2022EF003109.

4 Ullah, S., You, Q., Ullah, W. et al. (2019). Daytime and nighttime heat wave characteristics based on multiple indices over the China–Pakistan economic corridor. *Clim. Dyn.* 53 (9, 10): 6329–6349. http://link.springer.com/10.1007/s00382-019-04934-7.

5 Ullah, S., You, Q., Chen, D. et al. (2022). Future population exposure to daytime and nighttime heat waves in South Asia. *Earth's Future* 10 (5): https://onlinelibrary.wiley.com/doi/10.1029/2021EF002511.

6 Tahir, F., Ajjur, S.B., Serdar, M.Z. et al. (2021). *Qatar Climate Change Conference 2021: A Platform for Addressing Key Climate Change Topics Facing Qatar and the World* (ed. S.G. Al-Ghamdi and S.K. Al-Thani), 1–38. Doha, Qatar: Hamad bin Khalifa University Press (HBKU Press).

7 Trenberth, K.E. (2001). Climate variability and global warming. *Science (80-)* 293 (5527): 48–49. https://www.science.org/doi/10.1126/science.293.5527.48.

8 IPCC (2001). *Synthesis Report, Third Assessment Report.* Cambridge University Press.

9 McMichael, A.J. (2001). Impact of climatic and other environmental changes on food production and population health in the coming decades. *Proc. Nutr. Soc.* 60 (2): 195–201. https://www.cambridge.org/core/product/identifier/S0029665101000234/type/journal_article.

10 Tahir, F., Bansal, D., Rehman, A.U. et al. (2023). Assessing the impact of climate conditions on the distribution of mosquito species in Qatar. *Front. Public Heal.* 10. https://www.frontiersin.org/articles/10.3389/fpubh.2022.970694/full.

11 Pascual, M. and Dobson, A. (2005). Seasonal patterns of infectious diseases. *PLoS Med.* 2 (1): e5. https://dx.plos.org/10.1371/journal.pmed.0020005.

12 Hales, S., Kovats, S., and Woodward, A. (2000). What El Niño can tell us about human health and global climate change. *Glob. Chang. Hum. Health* 1 (1): 66–77.

13 McMichael, A.J. (2013). Globalization, climate change, and human health. *N. Engl. J. Med.* 368 (14): 1335–1343. http://www.nejm.org/doi/10.1056/NEJMc1305749.

14 Watts, N., Adger, W.N., Agnolucci, P. et al. (2015). Health and climate change: policy responses to protect public health. *Lancet* 386 (10006): 1861–1914. https://linkinghub.elsevier.com/retrieve/pii/S0140673615608546.

15 WHO (2021). Climate change and health. Fact Sheets [cited 2 Jul 2022]. https://www.who.int/news-room/fact-sheets/detail/climate-change-and-health.

16 Rana, M.S., Alam, M.M., Ikram, A. et al. (2021). Cocirculation of COVID-19 and dengue: a perspective from Pakistan. *J. Med. Virol.* 93 (3): 1217–1218. https://onlinelibrary.wiley.com/doi/10.1002/jmv.26567.

17 do Rosário, M.S. and de Siqueira, I.C. (2020). Concerns about COVID-19 and arboviral (chikungunya, dengue, zika) concurrent outbreaks. *Brazilian J. Infect. Dis.* 24 (6): 583–584. https://linkinghub.elsevier.com/retrieve/pii/S1413867020301239.

18 Shragai, T., Tesla, B., Murdock, C., and Harrington, L.C. (2017). Zika and chikungunya: mosquito-borne viruses in a changing world. *Ann. N.Y. Acad. Sci.* 1399 (1): 61–77. https://onlinelibrary.wiley.com/doi/10.1111/nyas.13306.

19 Ryan, S.J., Carlson, C.J., Tesla, B. et al. (2021). Warming temperatures could expose more than 1.3 billion new people to Zika virus risk by 2050. *Glob. Change Biol.* 27 (1): 84–93. https://onlinelibrary.wiley.com/doi/10.1111/gcb.15384.

20 Cardona-Ospina, J.A., Arteaga-Livias, K., Villamil-Gómez, W.E. et al. (2021). Dengue and COVID-19, overlapping epidemics? An analysis from Colombia. *J. Med. Virol.* 93 (1): 522–527. https://onlinelibrary.wiley.com/doi/10.1002/jmv.26194.

21 Gratz, N.G. (1999). Emerging and resurging vector-borne diseases. *Annu. Rev. Entomol.* 44 (1): 51–75. https://www.annualreviews.org/doi/10.1146/annurev.ento.44.1.51.

22 Harvell, C.D., Mitchell, C.E., Ward, J.R. et al. (2002). Climate warming and disease risks for terrestrial and marine biota. *Science (80-)* 296 (5576): 2158–2162. https://www.science.org/doi/10.1126/science.1063699.

23 Tonnang, H.E., Kangalawe, R.Y., and Yanda, P.Z. (2010). Predicting and mapping malaria under climate change scenarios: the potential redistribution of malaria vectors in Africa. *Malar J.* 9 (1): 111. https://malariajournal.biomedcentral.com/articles/10.1186/1475-2875-9-111.

24 Chapungu, L. and Nhamo, G. (2021). Interfacing vector-borne disease dynamics with climate change: implications for the attainment of SDGs in Masvingo city, Zimbabwe. *Jàmbá J. Disaster Risk Stud.* 13 (1): https://jamba.org.za/index.php/jamba/article/view/1175.

25 Ren, Z., Wang, D., Ma, A. et al. (2016). Predicting malaria vector distribution under climate change scenarios in China: challenges for malaria elimination. *Sci. Rep.* 6 (1): 20604. http://www.nature.com/articles/srep20604.

26 Mac Kenzie, W.R., Hoxie, N.J., Proctor, M.E. et al. (1994). A massive outbreak in milwaukee of Cryptosporidium infection transmitted through the public water supply. *N. Engl. J. Med.* 331 (3): 161–167. http://www.nejm.org/doi/abs/10.1056/NEJM199407213310304.

27 Wayne, P., Foster, S., Connolly, J. et al. (2002). Production of allergenic pollen by ragweed (*Ambrosia artemisiifolia* L.) is increased in CO_2-enriched atmospheres. *Ann. Allergy, Asthma Immunol.* 88 (3): 279–282. https://linkinghub.elsevier.com/retrieve/pii/S1081120610620091.

28 Ziska, L.H. and Caulfield, F.A. (2000). Rising CO_2 and pollen production of common ragweed (*Ambrosia artemisiifolia* L.), a known allergy-inducing species: implications for public health. *Funct. Plant Biol.* 27 (10): 893. http://www.publish.csiro.au/?paper=PP00032.

29 Patz, J.A., Campbell-Lendrum, D., Holloway, T., and Foley, J.A. (2005). Impact of regional climate change on human health. *Nature* 438 (7066): 310–317. http://www.nature.com/articles/nature04188.

30 Patz, J.A. and Olson, S.H. (2006). Climate change and health: global to local influences on disease risk. *Ann. Trop. Med. Parasitol.* 100 (5, 6): 535–549. http://www.tandfonline.com/doi/full/10.1179/136485906X97426.

31 Negev, M., Paz, S., Clermont, A. et al. (2015). Impacts of climate change on vector borne diseases in the Mediterranean Basin—implications for preparedness and adaptation policy. *Int. J. Environ. Res. Public Health* 12 (6): 6745–6770. http://www.mdpi.com/1660-4601/12/6/6745.

32 Qadir, S.A., Al-Motairi, H., Tahir, F., and Al-Fagih, L. (2021). Incentives and strategies for financing the renewable energy transition: a review. *Energy Rep.* 7: 3590–3606. https://linkinghub.elsevier.com/retrieve/pii/S2352484721004066.

33 Zierold, K.M. and Odoh, C. (2020). A review on fly ash from coal-fired power plants: chemical composition, regulations, and health evidence. *Rev. Environ. Health* 35 (4): 401–418. https://www.degruyter.com/document/doi/10.1515/reveh-2019-0039/html.

34 Climate: Qatar. Climate-Data [cited 29 September 2022]. https://en.climate-data.org/asia/qatar-183/.

35 Semenza, J.C., Suk, J.E., Estevez, V. et al. (2012). Mapping climate change vulnerabilities to infectious diseases in Europe. *Environ. Health Perspect.* 120 (3): 385–392. https://ehp.niehs.nih.gov/doi/10.1289/ehp.1103805.

36 Campbell-Lendrum, D., Corvalán, C., and Neira, M. (2007). Global climate change: implications for international public health policy. *Bull. World Health Organ.* 85 (3): 235–237.

37 Wilbanks, T.J., Leiby, P., Perlack, R. et al. (2007). Toward an integrated analysis of mitigation and adaptation: some preliminary findings. *Mitig. Adapt. Strateg. Glob. Chang.* 12 (5): 713–725. http://link.springer.com/10.1007/s11027-007-9095-4.

38 Feiock, R.C. (2009). Metropolitan governance and institutional collective action. *Urban Aff. Rev.* 44 (3): 356–377. http://journals.sagepub.com/doi/10.1177/1078087408324000.

39 Campbell-Lendrum, D., Manga, L., Bagayoko, M., and Sommerfeld, J. (2015). Climate change and vector-borne diseases: what are the implications for public health research and policy? *Philos. Trans. R. Soc. B Biol. Sci.* 370 (1665): 20130552. https://royalsocietypublishing.org/doi/10.1098/rstb.2013.0552.

40 Purse, B.V., Masante, D., Golding, N. et al. (2017). How will climate change pathways and mitigation options alter incidence of vector-borne diseases? A framework for leishmaniasis in South and Meso-America. Dowdy DW, editor. *PLoS One* 12 (10): e0183583. https://dx.plos.org/10.1371/journal.pone.0183583.

16

Enhancing Resilience: Surveillance Strategies for Monitoring the Spread of Vector-Borne Diseases

Furqan Tahir[1,2], Muhammed G. Madandola[3] and Sami G. Al-Ghamdi[1,2,3]

[1] Environmental Science and Engineering Program, Biological and Environmental Science and Engineering Division, King Abdullah University of Science and Technology (KAUST), Thuwal, Saudi Arabia
[2] KAUST Climate and Livability Initiative, King Abdullah University of Science and Technology (KAUST), Thuwal, Saudi Arabia
[3] Division of Sustainable Development, College of Science and Engineering, Hamad Bin Khalifa University, Qatar Foundation, Doha, Qatar

Introduction

The earth's climate significantly impacts the energy–water–food (EWF) nexus and human health [1–4]. Climate change affects the environment and has social and economic implications [5, 6]. Due to the increase in greenhouse gases in the atmosphere due to human activities, the earth's surface is getting warmer. Temperatures will rise along with climate-related thermal stress, infectious diseases, and floods, threatening human health [7–9]. Such events propel human beings beyond the limits they can adapt to, which affects the rise in mortality [10]. The relationship between temperature and mortality varies depending on the area where a population lives. Temperature changes control how vectors are redistributed [11]. One example of this phenomenon is the deviations that influence the presence of West Nile fever in Europe in vector distributions [12]. Such changes contribute to the burden of vector-borne diseases (VBDs) and the deterioration of human health [13].

Urban populations are more vulnerable to VBDs than nonurban populations, causing an increase in the death rate due to the lack of adequate housing equipped to deal with thermal stress. Climate change also affects the amount of annual rainfall in a region. Disproportionate amounts of rainfall cause human and animal waste to enter drinking water supplies and waterways, thus increasing the burden of water-borne diseases [14]. Climatic factors also influence the burden of infectious diseases via their interactions with vectors, nonhuman reservoir species, and the rate at which pathogens multiply.

An example is when climatic conditions change in a low-temperature area, causing the ecological balance to slant and create an epidemic [15]. Climate change is also predicted to influence the seasonal transmission of VBDs as well as increase the burden of infectious

Sustainable Cities in a Changing Climate: Enhancing Urban Resilience, First Edition.
Edited by Sami G. Al-Ghamdi.
© 2024 John Wiley & Sons Ltd. Published 2024 by John Wiley & Sons Ltd.

diseases due to human migration. Particularly, the spread of infectious diseases accelerates due to the displacement of populations when temperatures increase and cause a rise in sea levels [16]. The combination of extreme weather events, thermal stress, and the burden of infectious diseases resulting from climate change threatens the health and well-being of humans [17]. Research and global health initiatives that combat these issues must be prioritized to lessen their negative impact on human health.

In addition to the loss of quality of life brought on by these diseases, more than a million people are killed each year by more than 112 genera that branch out into the roughly 4,000 species of mosquitoes that transmit VBDs. Malaria, dengue fever, chikungunya, zika, yellow fever, and the West Nile virus are only a few of these ailments. *Aedes* mosquitoes carry dengue, chikungunya, and yellow fever, and 40 genera of *Anopheles* mosquitoes transmit malaria. Many locals and visitors to the Gulf region are from endemic mosquito-borne illness hotspots like West Africa, East Africa, and the Indian subcontinent. Many of these areas are potentially endangered by the spread of mosquito-borne diseases. There is also a larger risk of shifting the burden to illnesses with limited testing, diagnosis, treatment, and immunization capabilities [18–20].

Climatic variables such as temperature, rainfall, and humidity have always been studied to significantly influence mosquitoes' pathogenic activities, geographical distributions, population dynamics, and the spread of VBDs [21–27]. Many laboratories and natural field research have highlighted the diversity of mosquito species and their various relationships with climatic parameters. Malaria, zika, chikungunya, and dengue fever proliferate by interacting with mosquitoes and meteorological factors [28, 29]. While it is difficult to anticipate how global warming and atmospheric conditions will affect disease transmission, climate change is widely predicted to impact mosquito distribution as a vector for parasites and illnesses [30–34]. Figure 16.1 summarizes the climatic change factors that participate in the spread of VBDs, such as weather conditions, changes in habitat conditions due to rainfall and floods, water levels and irrigation, urbanization, and species migrations.

Finding and containing foreign hazards detrimental to human and animal health is a recurring issue for the public health sector. Given that most countries have limited capabilities for entomological surveillance, this is particularly crucial. Because of this, the scientific,

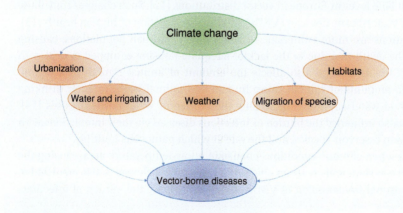

Figure 16.1 Climate change factors affecting VBDs.

public health, and funding communities support in-depth research and capacity-building initiatives, particularly emphasizing a local and global perspective for public health preparation. This global surveillance view adopts a preventative strategy to prepare for and lessen these threats. This chapter summarizes and integrates the surveillance techniques used to track the spread of VBDs as resilience measures. The investigation of human case monitoring, pathogen species identification, vector distribution, and behavior, the impact of climatic and environmental conditions, and control strategies are all crucial components. Conceptual frameworks for more effective modeling, design, and development regulations for resilient healthcare infrastructure are also discussed.

Vector-Borne Diseases

Environmental Factors and Vector-Borne Diseases

Compared to other variables, temperature has a more decisive influence on vectors such as adult *Aedes*, *Anopheles*, and *Culex* species' abundance than precipitation [27, 35, 36]. Several empirical and mathematical modeling studies focusing on epidemiology transmission have demonstrated that ambient temperature constitutes one of the most important abiotic criteria that control mosquito survival, egg-laying rate, and the development of the aquatic stages (eggs, larvae, and pupae), correlating to mosquitoes' ectothermic nature [37]. It is responsible for the biology, physiology, behavior, ecology, and fecundity of mosquitoes [38, 39]. Insects can only survive and operate within certain minimum and maximum temperature thresholds [40]. Thermal generalist mosquitoes have a broader spectrum of temperature performance range, but thermal specialists perform only at a narrow temperature range [41].

The parasite's growth within the mosquito (sporogonic stage) is temperature-dependent [28, 42], with a peak at 28 °C and a minimum of 16 °C [43]. Several research studies have shown that temperature influences the length and duration of the extrinsic incubation period (EIP) [34]. The blood-feeding rate and gonotrophic activities in adult female mosquitoes also depend on temperature, with faster and more aggressive feeding occurring at higher environmental temperatures [33, 44]. As a result, temperature changes can significantly influence disease transmission, mosquito lifespan, and vectorial competence [37, 45]. Additionally, temperature sensitivity varies throughout mosquito stages, with the juvenile stage's dependency significantly impacting adult recruitment and abundance while resulting in nonlinear population responses [46–48]. Since only 1–10% of mosquito larvae make it to the adult stage [49], tackling the larval stage presents a viable way of controlling mosquito population dynamics.

Studies of the effect of rainfall on mosquito abundance showed varying results that ranged from positive correlation to negative association or no effect [50, 51]. Yet, mosquito species use precipitation as a breeding ground, especially standing water, where the females lay their eggs [52, 53]. Pools, hydrology, and irrigation are artificial water sources that might serve as possible host sites. Global warming and abundant water from thawing ice would increase pathogenic transmission [54, 55]. On the other hand, several studies projected that significant rains would wash away the breeding grounds, resulting in the death of larvae [56].

Climate Change Impacts on Vector-Borne Diseases

Despite the general success of attempts in the twentieth century to eradicate infectious diseases, there has been a reversal in recent years due to climate change, the degradation of natural ecosystems, and increased urbanization. The co-circulation of mosquito-borne diseases increased the strain on the healthcare system in endemic locations, in 2020, with the COVID-19 pandemic [57, 58]. Climate change is anticipated to shift transmission seasons and affect the geographic distribution of different VBDs [59]. Furthermore, when variables like urbanization, international trade, and travel abroad rise, the impact of climate on VBDs will also rise [60]. Several studies were conducted to investigate the effects of climatic conditions on VBDs.

Liu-Helmersson et al. [61] concentrated on the possible spread of disease burdens in previously unaffected areas. This study projected the dengue epidemic potential using vectorial capacity. To determine how dengue affects these cities, three tropical cities and seven European cities were studied. The study concluded that higher temperatures and more significant temperature variance could raise the risk of a dengue pandemic in Europe. With Representative Concentration Pathway (RCP) scenarios, Lee et al. [62] evaluated the effects of dengue in South Korea. South Korea's temperature has changed from warm to subtropical, improving the mosquito life cycle environment and, in particular, the prevalence of dengue. The study also looked at the effects of vector-control techniques, concluding that South Korea should intensify its efforts to control mosquitoes and the mobility of affected people. In recent decades, monitoring diseases through surveillance systems, vaccine research, and quantitative risk modeling/mapping has become crucial for safeguarding human health [63].

Surveillance Strategies

Surveillance of VBDs is essential to preventing and controlling global health catastrophes. The extensive geographical distribution of diseases has necessitated that governmental health bodies monitor diseases within the country as well as beyond their boundaries [64]. In recent years, the world has witnessed an increase in the movement of people across borders due to global travel, migration, and the effects of climate change. While this mobility has brought various benefits, it has also led to the introduction of invasive vectors carrying unfamiliar diseases into local communities. These diseases pose significant threats to the well-being and health of affected populations, as they often lack effective technical countermeasures.

To adequately address this emerging challenge, it is crucial to develop a comprehensive understanding of the diverse factors that influence disease transmission and the associated health risks. By unraveling the complexities of these determinants, we can better comprehend how invasive vectors establish themselves in new environments and exploit local vulnerabilities. The impact of invasive diseases goes beyond mere health implications. Communities facing such threats often experience incapacitation on multiple levels, including economic, social, and psychological dimensions. Introducing unfamiliar diseases disrupts established healthcare systems, which may struggle to respond effectively due to limited knowledge, inadequate resources, and the absence of targeted preventive measures.

Moreover, the globalization of trade and travel has created interconnected networks that facilitate the rapid spread of infectious agents across geographical boundaries. Climate change further exacerbates this issue by altering ecosystems, favoring the proliferation of invasive vectors, and expanding their range. Consequently, local communities face the dual challenge of contending with the direct health risks these diseases pose and the indirect consequences of disrupted well-being.

To safeguard local well-being and mitigate the impacts of invasive diseases, a multidisciplinary approach is required. This approach should encompass research from various fields, such as epidemiology, entomology, environmental science, and social sciences, to comprehensively investigate the determinants and complexities of disease transmission. By identifying key factors influencing the spread and establishment of invasive vectors, we can develop targeted strategies to mitigate their impact on vulnerable populations.

VBD surveillance systems can come in various forms to accommodate the needs of people in a particular region. Modern technologies such as geographic information system (GIS) mapping, mobile health (mHealth), or handheld systems transform field surveillance and laboratory study reporting with efficient and cost-effective solutions. In one study by Rodriguez et al. [65], a mHealth surveillance application called FeverDx was evaluated to gauge its efficiency related to VBD surveillance, particularly in arboviral endemic areas in Columbia. The application includes the collection and analysis of data as well as notifications. However, epidemiological surveillance mobile applications are generally limited in scope, location, language, and operating systems. This necessitates a collective push toward developing more advanced surveillance systems.

Since vectors and pathogens depend on land use, temperature, rainfall, humidity, and vegetation, satellite and geospatial applications provide novel approaches to monitoring the environmental parameters important for disease growth. Several studies have utilized GIS, global positioning systems (GPSs), and remote sensing (RS) for the identification of disease hotspots and risk mapping through spatial analysis to facilitate the early detection, management, and processing of epidemiological data [66]. For example, the ECOSTRESS instrument onboard the international space station (ISS) allows for insights into the interaction between vector abundance, climate, and ecosystems. To maximize the effectiveness of data from earth-observing satellites, surveillance requires interdisciplinary collaboration between medical and other specialists such as environmental scientists, epidemiologists, and urban planners. Furthermore, granting free access to data that allows for interoperability between the two spheres—medical and environmental—can facilitate the eradication process of VBDs with appropriate decision-making.

Adopting a more global approach to VBD surveillance by studying emerging pathogens rather than focusing on specific endemic instances allows for greater preparedness when facing VBD threats. Particularly in developing countries, initiatives, such as the Emerging Pandemic Threats Program, enable researchers to make predictions about VBD presence and respond to them promptly. Implementing effective VBD surveillance methods relies on two main factors: (1) detailed cost–benefit analysis and (2) prioritization of resources. In conditions where both elements function successfully, the collaboration between policymakers, funding agencies, and public health stakeholders thrives [64].

Continuing monetary, political, and public support for surveillance programs run by public health in collaboration with vector control specialists is necessary to identify VBDs promptly. An experienced cadre of professionals with regional bases, intimately familiar with the local environment, recognized vectors and infections, and their natural habitats, are needed for effective entomologic (vector) surveillance. For effective surveillance of VBDs, models are needed to evaluate and forecast community risk of disease transmission informed by data on vector ranges, habitat associations, seasonal abundance, and blood-feeding habits. The epidemiological importance of field observations, and the identification of vectors and hosts capable of supporting disease invasion, depends significantly on the supporting data regarding the relative competence of various vectors and vertebrate hosts for pathogen transmission. Figure 16.2 exhibits some of the surveillance strategies to control and prevent the spread of VBDs and develop healthcare policies.

Monitoring of Human Cases

Monitoring human cases in the context of VBDs can greatly benefit from using cell phones and cloud-based technologies, which offer an efficient means of data collection. In particular, the integration of GPS technology allows for the identification and mapping of VBD hotspots. This can be exemplified by applying mHealth solutions that enable the collection of such data [67]. With the widespread adoption of smartphones, the potential for bias arises if certain subpopulations have greater smartphone ownership or access than others, thus limiting the overall utility. However, the implementation of mobile technologies designed to facilitate data collection in specific geographic areas has the potential to enhance the level of detail, standardize datasets, and enable seamless data exchange and comparison.

Through the deployment of user-friendly mHealth applications, individuals can actively participate in data collection, providing real-time information on their health status and potential exposure to vectors. This dynamic approach not only enables timely data acquisition

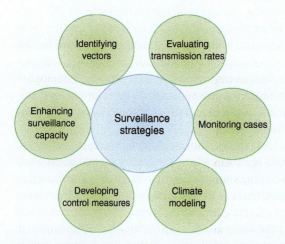

Figure 16.2 Surveillance strategies to control and prevent the spread of VBDs and develop healthcare policies.

but also reduces the reliance on traditional, resource-intensive methods of data collection. Additionally, the integration of GPS technology holds immense value in identifying and locating VBD hotspots. By combining geolocation data with information on reported cases, health professionals can effectively pinpoint areas of heightened disease prevalence.

However, it is important to acknowledge the potential biases that may arise when relying on mobile technologies for data collection. The widespread use of smartphones may not be evenly distributed across all populations, resulting in an unequal representation of certain subpopulations in the collected data. This can introduce limitations and challenges in interpreting and generalizing findings. To address this issue, it is crucial to employ strategies that account for potential biases and ensure the inclusion of diverse populations in data collection efforts. Collaborative partnerships with community organizations, targeted outreach initiatives, and the provision of alternative data collection methods can help mitigate these biases and promote equitable representation.

Identification of Pathogen Species

It is possible to quantitatively anticipate likely introduction channels using information on worldwide migration. Vectors, contaminated items, or vertebrate hosts, like people, can all spread pathogens. Vectors can occasionally enter through air, ground, or ship transportation while still connected to a host. Researchers have looked at global, human, and animal migration data in the context of pathogen and vector invasions for various pathogen systems; however, these data are occasionally expensive or difficult to gather, posing impediments to critical analysis. Surveillance and preventative measures can be employed to stop these invasions, at the port of entry, by predicting the likelihood that a pathogen would enter through a specific pathway. Examples of such measures include the examination of travelers for sickness at airports, the sanitization of airplanes, the requirement of immunizations and health certificates, and the imposition of quarantine periods for imported animals.

Distribution and Behavior of Vectors

Regional variations in the burden of VBDs and the associated abundance and dispersion of vectors may arise due to a complex interplay of climatic and non-climatic factors. Understanding these variations and their underlying determinants is essential for comprehensively assessing disease risks and population vulnerabilities. To gain valuable insights into the adaptive capacities of communities, it is crucial to examine and map these factors from a regional perspective. The non-climatic factors, including land use changes, urbanization, socioeconomic factors, and human behavior, can also contribute to regional variations in VBD burdens. These factors influence vector habitat availability, human exposure patterns, access to healthcare, and overall community resilience. By examining and mapping disease risks and population vulnerabilities at a regional level, researchers can better understand the complex interactions between climatic and non-climatic factors. This approach enables the identification of high-risk areas where vectors are abundant and diseases are prevalent. It also provides insights into the underlying determinants that contribute to the adaptive capacities of communities in the face of VBDs.

By identifying regions with higher burdens of VBDs, healthcare providers can tailor prevention and control strategies to address specific challenges. Additionally, mapping adaptive capacities provides valuable information for assessing the resilience of communities to cope with VBDs. This knowledge can guide the development of context-specific interventions to strengthen community capacities and mitigate the impact of diseases. Regional assessments allow for proactive planning and the development of adaptive strategies to minimize the health risks posed by VBDs.

Climatic and Environmental Changes

To identify the hotspots for the spread of VBDs in the coming years, detailed climate modeling is required to quantify the temperature, rainfall, and humidity distribution daily, monthly, and yearly. This will help develop control and mitigation measures that can be embedded in health-related policies. Furthermore, computational tools such as artificial intelligence can be employed in conjunction with climate models to deal with vast amounts of data and make predictions about VBDs.

Control Measures

An army of experts is needed for effective surveillance and response to identify invasive species and disease introductions, gather long-term entomological and virus infection data, and offer local communities the ability to respond to future epidemic occurrences. The devastation that insufficient public health infrastructure and surveillance systems can create on a local, regional, and global level was demonstrated by the West African Ebola outbreak in 2014 [64]. If there is no method to detect and respond to new diseases quickly, outbreaks can surpass existing capabilities. There is an urgent need for trained personnel at all levels of government and collaborations with industry, nongovernmental organizations, academia, and other groups focusing on public health. It has become clear that effective technical know-how, leadership, coordination between the government and other partner organizations, and improved pathogen testing and response capabilities are all necessary for controlling epidemics. If effective and prompt action is to be taken in future crises, it is crucial to understand and reduce delays in political action during an epidemic reaction.

Policy Development

Surveillance systems must be prioritized and include policy development to holistically address VBDs by targeting pathogens, hosts, and vectors within humans and animals. Surveillance provides a proactive, evidence-based approach to the diseases in circulation through data collection, analysis and management, and real-time sharing of information [68].

A suitable surveillance strategy includes early detection of the spread of infectious diseases, the interpretation of contributing variables in a new region, the modeling of pathogen transmission, and the active use of available information to forecast, discern, and mitigate worldwide hazards [64]. A better understanding of pathogen patterns, host sensitivity, seasonal abundance, interspecies differences, and vector competence in disease transmission would aid decision-making, as would interventions like mandatory health screening, vaccination, and quarantine for travelers from endemic areas.

Climate change will make many health and social inequities worse, and VBDs will disproportionately impact disadvantaged communities negatively. To increase climate resilience, planning VBD prevention and response initiatives locally requires collaboration between public health and climate science groups. Recognizing that not all people and communities will be impacted equally by climate change, local health professionals and public health organizations are ideally positioned to build resilience to climate-related stressors. Public health professionals can support local and state policymaking that simultaneously improves health while lowering health inequities by using a health equity lens to frame, explain, and measure the co-benefits of climate resilience action.

Conclusion

This chapter outlines the importance of surveillance techniques for VBDs in developing robust healthcare infrastructure. Some of the key highlights are as follows:

- The temperature has a more decisive influence on vectors compared to other variables.
- Due to the lack of adequate infrastructure, populations in urban areas are more vulnerable to VBDs than populations living in non-urban areas.
- Surveillance systems must be prioritized and included in policy development to address VBDs by targeting pathogens, hosts, and vectors.
- Satellite and geospatial applications provide novel approaches to monitoring the environmental parameters important for disease growth since vectors and pathogens depend on land use, temperature, rainfall, humidity, and vegetation.
- Human cases can be monitored by the use of cell phones and cloud-based technologies in an efficient way.
- Modern technologies like GIS mapping, mHealth, and handheld systems allow for efficient and cost-effective solutions that transform field surveillance and laboratory study reporting.
- A suitable surveillance strategy includes early detection of spreading infectious diseases, interpreting contributing variables, modeling pathogen transmission, and actively using available information to forecast and mitigate hazards.
- Given the relationship between health and climate, providing academic medical programs and environmental and public health education may help strengthen the healthcare infrastructure.

References

1 Tahir, F., Ajjur, S.B., Serdar, M.Z. et al. (2021). *Qatar Climate Change Conference 2021: A Platform for Addressing Key Climate Change Topics Facing Qatar and the World* (ed. S.G. Al-Ghamdi and S.K. Al-Thani), 1–38. Doha, Qatar: Hamad bin Khalifa University Press (HBKU Press). https://www.qscience.com/content/book/9789999999993.

2 Schaffner, F., Bansal, D., Mardini, K. et al. (2021). Vectors and vector-borne diseases in Qatar: current status, key challenges and future prospects. *J. Eur. Mosq. Control Assoc.* 19: 1–12. https://www.wageningenacademic.com/doi/10.52004/JEMCA2021.x001.

3 Tahir, F. and Al-Ghamdi, S.G. (2022). Integrated MED and HDH desalination systems for an energy-efficient zero liquid discharge (ZLD) system. *Energy Rep.* 8: 29–34. https://linkinghub.elsevier.com/retrieve/pii/S2352484722000282.

4 Al Huneidi, D.I., Tahir, F., and Al-Ghamdi, S.G. (2022). Energy modeling and photovoltaics integration as a mitigation measure for climate change impacts on energy demand. *Energy Rep.* 8: 166–171. https://linkinghub.elsevier.com/retrieve/pii/S2352484722001056.

5 Salimi, M. and Al-Ghamdi, S.G. (2020). Climate change impacts on critical urban infrastructure and urban resiliency strategies for the Middle East. *Sustain. Cities Soc.* 54: 101948. https://linkinghub.elsevier.com/retrieve/pii/S2210670719315872.

6 Afzal, M.S., Tahir, F., and Al-Ghamdi, S.G. (2022). Recommendations and strategies to mitigate environmental implications of artificial island developments in the Gulf. *Sustainability* 14 (9): 5027. https://www.mdpi.com/2071-1050/14/9/5027.

7 Trenberth, K.E. (2001). Climate variability and global warming. *Science (80-)* 293 (5527): 48–49. https://www.science.org/doi/10.1126/science.293.5527.48.

8 Ullah, S., You, Q., Ullah, W. et al. (2023). Climate change will exacerbate population exposure to future heat waves in the China-Pakistan economic corridor. *Weather Clim. Extremes* 40: 100570. https://linkinghub.elsevier.com/retrieve/pii/S2212094723000233.

9 Khan, S., Zeb, B., Ullah, S. et al. (2023). Assessment and characterization of particulate matter during the winter season in the urban environment of Lahore, Pakistan. *Int. J. Environ. Sci. Technol.* https://link.springer.com/10.1007/s13762-023-05011-7.

10 IPCC (2001). *Synthesis Report, Third Assessment Report*. Cambridge University Press.

11 Tonnang, H.E., Kangalawe, R.Y., and Yanda, P.Z. (2010). Predicting and mapping malaria under climate change scenarios: the potential redistribution of malaria vectors in Africa. *Malar J.* 9 (1): 111. https://malariajournal.biomedcentral.com/articles/10.1186/1475-2875-9-111.

12 Hubálek, Z. and Halouzka, J. (1999). West Nile fever–a reemerging mosquito-borne viral disease in Europe. *Emerg. Infect. Dis.* 5 (5): 643–650. https://wwwnc.cdc.gov/eid/article/5/5/99-0505_article.

13 Tahir, F., Bansal, D., Rehman, A.u. et al. (2023). Assessing the impact of climate conditions on the distribution of mosquito species in Qatar. *Front. Public Heal.* 10: https://www.frontiersin.org/articles/10.3389/fpubh.2022.970694/full.

14 McMichael, A.J. (2001). Impact of climatic and other environmental changes on food production and population health in the coming decades. *Proc. Nutr. Soc.* 60 (2): 195–201. https://www.cambridge.org/core/product/identifier/S0029665101000234/type/journal_article.

15 Pascual, M. and Dobson, A. (2005). Seasonal patterns of infectious diseases. *PLoS Med.* 2 (1): e5. https://dx.plos.org/10.1371/journal.pmed.0020005.

16 Hales, S., Kovats, S., and Woodward, A. (2000). What El Niño can tell us about human health and global climate change. *Glob. Chang. Hum. Health* 1 (1): 66–77.

17 McMichael, A.J. (2013). Globalization, climate change, and human health. *N. Engl. J. Med.* 368 (14): 1335–1343. http://www.nejm.org/doi/10.1056/NEJMc1305749.

18 Shragai, T., Tesla, B., Murdock, C., and Harrington, L.C. (2017). Zika and chikungunya: mosquito-borne viruses in a changing world. *Ann. N.Y. Acad. Sci.* 1399 (1): 61–77. https://onlinelibrary.wiley.com/doi/10.1111/nyas.13306.

19 Ryan, S.J., Carlson, C.J., Tesla, B. et al. (2021). Warming temperatures could expose more than 1.3 billion new people to Zika virus risk by 2050. *Glob. Chang. Biol.* 27 (1): 84–93. https://onlinelibrary.wiley.com/doi/10.1111/gcb.15384.

20 Cardona-Ospina, J.A., Arteaga-Livias, K., Villamil-Gómez, W.E. et al. (2021). Dengue and COVID-19, overlapping epidemics? An analysis from Colombia. *J. Med. Virol.* 93 (1): 522–527. https://onlinelibrary.wiley.com/doi/10.1002/jmv.26194.

21 Elbers, A.R.W., Koenraadt, C., and Meiswinkel, R. (2015). Mosquitoes and Culicoides biting midges: vector range and the influence of climate change. *Rev. Sci. Tech l'OIE* 34 (1): 123–137. https://doc.oie.int/dyn/portal/index.xhtml?page=alo&aloId=32331.

22 Ostfeld, R.S. (2009). Climate change and the distribution and intensity of infectious diseases. *Ecology* 90 (4): 903–905. http://doi.wiley.com/10.1890/08-0659.1.

23 Sutherst, R. (1998). Implications of global change and climate variability for vector-borne diseases: generic approaches to impact assessments. *Int. J. Parasitol.* 28 (6): 935–945. https://linkinghub.elsevier.com/retrieve/pii/S0020751998000563.

24 Patz, J.A., Martens, W.J., Focks, D.A., and Jetten, T.H. (1998). Dengue fever epidemic potential as projected by general circulation models of global climate change. *Environ. Health Perspect.* 106 (3): 147–153. https://ehp.niehs.nih.gov/doi/10.1289/ehp.98106147.

25 Epstein, P.R., Diaz, H.F., Elias, S. et al. (1998). Biological and physical signs of climate change: focus on mosquito-borne diseases. *Bull. Am. Meteorol. Soc.* 79 (3): 409–417. http://journals.ametsoc.org/doi/10.1175/1520-0477(1998)079%3C0409:BAPSOC%3E2.0.CO;2.

26 Benedict, M.Q., Levine, R.S., Hawley, W.A., and Lounibos, L.P. (2007). Spread of the tiger: global risk of invasion by the mosquito *Aedes albopictus*. *Vector-Borne Zoonotic Dis.* 7 (1): 76–85. https://www.liebertpub.com/doi/10.1089/vbz.2006.0562.

27 Abiodun, G.J., Maharaj, R., Witbooi, P., and Okosun, K.O. (2016). Modelling the influence of temperature and rainfall on the population dynamics of *Anopheles arabiensis*. *Malar J.* 15 (1): 364. http://malariajournal.biomedcentral.com/articles/10.1186/s12936-016-1411-6.

28 Reinhold, J., Lazzari, C., and Lahondère, C. (2018). Effects of the environmental temperature on *Aedes aegypti* and *Aedes albopictus* mosquitoes: a review. *Insects* 9 (4): 158. https://www.mdpi.com/2075-4450/9/4/158.

29 Couret, J. and Benedict, M.Q. (2014). A meta-analysis of the factors influencing development rate variation in *Aedes aegypti* (Diptera: Culicidae). *BMC Ecol.* 14 (1): 3. http://bmcecol.biomedcentral.com/articles/10.1186/1472-6785-14-3.

30 Craig, M.H., Snow, R.W., and le Sueur, D. (1999). A climate-based distribution model of malaria transmission in sub-saharan Africa. *Parasitol. Today* 15 (3): 105–111. https://linkinghub.elsevier.com/retrieve/pii/S0169475899013964.

31 Gething, P.W., Smith, D.L., Patil, A.P. et al. (2010). Climate change and the global malaria recession. *Nature* 465 (7296): 342–345. http://www.nature.com/articles/nature09098.

32 Hay, S.I., Cox, J., Rogers, D.J. et al. (2002). Climate change and the resurgence of malaria in the East African highlands. *Nature* 415 (6874): 905–909. http://www.nature.com/articles/415905a.

33 Paaijmans, K.P., Blanford, S., Bell, A.S. et al. (2010). Influence of climate on malaria transmission depends on daily temperature variation. *Proc. Natl. Acad. Sci.* 107 (34): 15135–15139. https://pnas.org/doi/full/10.1073/pnas.1006422107.

34 Paaijmans, K.P., Read, A.F., and Thomas, M.B. (2009). Understanding the link between malaria risk and climate. *Proc. Natl. Acad. Sci.* 106 (33): 13844–13849. http://www.pnas.org/cgi/doi/10.1073/pnas.0903423106.

35 Alto, B.W. and Juliano, S.A. (2001). Precipitation and temperature effects on populations of *Aedes albopictus* (Diptera: Culicidae): implications for range expansion. *J. Med. Entomol.* 38 (5): 646–656. https://academic.oup.com/jme/article-lookup/doi/10.1603/0022-2585-38.5.646.

36 Ciota, A.T., Matacchiero, A.C., Kilpatrick, A.M., and Kramer, L.D. (2014). The effect of temperature on life history traits of culex mosquitoes. *J. Med. Entomol.* 51 (1): 55–62. https://academic.oup.com/jme/article-lookup/doi/10.1603/ME13003.

37 Beck-Johnson, L.M., Nelson, W.A., Paaijmans, K.P. et al. (2013). The effect of temperature on *Anopheles* mosquito population dynamics and the potential for malaria transmission. Costa FTM, editor. *PLoS One* 8 (11): e79276. https://dx.plos.org/10.1371/journal.pone.0079276.

38 Denlinger, D.L. and Yocum, G.D. (2019). Physiology of heat sensitivity. In: *Temperature Sensitivity in Insects and Application in Integrated Pest Management* (ed. G.J. Hallman and D.L. Denlinger), 7–53. Boca Raton, FL, USA: CRC Press. https://www.taylorfrancis.com/books/9781000242430/chapters/10.1201/9780429308581-2.

39 Heinrich, B. (1993). *The Hot-Blooded Insects*. Berlin, Heidelberg: Springer Berlin Heidelberg. http://link.springer.com/10.1007/978-3-662-10340-1.

40 Mpho, M., Callaghan, A., and Holloway, G.J. (2002). Temperature and genotypic effects on life history and fluctuating asymmetry in a field strain of *Culex pipiens*. *Heredity (Edinb)* 88 (4): 307–312. http://www.nature.com/articles/6800045.

41 Angilletta, M.J. Jr. (2009). *Thermal Adaptation: A Theoretical and Empirical Synthesis*. Oxford University Press. https://oxford.universitypressscholarship.com/view/10.1093/acprof:oso/9780198570875.001.1/acprof-9780198570875.

42 Yé, Y., Louis, V.R., Simboro, S., and Sauerborn, R. (2007). Effect of meteorological factors on clinical malaria risk among children: an assessment using village-based meteorological stations and community-based parasitological survey. *BMC Public Health* 7 (1): 101. https://bmcpublichealth.biomedcentral.com/articles/10.1186/1471-2458-7-101.

43 Ermert, V. (2010). Risk assessment with regard to the occurrence of malaria in Africa under the influence of observed and projected climate change. PhD Thesis. Universität zu Köln. http://kups.ub.uni-koeln.de/id/eprint/3109.

44 Tompkins, A.M. and Ermert, V. (2013). A regional-scale, high resolution dynamical malaria model that accounts for population density, climate and surface hydrology. *Malar J.* 12 (1): 65. https://malariajournal.biomedcentral.com/articles/10.1186/1475-2875-12-65.

45 Carrington, L.B., Armijos, M.V., Lambrechts, L. et al. (2013). Effects of fluctuating daily temperatures at critical thermal extremes on *Aedes aegypti* life-history traits. Vasilakis N, editor. *PLoS One* 8 (3): e58824. https://dx.plos.org/10.1371/journal.pone.0058824.

46 Bayoh, M.N. and Lindsay, S.W. (2004). Temperature-related duration of aquatic stages of the Afrotropical malaria vector mosquito *Anopheles gambiae* in the laboratory. *Med. Vet. Entomol.* 18 (2): 174–179. https://onlinelibrary.wiley.com/doi/10.1111/j.0269-283X.2004.00495.x.

47 Bayoh, M.N. and Lindsay, S.W. (2003). Effect of temperature on the development of the aquatic stages of *Anopheles gambiae sensu stricto* (Diptera: Culicidae). *Bull. Entomol. Res.* 93 (5): 375–381. https://www.cambridge.org/core/product/identifier/S0007485303000440/type/journal_article.

48 Delatte, H., Gimonneau, G., Triboire, A., and Fontenille, D. (2009). Influence of temperature on immature development, survival, longevity, fecundity, and gonotrophic cycles of *Aedes albopictus*, vector of chikungunya and dengue in the Indian Ocean. *J. Med. Entomol.* 46 (1): 33–41. https://academic.oup.com/jme/article-lookup/doi/10.1603/033.046.0105.

49 Munga, S., Minakawa, N., Zhou, G. et al. (2007). Survivorship of immature stages of *Anopheles gambiae* s.l. (Diptera: Culicidae) in natural habitats in Western Kenya highlands. *J. Med. Entomol.* 44 (5): 758–764. https://academic.oup.com/jme/article-lookup/doi/10.1093/jmedent/44.5.758.

50 Tran, A., L'Ambert, G., Lacour, G. et al. (2013). A rainfall- and temperature-driven abundance model for *Aedes albopictus* populations. *Int. J. Environ. Res. Public Health* 10 (5): 1698–1719. http://www.mdpi.com/1660-4601/10/5/1698.

51 Bomblies, A. (2012). Modeling the role of rainfall patterns in seasonal malaria transmission. *Clim. Change* 112 (3–4): 673–685. http://link.springer.com/10.1007/s10584-011-0230-6.

52 Lunde, T.M., Korecha, D., Loha, E. et al. (2013). A dynamic model of some malaria-transmitting Anopheline mosquitoes of the Afrotropical region. I. Model description and sensitivity analysis. *Malar J.* 12 (1): 28. https://malariajournal.biomedcentral.com/articles/10.1186/1475-2875-12-28.

53 Soti, V., Tran, A., Degenne, P. et al. (2012). Combining hydrology and mosquito population models to identify the drivers of rift valley fever emergence in semi-arid regions of west Africa.Anyamba A, editor. *PLoS Negl. Trop. Dis.* 6 (8): e1795. https://dx.plos.org/10.1371/journal.pntd.0001795.

54 Bashar, K. and Tuno, N. (2014). Seasonal abundance of Anopheles mosquitoes and their association with meteorological factors and malaria incidence in Bangladesh. *Parasit. Vectors* 7 (1): 442. http://parasitesandvectors.biomedcentral.com/articles/10.1186/1756-3305-7-442.

55 Valdez, L.D., Sibona, G.J., and Condat, C.A. (2018). Impact of rainfall on *Aedes aegypti* populations. *Ecol. Modell.* 385: 96–105. https://linkinghub.elsevier.com/retrieve/pii/S0304380018302382.

56 Paaijmans, K.P., Wandago, M.O., Githeko, A.K., and Takken, W. (2007). Unexpected high losses of *Anopheles gambiae* larvae due to rainfall.Carter D, editor. *PLoS One* 2 (11): e1146. https://dx.plos.org/10.1371/journal.pone.0001146.

57 Rana, M.S., Alam, M.M., Ikram, A. et al. (2021). Cocirculation of COVID-19 and dengue: a perspective from Pakistan. *J. Med. Virol.* 93 (3): 1217–1218. https://onlinelibrary.wiley.com/doi/10.1002/jmv.26567.

58 do Rosário, M.S. and de Siqueira, I.C. (2020). Concerns about COVID-19 and arboviral (chikungunya, dengue, zika) concurrent outbreaks. *Brazilian J. Infect. Dis.* 24 (6): 583–584. https://linkinghub.elsevier.com/retrieve/pii/S1413867020301239.

59 Woodward, A., Smith, K.R., Campbell-Lendrum, D. et al. (2014). Climate change and health: on the latest IPCC report. *Lancet* 383 (9924): 1185–1189. https://linkinghub.elsevier.com/retrieve/pii/S0140673614605766.

60 Campbell-Lendrum, D., Manga, L., Bagayoko, M., and Sommerfeld, J. (2015). Climate change and vector-borne diseases: what are the implications for public health research and policy? *Philos. Trans. R. Soc. B Biol. Sci.* 370 (1665): 20130552. https://royalsocietypublishing.org/doi/10.1098/rstb.2013.0552.

61 Liu-Helmersson, J., Quam, M., Wilder-Smith, A. et al. (2016). Climate change and aedes vectors: 21st century projections for dengue transmission in Europe. *EBioMedicine* 7: 267–277. https://linkinghub.elsevier.com/retrieve/pii/S2352396416301335.

62 Lee, H., Kim, J.E., Lee, S., and Lee, C.H. (2018). Potential effects of climate change on dengue transmission dynamics in Korea. Sekaran SD, editor. *PLoS One* 13 (6): e0199205. https://dx.plos.org/10.1371/journal.pone.0199205.

63 Caminade, C., McIntyre, K.M., and Jones, A.E. (2019). Impact of recent and future climate change on vector-borne diseases. *Ann. N.Y. Acad. Sci.* 1436 (1): 157–173. https://onlinelibrary.wiley.com/doi/10.1111/nyas.13950.

64 Kading, R.C., Golnar, A.J., Hamer, S.A., and Hamer, G.L. (2018). Advanced surveillance and preparedness to meet a new era of invasive vectors and emerging vector-borne diseases. Bartholomay LC, editor. *PLoS Negl. Trop. Dis.* 12 (10): e0006761. https://dx.plos.org/10.1371/journal.pntd.0006761.

65 Rodríguez, S., Sanz, A.M., Llano, G. et al. (2020). Acceptability and usability of a mobile application for management and surveillance of vector-borne diseases in Colombia: an implementation study. Samy AM, editor. *PLoS One* 15 (5): e0233269. https://dx.plos.org/10.1371/journal.pone.0233269.

66 Malone, J., Bergquist, R., Martins, M., and Luvall, J. (2019). Use of geospatial surveillance and response systems for vector-borne diseases in the elimination phase. *Trop. Med. Infect. Dis.* 4 (1): 15. http://www.mdpi.com/2414-6366/4/1/15.

67 WHO (2011). mHealth New horizons for health through mobile technologies. [cited 2022 Jun 30]. https://apps.who.int/iris/bitstream/handle/10665/44607/9789241564250_eng.pdf?sequence=1&isAllowed=y.

68 Fournet, F., Jourdain, F., Bonnet, E. et al. (2018). Effective surveillance systems for vector-borne diseases in urban settings and translation of the data into action: a scoping review. *Infect. Dis. Poverty* 7 (1): 99. https://idpjournal.biomedcentral.com/articles/10.1186/s40249-018-0473-9.

Glossary

The following glossary encompasses terms relevant to this book's content, enhancing the readers' understanding of the key concepts and facilitating their engagement with the topics discussed.

Adaptability The capacity to adjust to changing circumstances during and after the disruption.
Adaptation measures Specific actions and interventions implemented to adjust or modify systems, infrastructure, policies, and practices in response to climate change impacts and to reduce vulnerability.
Adaptation strategies Actions taken to adjust and respond to changing environmental conditions caused by climate change, aiming to reduce vulnerability and enhance resilience.
Blue infrastructure Includes water-related features within urban areas, such as rivers, lakes, and stormwater systems, used to manage and improve water quality and supply.
Built environment Refers to the human-made physical surroundings, including buildings, infrastructure, public spaces, and transportation systems, which form the urban environment.
Carbon footprint The total amount of greenhouse gases, particularly carbon dioxide, emitted directly or indirectly by an individual, organization, event, or product throughout its life cycle, contributing to climate change.
Climate change mitigation Actions and policies aimed at reducing greenhouse gas emissions or removing them from the atmosphere, contributing to the global effort to limit the magnitude of climate change.
Climate change Refers to long-term shifts in temperature, precipitation patterns, and other aspects of the Earth's climate system, primarily caused by human activities such as the burning of fossil fuels and deforestation.
Climate resilience assessment The process of evaluating the capacity of a system or community to withstand climate change impacts and identifying areas for improvement or intervention.

Climate resilience The ability of a system or community to anticipate, absorb, adapt to, and recover from climate change impacts while minimizing disruption and maintaining essential functions and structures.

Community engagement The process of involving and empowering community members, residents, and stakeholders in decision-making, planning, and implementation processes to ensure their needs and perspectives are considered.

Critical Infrastructure Refers to the systems, networks, and public works that a government deems vital for its functioning and the safety of its citizens.

Disaster resilience The ability of a system or community to resist, absorb, respond to, and recover from the impacts of disasters, including natural disasters intensified by climate change.

Disaster risk reduction Measures taken to minimize the impacts of natural and human-induced disasters, including climate-related events, through prevention, preparedness, and response strategies.

Distributed Generation Electric power generation within distribution networks or on the customer site of the meter.

Ecosystem services The benefits that ecosystems provide to humans, including the provision of clean air and water, soil fertility, biodiversity conservation, climate regulation, and recreational opportunities.

Energy efficiency Reductions in the amount of energy needed to provide an energy service, thanks to more efficient technologies and measures.

Energy resilience The ability of an energy system to anticipate, prepare for, adapt to, withstand, respond to, and recover rapidly from disruptions.

Energy security The uninterrupted availability of energy sources at an affordable price. Energy security has many aspects: long-term energy security mainly deals with timely investments to supply energy in line with economic developments and sustainable environmental needs. Short-term energy security focuses on the ability of the energy system to react promptly to sudden changes in the supply-demand balance.

Energy system An interconnected network of physical infrastructure, or organizational arrangements related to the production, conversion, delivery, management, and utilization of energy.

Environmental sustainability The practice of using natural resources in a way that preserves and protects the environment for current and future generations, avoiding depletion or irreversible damage.

Gray infrastructure Comprises traditional, human-made elements like roads, bridges, and pipelines, which are essential for urban development and can be adapted to enhance sustainability.

Green infrastructure Involves natural or nature based elements such as parks, wetlands, and vegetation designed to manage environmental issues like flooding, while also providing recreational and ecological benefits.

Green–blue–gray infrastructure The approach that combines three types of infrastructure to provide a comprehensive solution for adapting to and mitigating climate change impacts.

Health policy Strategies, plans, and actions implemented by governments and organizations to promote and protect public health, including measures to address climate change impacts on health.

Knowledge sharing The exchange and dissemination of information, research findings, best practices, and lessons learned among stakeholders, policymakers, researchers, and practitioners to foster collaboration and informed decision-making.

Microgrid A local grid area network that can disconnect from the traditional grid to operate autonomously. Microgrids enhance local energy supply by locally generating, storing, and controlling the flow of energy.

Mitigation measures Actions and strategies aimed at reducing the severity, impact, or occurrence of climate change hazards, such as greenhouse gas emission reduction or implementing nature-based solutions.

Multi-level governance Collaborative decision-making and coordination among different levels of government, stakeholders, and communities to address complex challenges such as climate change.

Rapidity The capacity to meet priorities and achieve goals in a timely manner in order to contain losses and avoid future disruption.

Redundancy The extent to which elements, systems, or other units of analysis exist that are substitutable, i.e., capable of satisfying functional requirements in the event of disruption, degradation, or loss of functionality.

Resilience The ability of a system or community to withstand, adapt to, and recover from disturbances or shocks while maintaining its essential functions, structures, and identity.

Resilience indicators Quantitative or qualitative metrics used to measure and assess the resilience of systems or communities, providing insights into their adaptive capacity, vulnerability, and overall resilience level.

Resilient infrastructure Physical systems and networks that are designed and built to withstand climate change impacts, maintain functionality during disruptions, and recover quickly from disturbances.

Resourcefulness The capacity to identify problems, establish priorities, and mobilize resources when conditions exist that threaten to disrupt some element, system, or other unit of analysis. Resourcefulness can be further conceptualized as consisting of the ability to apply material (i.e., monetary, physical, technological, and informational) and human resources to meet priorities and achieve goals.

Risk assessment The process of evaluating potential hazards, vulnerabilities, and risks to determine their likelihood and potential consequences, aiding in the development of strategies to mitigate or manage those risks.

Robustness The ability of elements, systems, and other units of analysis to withstand a given level of stress or demand without suffering degradation or loss of function.

Smart grid An electricity network that uses digital and other advanced technologies to monitor and manage the transport of electricity from all generation sources to meet the varying electricity demands of end-users.

Stakeholders Individuals, organizations, or groups that have an interest, involvement, or influence in a particular issue, project, or decision-making process related to urban development and climate change resilience.

Sustainable development Development that meets the needs of the present generation without compromising the ability of future generations to meet their own needs, balancing economic, social, and environmental factors.

Sustainable urban development The approach of planning and managing cities to achieve social equity, economic prosperity, and environmental sustainability while considering the long-term well-being of communities.

Undergrounding Installation of overhead electric distribution lines underground, eliminating exposure to weather-related damage and other risks.

Urban planning The practice of designing and organizing urban areas to optimize land use, transportation, infrastructure, and the overall quality of life of residents.

Urban systems The interconnected physical, social, economic, and environmental elements that make up urban areas, including infrastructure, transportation networks, housing, and community services.

Urbanization The process of population growth and the expansion of urban areas, resulting in the transformation of rural or undeveloped land into urbanized spaces.

Vector-borne diseases Illnesses transmitted to humans through the bites of infected arthropods, such as mosquitoes and ticks. These diseases include malaria, dengue fever, and Lyme disease, which can be influenced by climate change.

Index

Note: *Italicized* and **bold** page numbers refer to figures and tables, respectively.

a

AAO *see* Antarctic Oscillation (AAO)
ABM *see* agent-based modeling (ABM)
absolute global warming potential (AGWP)   65
absorptive capacity   123
adaptive capacity   123, 127
adaptive resilience   **239**
Aedes mosquitoes   264
aerobic sludge age control   75
aerosols   4, 6
agent-based modeling (ABM)   84, 103
AGWP *see* absolute global warming potential (AGWP)
AHP *see* analytic hierarchy process (AHP)
AIACC *see* Assessments of Impacts and Adaptation of Climate Change (AIACC) sustainable livelihood approach
air pollution   20, 22, 24, 255
Alliance for National and Community Resilience Benchmarking System   **170**
Ambrosia artemisiifolia   255
analytic hierarchy process (AHP)   166
Anopheles mosquitoes   264
Antarctic Oscillation (AAO)   8
anthropogenic climate change   5–6, 212, 258, 259
anthropogenic emissions data   24
anthropogenic global warming   187
anthropogenic stressors   145
AO *see* Arctic Oscillation (AO)
Arabian Peninsula   37, 97, 187
Arctic Oscillation (AO)   8
Arup's City Resilience Framework | Rockefeller Foundation   **167**
asphalt industry   45
asphalt production, embodied carbon from   44–45
ASPIRE framework   177
Assessments of Impacts and Adaptation of Climate Change (AIACC) sustainable livelihood approach   **168**
Atlantic equatorial mode   8
Atlantic meridional mode   8
Atlantic Niño   8
Atlantic zonal mode   8
atmosphere   4

b

Baseline Resilience Indicators for Communities (BRIC)   **170**
Beddington Zero Energy Development (BedZED)   200, 201, 202
BedZED *see* Beddington Zero Energy Development (BedZED)
BEM *see* building energy model (BEM)
BEP scheme *see* building effect parameterization (BEP) scheme

Sustainable Cities in a Changing Climate: Enhancing Urban Resilience. First Edition.
Edited by Sami G. Al-Ghamdi.
© 2024 John Wiley & Sons Ltd. Published 2024 by John Wiley & Sons Ltd.

betweenness centrality 100–101
BFS *see* blast furnace slag (BFS)
BI *see* blue infrastructure (BI)
blast furnace slag (BFS) 48
blower technology 75
blue infrastructure (BI) 211, 212
BRACED framework *see* Building Resilience and Adaptation to Climate Extremes and Disasters (BRACED) framework
BREEAM assessment *see* British Building Research Establishment Environmental Assessment Method (BREEAM) assessment
BRIC *see* Baseline Resilience Indicators for Communities (BRIC)
British Building Research Establishment Environmental Assessment Method (BREEAM) assessment 194
BRT systems *see* bus rapid transit (BRT) systems
building earthquake resilience 113, *113*
building effect parameterization (BEP) scheme 22
building energy model (BEM) 24
building environment 39
  carbon emissions in 51–52
  carbon mitigation strategies 52
    heating, ventilation, and air conditioning (HVAC) systems 53
    lighting,. use of 54–55
    renewable resources integration 53
    water use, strategy for 54
  embodied carbon (EC) emission 40
    asphalt production and construction 44–45
    from cement and concrete manufacturing 42–44
    of limestone quarrying 41–42
    of steel production 45–46
  embodied carbon (EC) mitigation strategies 46
    carbon sequestration 51
    design phase and efficient construction 49–51
    extending the building's life 51
    heavy building materials 49
    lower embodied carbon, using materials with 46–49
building resilience, enhancing 111, 240
  performance-based design (PBD) 114–115
  structural resilience representation 112–114
  supporting systems 115
    building limits, beyond 116–117
    within the building 116
Building Resilience and Adaptation to Climate Extremes and Disasters (BRACED) framework **167**
built environment resilience, enhancing 121
  resilience, types of 124
    community resilience 127–128
    critical infrastructure resilience 128–129
    ecological resilience 125–127, *126*
    engineering resilience 125–127
    social resilience 127–128
    specified and general resilience 128
    technical systems, products, and production resilience 129
  resilience capacities 123–124
  resilience components 124
  resilience dimensions and capitals 129–130, *131*
  resilience measuring 130–133
  risk identification and assessment 123
  uncertainty 122–123
built environment systems 134, 148, 149, 210, *211*, 212
  categorization of **149**
  climate change impacts on 140–144
bus rapid transit (BRT) systems 188

c

C&D materials *see* construction and demolition (C&D) materials
calcination process 43
CanESM2-r0i0p0 9
carbon budget 4

carbon dioxide emissions   4, 6, 40, 41, *41*, 42, 43, **43**, 44, 45, 46, 48, 52, *52*, 61, 65, **66**, 67, 68, 70, 71, 189, 194
carbon emissions   4
  in building environment   51–52
  embodied   39
    from asphalt production and construction   44–45
    in building environment   40
    from cement and concrete manufacturing   42–44
    of limestone quarrying   41–42
    of steel production   45–46
    strategies to tackle   47
  operational   39, 40
    in building environment   51–52
  reducing   256–257
carbon mitigation strategies   52
  embodied
    building's life, extending   51
    carbon sequestration   51
    design phase and efficient construction   49–51
    heavy building materials   49
    lower EC, using materials with   46–48
  opeartional
    heating, ventilation, and air conditioning (HVAC) systems   53
    lighting, use of   54–55
    renewable resources integration   53
    water use, strategy for   54
carbon sequestration   51
carbon tax   4–5, 199
CART *see* Community Assessment of Resilience Tool (CART)
CCRAM *see* Conjoint Community Resilience Assessment Measure (CCRAM)
CDO *see* Climate Data Operators (CDO)
CDRI *see* Community Disaster Resilience Index (CDRI)
cement production   42–44
  greenhouse gas emissions from   *44*
CFSR *see* Climate Forecast System Reanalysis (CFSR)
Characteristics of a Disaster-Resilient Community   168, **170**
chikungunya   254, 264
CHP processes *see* combined heat and power (CHP) processes
CI *see* critical infrastructure (CI)
City Resilience Framework (CRF) *see* City Resilience Index (CRI)
City Resilience Index (CRI)   **170**
climate change-resilient system   144, *145*
Climate Data Operators (CDO)   14, **14**
Climate Forecast System Reanalysis (CFSR)   **11**
climate models   8–10, 29
climate resilience   121, 134, 150, 242–243
Climate Resilience Screening Index (CRSI)   **170**
climate-responsive urban planning, case studies of   200–202
Climate Risk and Adaptation Framework and Taxonomy (CRAFT)   **171**
climate variability   3
  modes of   6–8
climatic data   8
  climate models   8–10, 29
  observations and reanalysis   10–12
  pathways and scenarios   10
  visualizing and processing   12–14, **14**
CLT *see* cross-laminated timber (CLT)
CMIP *see* Coupled Model Intercomparison Project (CMIP)
CNs *see* complex networks (CNs)
CO_2 equivalent (CO_2-eq) emission   4, 42, 43, 44
coal power plant with carbon capture technology in Czech Republic (case study)   67–68
Coastal Resilience Decision Support System   **171**
Coastal Resilience Index   **171**
CoBRA *see* Community-Based Resilience Analysis (CoBRA)
collapse prevention (CP)   113, *113*
combined heat and power (CHP) processes   75, 201

communication networks 83, 163
Community Assessment of Resilience Tool (CART) **171**
Community-Based Resilience Analysis (CoBRA) **169, 173**
Community Disaster Resilience Index (CDRI) **172**
community resilience 127–128, 157–158
 and climate change **132, 167**
 defined **142**
 and example programs **159**
 USAID measurement for **168**
Community Resilience: Conceptual Framework and Measurement Feed the Future Learning Agenda **173**
community resilience against climate change **132**
Community Resilience Indicators and National-Level Measures **172**
Community Resilience Manual **172**
Community Resilience Planning Guide **172**
Community Resilience System (CRS) **172**
community resilience to climate change frameworks 166, **167–169**
compact cities 190–191
Comparative Study on Urban Transport and the Environment (CUTE) matrix 189
complex networks (CNs) 82, 98
 advantages 101
 drawbacks 101
 graphs, types of 98
  directed and undirected graphs 99, *100*
  weighted and unweighted graphs 99–100
 resilience assessment, main applications in
  betweenness centrality 100
  graph percolation 101
conceptual/theoretical framework 157, 159, 161–163
concrete 42–44
Conference of the Parties (COP) 198
Conjoint Community Resilience Assessment Measure (CCRAM) **173**
construction and demolition (C&D) materials 49

content-related uncertainty 123
Coordinated Regional Climate Downscaling Experiment (CORDEX) 9–10, 30–31, 33, 35–36
coping capacity *see* absorptive capacity
COP *see* Conference of the Parties (COP)
CORDEX *see* Coordinated Regional Climate Downscaling Experiment (CORDEX)
cost–benefit analysis **196**
Coupled Model Intercomparison Project (CMIP) 31
 CMIP6 6
Coupled Ocean-Atmosphere Reanalysis of the Twentieth Century *11*
COVID-19 pandemic 65, 145, 254, 266
CP *see* collapse prevention (CP)
CRAFT *see* Climate Risk and Adaptation Framework and Taxonomy (CRAFT)
CRI *see* City Resilience Index (CRI)
critical infrastructure (CI) 90, 92, 116, 128–129, 186, 207
cross-laminated timber (CLT) 47
CRSI *see* Climate Resilience Screening Index (CRSI)
CRS *see* Community Resilience System (CRS)
Cryptosporidium parasites 255
Culex pipiens 254
cultural (institutional) capital 130
CUTE matrix *see* Comparative Study on Urban Transport and the Environment (CUTE) matrix

d

DECK *see* Diagnostic, Evaluation and Characterization of Klima (DECK)
Delphi method 157, **158**
dengue fever 264
Department of Energy (DOE) **228**, 229
"designing for disassembly," strategy of 50
deterministic performance-based approach **164**, 165
Diagnostic, Evaluation and Characterization of Klima (DECK) 10
diarrhea 255
digital urban climate twin (DUCT) 203

directed graph   99, *100*
disaster   82, 84, 97, 98, 105, 111, 112–113, 128, 144, 256
Disaster Resilience Scorecard   **173**
Disaster Resilience Scorecard for Cities   **174**
distributed generation   84, 85, 240–241
DOE *see* Department of Energy (DOE)
DUCT *see* digital urban climate twin (DUCT)

e

early warning system (EWS)   123, 160, **195**
earth masonry *see* unfired brick
Earthquake Recovery Model   **174**
earthquakes   111–116, *113*
Earth System Grid Federation (ESGF) nodes   31
EC emissions *see* embodied carbon (EC) emissions
ECMWF *see* European Centre for Medium-Range Weather Forecasts (ECMWF)
ecological resilience   125–127, *126*, **239**
  defined   125
ecological systems   81, 121
  defined   **141**
ECOSTRESS instrument   267
ecosystem resilience   125
  defined   **142**
EHEs *see* extreme heat events (EHEs)
EIP *see* extrinsic incubation period (EIP)
electrical and water distribution infrastructures   102
electrical network resilience   84–85, 88
electricity generation in Turkey (case study)   65–67
Electric Power Research Institute (EPRI) guide   243
El Niño-Southern Oscillation (ENSO)   8
embodied carbon (EC) emissions   39
  from asphalt production and construction   44–45
  in building environment   40
  from cement and concrete manufacturing   42–44
  of limestone quarrying   41–42
  of steel production   45–46
  strategies to tackle   47
embodied carbon (EC) mitigation strategies   46
  building's life, extending   51
  carbon sequestration   51
  design phase and efficient construction   49–51
  heavy building materials   49
  lower EC, using materials with   46
    ethylene tetrafluoroethylene (ETFE)   49
    precast hollow-core slabs   48
    steel framework system   48
    unfired brick, use of   48
Emerging Pandemic Threats Program   267
emissions savings
  from energy sector   69
    energy efficiency, increase in   70–71
    nuclear plants, running   72
    wind and solar plant installation   71
  from water sector
    energy management in water system   75
    groundwater management   74
    smart wastewater treatment technology   75
energy efficiency   18, 52, 55, 70–71
  enhancing   242
  resilience benefits of   **243**
energy management in water system   75
energy sector impact on climate change   65
  coal power plant with carbon capture technology in Czech Republic (case study)   67–68
  electricity generation in Turkey (case study)   65–67
  solar power with energy storage (case study)   68–69
energy storage   241
energy systems   227
  adapting the theory of resilience to   229
  climate resilience and   242–243
  climatic changes and   236, *237*, **238**
  components of   *228*
  conventional   227
  damages to, by natural disasters   *228*

energy systems (*continued*)
  domains of resilience approaches to   237
  failure scenarios in   *236*
  incorporating resilience into   234–235
  paradigms of resilience applicable to   **239**
  resilience enhancement
    approaches for   240
    climate resilience   242–243
    distributed generation   240–241
    energy efficiency, enhancing   242
    energy storage   241
    smart grid technology   241–242, *241*
    system hardening   240
  resilience of   229
  threats to   235–237
energy–water–food (EWF) nexus   263
engineering resilience   125–127, *126*, **239**
  defined   **142**
ENSO *see* El Niño-Southern Oscillation (ENSO)
environmental impact   39, 44, 50
  of floods and green climate change adaptation   210–211
  life cycle assessment for   63–64
Environmental Protection Agency (EPA)   19, 44, 198
EPA *see* Environmental Protection Agency (EPA)
EPRI guide *see* Electric Power Research Institute (EPRI) guide
ERA5 atmospheric reanalysis   **11**, 12
ESGF nodes *see* Earth System Grid Federation (ESGF) nodes
ETFE *see* ethylene tetrafluoroethylene (ETFE)
ethylene tetrafluoroethylene (ETFE)   49
European Centre for Medium-Range Weather Forecasts (ECMWF)   12
Evaluating Urban Resilience to Climate Change   **174**
EWF nexus *see* energy–water–food (EWF) nexus
EWS *see* early warning system (EWS)
extreme heat events (EHEs)   191
extreme natural events   111
extrinsic incubation period (EIP)   265

f

FA *see* fly ash (FA)
Fifteen Minutes City (FMC)   190
financial capital   130
flooding
  climate change-induced   187
  environmental impacts of   210
Flood Resilience Measurement Framework   **174**
flood risk management
  green–blue–gray infrastructure (GBGI) for   210, *211*
  resilience   212
fluorinated gases   4
fly ash (FA)   48
FMC *see* Fifteen Minutes City (FMC)
fossil fuels   4, 5, 53, 61, *62*, 63, 67, 242
fragmentation   101, 198, 258
freshwater sector impact on climate change   72
  multistage flash (MSF) desalination in Qatar (case study)   73–74, *74*
  seawater desalination in South Africa (case study)   73
  water supply in Singapore (case study)   72
FTOPSIS *see* fuzzy technique for order preference by similarity to ideal solution (FTOPSIS)
fuzzy logic models   **164**, 166
fuzzy technique for order preference by similarity to ideal solution (FTOPSIS)   166

g

game theory   105
GBGI systems *see* green–blue–gray infrastructure (GBGI) systems
GBI *see* green and blue infrastructure (GBI)
GCMs *see* general circulation models (GCMs)
general circulation models (GCMs)   8–9, 14, 29–30
  vs. regional climate models (RCMs) in MENA region   32–33
geographic information system (GIS)   103, 117, 145, 191, 193, **196**, 267

GHGs *see* greenhouse gases (GHGs)
GI *see* green infrastructure (GI)
GIS *see* geographic information system (GIS)
Global Positioning Systems (GPSs)   104, 267, 268–269
global warming   3, 4, 6, 208, 255, 264, 265
  anthropogenic   187
  in CMIP6 models   7
  spatial distribution of   6
Global Warming Potential (GWP) index   4
GPSs *see* Global Positioning Systems (GPSs)
GrADS *see* Grid Analysis and Display System (GrADS)
GRAI *see* gray infrastructure (GRAI)
graph percolation   101
graphs, types of   98
  directed and undirected graphs   99, *100*
  weighted and unweighted graphs   99–100
gray infrastructure (GRAI)   208, 209
green and blue infrastructure (GBI)   208, 209, 210, 211, 221
green–blue–gray infrastructure (GBGI) systems   207, 208, *212*
  benefits of combining   209
  flood risk management   210, *211*
  flood risk management resilience   212
  floods, environmental impacts of   210
  gray infrastructure (GRAI)   209
  green climate change adaptation   210
  green infrastructure (GI)   208
  regional progress in GBGI nexus research   211
green climate change adaptation   209, 210
greenhouse gases (GHGs)   4, 5, 29, 139
  emission of   6, 18, 40, 42, 45, 54, 64, 67, 72, 73, 139, 188, 190
    from cement production   44
    evaluation system boundary for   46
Greenhouse Gas Protocol methodology   41, 42
green infrastructure (GI)   18, 192, 193, 197, 208–210, 212
  and urban greening initiatives   193–194
Green New Deal   198
GRIB *see* GRIdded Binary (GRIB)

Grid Analysis and Display System (GrADS)   **14**
GRIdded Binary (GRIB)   12, 14
groundwater management   74
GWP index *see* Global Warming Potential (GWP) index

h

HadEX3   12, *13*
handheld systems   267
health policy development   256
  carbon emissions, reducing   256–257
  healthy lifestyle   257
  institutions, strengthening   258–259
  medical interventions   257
  monitoring   257–258
  proactive approaches   258
heating, ventilation, and air conditioning (HVAC) system   52, 53, 55, 84, 112
heatwave   97, 111–112
heavy building materials   49
HGBGIs *see* hybrid green–blue–gray infrastructures (HGBGIs)
High Latitude Mode   8
high-speed turbo blowers   75
human capital   130
human influence on climate system   61
  emission savings from water sector
    energy management in water system   75
    groundwater management   74
    smart wastewater treatment technology   75
  emissions savings from energy sector   69
    energy efficiency, increase in   70–71
    nuclear plants, running   72
    wind and solar plant installation   71
  energy sector impact on climate change   65
    coal power plant with carbon capture technology in Czech Republic (case study)   67–68
    electricity generation in Turkey (case study)   65–67
    solar power with energy storage (case study)   68–69

human influence on climate system (*continued*)
  freshwater sector impact on climate change 72
    multistage flash (MSF) desalination in Qatar (case study) 73–74, *74*
    seawater desalination in South Africa (case study) 73
    water supply in Singapore (case study) 72
  life cycle assessment (LCA) for environmental impact 63–64
  ReCiPe impact category 64–65
HVAC system *see* heating, ventilation, and air conditioning (HVAC) system
hybrid green–blue–gray infrastructures (HGBGIs) 211–212, 221
hydrological model 191

i

IEA *see* International Energy Agency (IEA)
IFRC Framework for Community Resilience **174**
immediate occupancy (IO) 113, *113*, 116
IMPACT 2002+ analysis 41–42
indicator method 157–158, **157**
Indicators of Disaster Risk and Risk Management **175**
individual network resilience 83
  case study about 86–88
  electrical network resilience 84–85
  transportation network resilience 84
  water network resilience 85–86
infectious diseases 252, 254–255, 263–264, 266, 270
infrastructure resilience 97
  complex networks theory 98
    betweenness centrality 100
    graph percolation 101
    graphs, types of 98–100
    strengths and limitations of 101
  infrastructures interdependencies and 88–90
  optimization approaches 104–105
  simulation approaches 101

  advantages 103
  agent-based modeling (ABM) 103
  drawbacks 104
  Geographic Information Systems (GIS)-based approaches 103
  system simulation 102–103
statistical approaches 104
interacting stresses 144–146
interdependent systems resilience, case study about 90–92
Intergovernmental Panel on Climate Change (IPCC) 61, 72, 139, 198
International Energy Agency (IEA) 65, 70, 229
International Space Station (ISS) 267
interview method 157, 158, **158**
Inter-decadal Pacific Oscillation (IPO) 8
IO *see* immediate occupancy (IO)
IPCC *see* Intergovernmental Panel on Climate Change (IPCC)
IPO *see* Inter-decadal Pacific Oscillation (IPO)
irreducible uncertainty 125
ISS *see* International Space Station (ISS)

j

JRA-55 reanalysis **11**, 12

l

land cover and land use change (LCLUC) models 191
LCA *see* life cycle assessment (LCA)
LCCA *see* life cycle cost analysis (LCCA)
LCIA *see* life cycle impact assessment (LCIA)
LCLUC models *see* land cover and land use change (LCLUC) models
Leadership in the Energy and Environmental Design (LEED) system 194
LEED system *see* Leadership in the Energy and Environmental Design (LEED) system
LID *see* low-impact development (LID)
life cycle assessment (LCA) 41, 50
  for environmental impact 63–64
life cycle cost analysis (LCCA) 63
life cycle impact assessment (LCIA) 63–64

life safety (LS)   113, *113*
life support systems   143
lighting, use of   54–55
limestone quarrying, embodied carbon emission of   41–42
Little Stringybark Creek (LSC) catchment   210
livelihood vulnerability analysis   **196**
low-impact development (LID)   191–192
LS *see* life safety (LS)
LSC catchment *see* Little Stringybark Creek (LSC) catchment

m

Machine Learning (ML)   19, 145
malaria   254–255, 264
mapping method   157, **158**
materials/physical systems, defined   **141**
MCA *see* multi-criteria analysis (MCA)
mechanical, electrical and plumbing (MEP)   113, *113*, 116
medical interventions   257
mega sport events (MSEs)
  multilevel resilience assessment framework developed for   87
  transportation resilience during   86–88
Mekong River runoff   187
MENA region *see* Middle East and North Africa (MENA) region
MEP *see* mechanical, electrical and plumbing (MEP)
methane   4
mHealth *see* mobile health (mHealth)
Middle East and North Africa (MENA) region   29
  Coordinated Regional Climate Downscaling Experiment (CORDEX)   31, 35–36
  natural disasters in   144
  regional climate models (RCMs)
    general circulation models (GCMs) vs.   32–33
    mean annual precipitation in   *33*
    mean surface air temperature in   *32*
  performance in simulating MENA climatic changes   34–35
MIPs *see* Model Intercomparison Projects (MIPs)
ML *see* Machine Learning (ML)
MME *see* multi-model ensemble (MME)
mobile health (mHealth)   267, 268
Model Intercomparison Projects (MIPs)   10
mosquito-borne diseases   254, 264, 266
movement systems related to transportation infrastructure   143
MSEs *see* mega sport events (MSEs)
multiple stable states, existence of   125
multistage flash (MSF) desalination in Qatar (case study)   73–74, *74*
multi-criteria analysis (MCA)   **195**
multi-model ensemble (MME)   14

n

NAM *see* Northern Hemisphere Annular Mode (NAM)
NAO *see* North Atlantic Oscillation (NAO)
NASA MERRA   **11**
National Center for Atmospheric Research (NCAR)   19, 31
National Health Security Preparedness Index   **175**
National Oceanic and Atmospheric Administration (NOAA)   31
natural (ecological) capital   130
natural disasters   111, 256, 258
  critical infrastructure and   128
  damages to energy systems by   *228*
  frequency of   144
NCAR *see* National Center for Atmospheric Research (NCAR)
NCAR Command Language (NCL)   **14**
NCEP-DOE Reanalysis 2   **11**
network Common Data Form (netCDF) format   12, 14
network resilience   83
  electrical   84–85
  transportation   84
  water   85–86
nitrogen removal process   75

nitrous oxide 4
NOAA *see* National Oceanic and Atmospheric Administration (NOAA)
NOAA-CIRES Twentieth Century Reanalysis **11**
non-revenue water 54
North Atlantic Oscillation (NAO) 8
Northern Hemisphere Annular Mode (NAM) 6–8
nuclear plants 72
numerical method 157, **157**
numerical modeling 17–18, 21, 24

o

open space systems 143
operational carbon emissions 39, 40
  in building environment 51–52
operation carbon mitigation strategies 52
  efficient HVAC systems in buildings 53
  lighting, use of 54–55
  renewable resources integration 53
  water use, strategy for 54
optimization models 104–105, **164**, 165
organizational resilience, defined **141**

p

PACE *see* Property Assessed Clean Energy (PACE)
Pacific Decadal Oscillation (PDO) index 8
Panoply 14, **14**
Paris Climate Accords (2015) 45
participatory risk assessment **196**
participatory scenario planning (PSP) **196**
PBD *see* performance-based design (PBD)
PDO index *see* Pacific Decadal Oscillation (PDO) index
PEG *see* polyethylene glycol (PEG)
PEOPLES Resilience Framework **175**
Peramin SRA 40 48
percolation 101
percolation threshold 101
performance-based design (PBD) 112–116
physical (built) capital 130
political capital 130
polyethylene glycol (PEG) 48

polypropylene glycol (PPG) 48
polytetrafluoroethylene (PTFE) 49
post-traumatic stress disorder (PTSD) 256
PPG *see* polypropylene glycol (PPG)
precast hollow-core slabs 48
probabilistic performance-based approach **164**, 165
Program for Predicting Polluting Particle Passage Through Pits, Puddles, and Ponds (P8 Urban Catchment Model) 193
Property Assessed Clean Energy (PACE) 56
PSP *see* participatory scenario planning (PSP)
PTFE *see* polytetrafluoroethylene (PTFE)
PTSD *see* post-traumatic stress disorder (PTSD)
public health, climate change impacts on 253
  air pollution 255
  extreme events 256
  infectious diseases 254–255

q

qualitative assessment 157, 159, 160, 161, 177
qualitative resilience assessment 156, 160
  conceptual frameworks 161–163
  semiquantitative indices 163
quantitative assessment 157, 159, 177
quantitative indicators 69, 128, 160
quantitative resilience assessment 163, **164**

r

radiative forcing 4, 10, 64–65
ragweed (*Ambrosia artemisiifolia*) plants 255
rainwater harvesting 54
RCD *see* regional climate downscaling (RCD)
RCI *see* Resilience Capacity Index (RCI)
RCMs *see* regional climate models (RCMs)
RCPs *see* Representative Concentration Pathways (RCPs)
RCSMs *see* regional climate system models (RCSMs)
RECARGA model 193

ReCiPe methodology   64–65, 68, 69
recovery capacity *see* restorative capacity
recycling   50, 194–195
reducible and irreducible uncertainty   125
reflectiveness   104
RegCM4-3 simulations   31–37
regional climate downscaling (RCD)   30
regional climate models (RCMs)   9, 29, 32
   mean annual precipitation in   *33*
   mean surface air temperature in   *32*
   performance in simulating MENA climatic changes   34–35
   vs. general circulation models (GCMs) in Middle East and North Africa (MENA) region   32–33
regional climate system models (RCSMs)   30
remote sensing (RS)   103, 267
renewable resources integration   53
Representative Concentration Pathways (RCPs)   31, 266
resilience   81, 139, 150, 263
   -based design thinking   82, 112, 116
   capacities   123–124
   characteristics   **233–234**
   components   124
   concept of   121, **122**
   defined   **142**
   dimensions and capitals   129–130, *131*
   in ecology   125
   of energy system   229
   individual network resilience   83
     case study about   86–88
     electrical network resilience   84–85
     transportation network resilience   84
     water network resilience   85–86
   infrastructures interdependencies and   88–90
   interdependent systems resilience, case study about   90–92
   measuring   130–133
   mega sport events (MSEs), transportation resilience during   86–88
   qualities   146
     flexibility   147
     inclusivity   148
     integration   148–149
     interrelation of   149
     rapidity of recovery   148
     redundancy   147
     reflectivity   146–147
     resourcefulness   148
     robustness   147
   surveillance strategies   266, *268*
     climatic and environmental changes   270
     control measures   270
     distribution and behavior of vectors   269–270
     identification of pathogen species   269
     monitoring of human cases   268–269
     policy development   270–271
   of transportation systems   163
   types of   124
     community resilience   127–128
     critical infrastructure resilience   128–129
     ecological resilience   125–127, *126*
     engineering resilience   125–127
     social resilience   127–128
     specified and general resilience   128
     technical systems, products, and production resilience   129
   vector-borne diseases (VBDs)
     climate change impacts on   266
     environmental factors and   265
Resilience Alliance   161
resilience assessment   159
   applications in
     betweenness centrality   100
     graph percolation   101
     strengths and limitations of complex networks   101
   classification of   *161*
   general resilience approaches   164
     deterministic performance-based approach   165
     probabilistic performance-based approach   165
   qualitative resilience assessment   160
     conceptual frameworks   161–163

resilience assessment (*continued*)
   semiquantitative indices   163
   quantitative resilience assessment   163
   structural-based models   165
      fuzzy logic models   166
      optimization models   **164**, 165
      simulation models   165–166
resilience benefits of energy efficiency   **243**
Resilience Capacity Index (RCI)   **175**
resilience energy system   *232*
resilience engineering, defined   **141**
resilience enhancement approaches for energy systems   240
   climate resilience   242–243
   distributed generation   240–241
   energy efficiency, enhancing   242
   energy storage   241
   smart grid technology   241–242, *241*
   system hardening   240
resilience-enhancing measures   **244**
resilience evolution   *141*
Resilience Index Measurement and Analysis (RIMA)   **175**
Resilience Inference Measurement (RIM)   **175**
resilience in systems   112
Resilience Measurement Index (RMI)   **176**
Resilience Scorecard   **176**
resilience strategies   156
resilience thinking   81
resilience tools   **170–176**
resilience triangle   113, **133**
Resilience United States (ResilUS)   **176**
resilient health policies, building   251
   climate change impacts on public health   253
      air pollution   255
      extreme events   256
      infectious diseases   254–255
   health policy development, considerations in   256
      carbon emissions, reducing   256–257
      healthy lifestyle   257
      institutions, strengthening   258–259
      medical interventions   257
      monitoring   257–258
      proactive approaches   258
resilient system   98, 144, *145*
ResilUS *see* Resilience United States (ResilUS)
resistive capacity *see* absorptive capacity
restorative capacity   124
reverse osmosis (RO)   73
RIMA *see* Resilience Index Measurement and Analysis (RIMA)
RIM *see* Resilience Inference Measurement (RIM)
risk assessment and adaptation in urban planning   195–199
risk identification and assessment   123
RMI *see* Resilience Measurement Index (RMI)
RO *see* reverse osmosis (RO)
RRI *see* Rural Resilience Index (RRI)
RS *see* remote sensing (RS)
Rural Resilience Index (RRI)   **176**

S

SAM *see* Southern Hemisphere Annular Mode (SAM)
scenario modeling   163, **196**, 258
SDGs *see* Sustainable Development Goals (SDGs)
sea level rise (SLR)   9, 144, 187, 210, 238
seawater desalination in South Africa (case study)   73
semiquantitative indices   159, 163
Shared Socioeconomic Pathways (SSPs)   10
shelter systems   143
simulation approaches   101
   advantages   103
   agent-based modeling (ABM)   103
   drawbacks   104
   Geographic Information Systems (GIS)-based approaches   103
   system simulation   102–103
simulation models   **164**, 165–166
SIPs *see* structural insulated panels (SIPs)
SLR *see* sea level rise (SLR)
smart grid technology   241–242, *241*
smart wastewater treatment technology   75

social capital   130
social resilience   127–128
  defined   **143**
social vulnerability index (SVI)   **196**
socio-ecological resilience, defined   **142**
socio-ecological systems, defined   **141**
SOI *see* Southern Oscillation Index (SOI)
solar power with energy storage (case study)   68–69
Southern Hemisphere Annular Mode (SAM)   6–8
Southern Oscillation Index (SOI)   8
Special Report on Emissions Scenarios (SRES)   10
specified and general resilience   128
sponge cities   192–193, 212
SRES *see* Special Report on Emissions Scenarios (SRES)
SSPs *see* Shared Socioeconomic Pathways (SSPs)
standard/unweighted graphs   99–100
statistical approaches   9, 104
steel framework system   48
steel production, embodied carbon emission of   45–46
stormwater management   192, 208, 210, 221
stormwater management model (SWMM)   193
strengths, weaknesses, opportunities, and threats (SWOT)   **195**
stresses, interacting   144–146
structural insulated panels (SIPs)   51
structural resilience   112–114
successful climate-responsive urban planning, case studies of   200–202
supporting systems   112, 113, 115
  beyond building limits   116–117
  within the building   116
surveillance strategies for monitoring the spread of vector-borne diseases   266, *268*
  climatic and environmental changes   270
  control measures   270
  human cases, monitoring of   268–269

pathogen species, identification of   269
policy development   270–271
vectors, distribution and behavior of   269–270
survey method   157, **158**
Sustainable Development Goals (SDGs)   139, 198
sustainable land use and development policies   191
SVI *see* social vulnerability index (SVI)
SWMM *see* stormwater management model (SWMM)
SWOT *see* strengths, weaknesses, opportunities, and threats (SWOT)
system hardening   240, 243
system simulation   102–103

t

TDM *see* transportation demand management (TDM)
technical systems, products, and production resilience   129
teleconnections   6, 8, 30
temperature rise   36, 144
Tensor Processing Unit (TPU)   71
3-30-300 rule   193
tipping points   3
TOD *see* transit-oriented development (TOD)
total suspended solid (TSS) removal   193
TPU *see* Tensor Processing Unit (TPU)
transformative capacity   124
transit-oriented development (TOD)   188–190, 191
transportation demand management (TDM)   **196**
transportation networks   82, 86, 100, 102, 103, 116
  resilience of   84
transportation sector   18
transportation systems   18, 84, 88
  resilience of   163, 165
TSS removal *see* total suspended solid (TSS) removal

U

UCAR *see* University Corporation for Atmospheric Research (UCAR)
UF *see* ultrafiltration (UF)
UHC *see* universal health coverage (UHC)
UHI effect *see* urban heat island (UHI) effect
ultrafiltration (UF) 73
uncertainty 122–123, 146, 150, 165
  content-related 123
  context-related 123
  irreducible 125
  reducible 125
undirected graph 99, *100*
UNDP *see* United Nations Development Programme (UNDP)
unfired brick 48
United Nations Climate Change Conferences 198
United Nations Development Programme (UNDP) **169**
United States Agency for International Development (USAID) **168**
universal health coverage (UHC) 253
University Corporation for Atmospheric Research (UCAR) 14
unweighted graph 99–100
urban air quality forecasting 20
urban and community resilience assessment 157–158
urban areas 17, 19, 22, 82, 85, 97
  built environment in 111
  climate change impacts on 186–188, 195, 197
  climate extremes on 18
  cost of disasters in 112
  energy supply in 227
  green–blue–gray infrastructures (GBGIs) in 209
  green infrastructure (GI) in 208
  healthy lifestyle in 257
  modernization of 143
  waterborne diseases in 18
urban climate change studies 20
urban climate resilience 122, 150
urban emergency response planning and management 21
urban energy modeling 21
Urban Footprint 191
urban heat island (UHI) effect 19, 20, 23, 193, 197
urban hydrological modeling 20
urban infrastructure flow types **83**
urban infrastructures 82, 83, 88, 105, 116
urbanization and climate change, nexus between 18–19
urban land use planning 21
urban planning 185
  challenges and opportunities 202–203
  risk assessment and adaptation in 195–199
  significance of 185
  strategies for climate change 188
    compact cities 190–191
    Fifteen Minutes City (FMC) 190
    green infrastructure and urban greening initiatives for cool cities 193–194
    low-impact development (LID) 191–192
    sponge cities 192–193
    sustainable land use and development policies 191
    transit-oriented development (TOD) 188–190
    waste management and recycling systems 194–195
  successful climate-responsive urban planning, case studies of 200–202
urban resilience 17, 139, 156, 185–186
  applied methods in the assessment of **157–158**
  characteristics of 124, **124**
  to climate change 140
    built environment systems, climate change impacts on 140–144
    major uncertainties and interrelations 146
    sea level rise (SLR) 144
    stresses, interacting 144–146
    temperature rise 144

defined 142
resilience qualities 146
  flexibility 147
  inclusivity 148
  integration 148–149
  interrelation 149
  rapidity of recovery 148
  redundancy 147
  reflectivity 146–147
  resourcefulness 148
  robustness 147
urban resilience, quantifying 155, 156
  frameworks and tools for measuring resilience 166–177
  general resilience approaches 164
    deterministic performance-based approach 165
    probabilistic performance-based approach 165
  qualitative resilience assessment 160
    conceptual frameworks 161–163
    semiquantitative indices 163
  quantitative resilience assessment 163
  resilience strategies 156
  structural-based models 165
    fuzzy logic models 166
    optimization models **164**, 165
    simulation models 165–166
  urban and community resilience assessment 157–158
urban sustainability 17–18, 24–25
urban system resilience theory **162**
urban weather forecasting 20, 22
USAID *see* United States Agency for International Development (USAID)

V

VBDs *see* vector-borne diseases (VBDs)
vector-borne diseases (VBDs) 252, 254–255, 263
  climate change impacts on *264*, 266
  environmental factors and 265
  surveillance strategies for monitoring the spread of 266, *268*
    climatic and environmental changes 270

control measures 270
human cases, monitoring of 268–269
pathogen species, identification of 269
policy development 270–271
vectors, distribution and behavior of 269–270

W

waste management and recycling systems 194–195
water distribution networks 82, 102–103
water management 54, 201–202
water network resilience 85–86, 102
water recycling systems 53
water sector
  emission savings from
    energy management in water system 75
    groundwater management 74
    smart wastewater treatment technology 75
  energy demand 72
water supply in Singapore (case study) 72
water use, strategy for 54
water vapor 4
WCRP *see* World Climate Research Programme (WCRP)
Weather Research and Forecasting (WRF) model 19
Weather Research and Forecasting Urban (WRF-Urban) Model 17, 20
  applications of 20
    urban air quality forecasting 20
    urban climate change studies 20
    urban emergency response planning and management 21
    urban energy modeling 21
    urban heat island (UHI) effect studies 20
    urban hydrological modeling 20
    urban land use planning 21
    urban weather forecasting 20
  case studies 21
    coastal-urban meteorology study in the metropolitan region of Vitória, Brazil 22

Weather Research and Forecasting Urban
(WRF-Urban) Model (*continued*)
  summertime air conditioning electric
    loads modeling in Beijing,
    China 21–22
  urban climate modeling in
    Singapore 21
  limitations of 22–23
    computational capabilities 22
    data requirements 22
    limited chemical mechanisms 23
    model evaluation and validation 23
    parameterization uncertainties 23
    sensitivity to initial and boundary
      conditions 23
    urban complexity representation 23
    user expertise 23
  urbanization and climate change, nexus
    between 18–19
  urban modeling through 19–21
  ways forward for improvement 23–24

weighted graph 99–100
West Nile virus 264
wgrib **14**
wind and solar plant installation 71
windstorms 134, 188
WMO *see* World Meteorological
    Organization (WMO)
World Climate Research Programme
    (WCRP) 10
World Meteorological Organization
    (WMO) 12
WRF model *see* Weather Research and
    Forecasting (WRF) model
WRF-Urban Model *see* Weather Research and
    Forecasting Urban (WRF-
    Urban) Model

y
yellow fever 264

z
zika virus 264